343
Topics in Current Chemistry

For further volumes:
http://www.springer.com/series/128

Aims and Scope

The series Topics in Current Chemistry presents critical reviews of the present and future trends in modern chemical research. The scope of coverage includes all areas of chemical science including the interfaces with related disciplines such as biology, medicine and materials science.

The goal of each thematic volume is to give the non-specialist reader, whether at the university or in industry, a comprehensive overview of an area where new insights are emerging that are of interest to larger scientific audience.

Thus each review within the volume critically surveys one aspect of that topic and places it within the context of the volume as a whole. The most significant developments of the last 5 to 10 years should be presented. A description of the laboratory procedures involved is often useful to the reader. The coverage should not be exhaustive in data, but should rather be conceptual, concentrating on the methodological thinking that will allow the non-specialist reader to understand the information presented.

Discussion of possible future research directions in the area is welcome.

Review articles for the individual volumes are invited by the volume editors.

Readership: research chemists at universities or in industry, graduate students.

Wei Li • Xumu Zhang
Editors

Stereoselective Formation of Amines

With contributions by

A.B. Charette · Q.-H. Fan · G.K. Friestad · Y.-M. He · K.C. Hultzsch · W. Li · V. Lindsay · A.J. Nawara-Hultzsch · A.L. Reznichenko · F.-T. Song · C. Wang · J. Xiao · J.-H. Xie · X. Zhang · Q.-L. Zhou

Springer

Editors
Wei Li
Institute of Chemical Biology and Drug Discovery
Stony Brook University
Stony Brook, NY
USA

Xumu Zhang
Department of Chemistry and Chemical Biology, Department of Medicinal Chemistry
Rutgers, The State University of New Jersey
Piscataway, NJ
USA

ISSN 0340-1022 ISSN 1436-5049 (electronic)
ISBN 978-3-642-53928-2 ISBN 978-3-642-53929-9 (eBook)
DOI 10.1007/978-3-642-53929-9
Springer Heidelberg New York Dordrecht London

Library of Congress Control Number: 2014933584

Printed on acid-free paper

Springer is part of Springer Science+Business Media (www.springer.com)

Preface

Chiral amines are widespread structural units in chiral auxiliaries and bioactive molecules, such as agrochemicals and active pharmaceutical ingredients. Historically, there has been interest in producing chiral amines, and such interest has promoted a variety of methodologies to synthesize chiral amines – ideally in a single isomer. Organic chemists, particularly organometallic chemistry researchers, have made tremendous strides toward developing novel chemical technologies to tackle the long-standing challenge of producing chiral amines via operationally simple, economic, and highly stereoselective methods.

Despite the fact that chiral amines have been recognized for their importance dating back several decades, the top long-existing challenges on the list have been narrow substrate scope, relatively low turnover numbers caused by catalyst inhibition/deactivation by amine products, and the necessity of protecting groups. Following the seminal studies in the 1990s, in more recent years a fast-growing body of creative research has emerged, yielding versatile and powerful transformations for introducing chiral amines stereoselectively. This array of methods has made this once challenging ambition easier to realize. These new methods have drastically shortened possible routes to a given natural product or pharmaceutical drug containing a chiral amine moiety. The increasingly versatile development of new methods will be implemented into novel synthetic strategies in the future.

This volume contains valuable and insightful contributions from several leading experts in this field, and the reviews within collectively provide an overview of a diverse set of valuable approaches as well as their implementation toward natural product and pharmaceutical synthesis. Asymmetric free radical addition to imino compounds is addressed in the first chapter, mainly dealing with the use of chiral hydrazones as radical acceptors. The second chapter presents an in-depth review of the nucleophilic addition of unstabilized carbanions to azomethine derivatives. With a biased emphasis on asymmetric hydrogenation methodology, the subsequent three chapters provide an in-depth overview of the most powerful and straightforward asymmetric transformations using hydrogen as the reducing reagent on three main substrate categories: enamide/enamine, imine, and heteroaromatics, respectively. A comprehensive contribution on asymmetric hydroamination displays solid examples of efficient catalyst systems that underline the usefulness of the direct addition of N–H bonds across an unsaturated carbon–carbon bond to formal chiral amines. The final section covers current methods and recent advances in asymmetric reductive amination by hydrogenation, transfer hydrogenation, organocatalytic reduction, and biocatalytic reduction, as a review of the state-of-the-art method of asymmetric reductive amination. Unfortunately, we were unable to include other important topics like enantioselective C–H amination, and recently-emerging enzymatic approaches via dynamic kinetic resolution or transaminase technology as chapters in this short volume.

We would like to express our sincere gratitude to those who contributed to this volume. We appreciate their valuable time and effort and expect that their views on future research trends will be useful to all researchers in the field.

Rutgers, The State University of New Jersey, Piscataway, NJ, USA Xumu Zhang
Stony Brook University, Stony Brook, NY, USA Wei Li
September 2013

Contents

Top Curr Chem (2014) 343: 1–32
DOI: 10.1007/128_2013_481

Published online: 28 September 2013

Control of Asymmetry in the Radical Addition Approach to Chiral Amine Synthesis

Gregory K. Friestad

Abstract The state-of-the-science in asymmetric free radical additions to imino compounds is presented, beginning with an overview of methods involving stereocontrol by various chiral auxiliary approaches. Chiral *N*-acylhydrazones are discussed with respect to their use as radical acceptors for Mn-mediated intermolecular additions, from design to scope surveys to applications to biologically active targets. A variety of aldehydes and ketones serve as viable precursors for the chiral hydrazones, and a variety of alkyl iodides may be employed as radical precursors, as discussed in a critical review of the functional group compatibility of the reaction. Applications to amino acid and alkaloid synthesis are presented to illustrate the synthetic potential of these versatile stereocontrolled carbon–carbon bond construction reactions. Asymmetric catalysis is discussed, from seminal work on the stereocontrol of radical addition to imino compounds by non-covalent interactions with stoichiometric amounts of catalysts, to more recent examples demonstrating catalyst turnover.

Keywords Asymmetric synthesis · Hydrazones · Imines · Oxime ethers · Radical reactions

Contents

G.K. Friestad (✉)
Department of Chemistry, University of Iowa, Iowa City, IA 52242, USA
e-mail: gregory-friestad@uiowa.edu

1 Background and Introduction

Chiral α-branched amine functionalities are present in a wide range of bioactive synthetic targets, both natural and unnatural. Accordingly, a variety of methods for asymmetric amine synthesis have been developed over recent years, many of which involve carbon–carbon bond constructions by addition to the C=N bond of carbonyl imino derivatives, represented in the retrosynthetic direction in Fig. 1 (Recent reviews: see [1–10]).

As is typical in most synthetic chemistry, methods accrue higher impact if they offer configurational control under mild conditions compatible with a range of spectator functional groups. Unfortunately, many methods involving nucleophilic additions of carbanion-type reagents to C=N bonds have limited compatibility with electrophilic or acidic functionality, or may result in competing deprotonation at the imine α-carbon due to the basicity of the organometallic reagent (For examples of aza-enolization of imino compounds by organometallic reagents see [11–14]). These limitations have spurred the development of free radical additions (Reviews of free radical reactions in synthesis: see [15–21]) to imino compounds (Fig. 2) as a C–C bond construction approach to chiral amines under mild conditions, offering a valuable complement to organometallic additions and expanding the scope of this α-C–C retrosynthetic disconnection (Reviews of radical additions to imines and related acceptors: see [2, 22–25]).

Seeking to probe the improved versatility which might be associated with the radical addition approach, we initiated a program to develop a variety of radical additions to imino compounds [26]. In the process we introduced several new modes of stereocontrol using hydrazones as the C = N radical acceptor functionality. Imino compounds have been extensively used for cyclizations [27], and initially we built upon this foundation by introducing a temporary tethering approach [28–34] which has been effective in establishing relative configuration in a predictable diastereoselective fashion via reactions of α-hydroxyaldehyde hydrazones (Reviews of temporary tethers: see [35–43]). In this review we will focus on the intermolecular variant of the reaction, for which we have introduced methodology to build chiral amines from achiral precursors using chiral auxiliaries or chiral catalysts for stereocontrol [44]; the design and experimental evaluation of these strategies will be described along with synthetic applications.

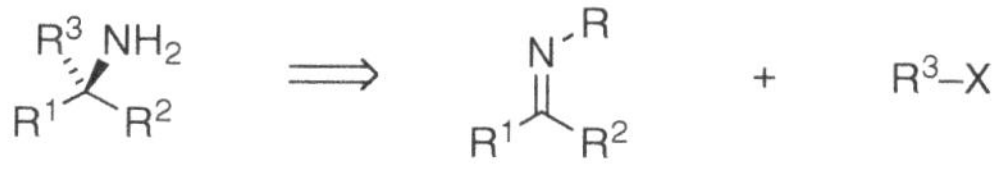

Fig. 1 Disconnecting a C–C bond of a chiral amine suggests an imino compound (e.g., imine, oxime, hydrazone, etc.) and an organic halide as precursors

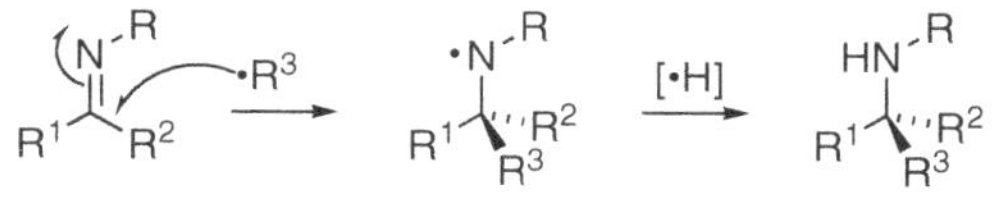

Fig. 2 Radical addition to imino compounds

2 Intermolecular Radical Addition to Chiral *N*-Acylhydrazones

2.1 Use of Chiral Auxiliaries in Radical Additions to Imino Compounds

In many areas of synthetic methods development, the reaction methodology has progressed to a point where it is quite well-understood, and the exciting new advances are mostly focused on extension to asymmetric catalysis. In contrast, in the area of radical additions to C=N bonds, only rare examples of intermolecular coupling reactions were available prior to the initial work on asymmetric methods [45–50] Thus, the development of asymmetric induction for radical addition to imino compounds has been slow, pending important advances that are still needed in the fundamental reaction chemistry. As this chemistry emerges, the challenges of adapting the new chemistry to all types of stereocontrol modes, including auxiliary, reagent and catalyst, continue to attract active investigation. An overview of some of the main approaches to chiral auxiliary-mediated stereocontrol is provided below.

Naito and coworkers reported the first general method for achieving reductive addition of carbon-centered radicals to imino compounds; by using Et_3B/oxygen initiation in the presence of BF_3•OEt_2, neutral radicals were added successfully to prochiral aldoximes [51, 52]. Although unhindered formaldoximes and activated glyoxylic oxime ethers did not require Lewis acid activation, the beneficial effect of BF_3•OEt_2 allowed general application to various oxime ethers. This enabled the use of **1** bearing a chiral auxiliary for asymmetric induction (Table 1) [53, 54]. The proposed stereocontrol model invokes steric blocking by one of the sultam oxygens (Fig. 3). Although difunctional alkyl iodides afforded lower yields, subsequent work has successfully employed more complex radicals [55].

Bertrand and coworkers reported the radical addition of various alkyl iodides to cyclic and acyclic chiral glyoxylate imines using either tin-mediated conditions initiated by AIBN, or triethylborane conditions initiated by oxygen [56]. Building on these initial studies, Bertrand discovered the combination of diethylzinc and air

Table 1 Radical addition to chiral glyoxylic oxime ethers

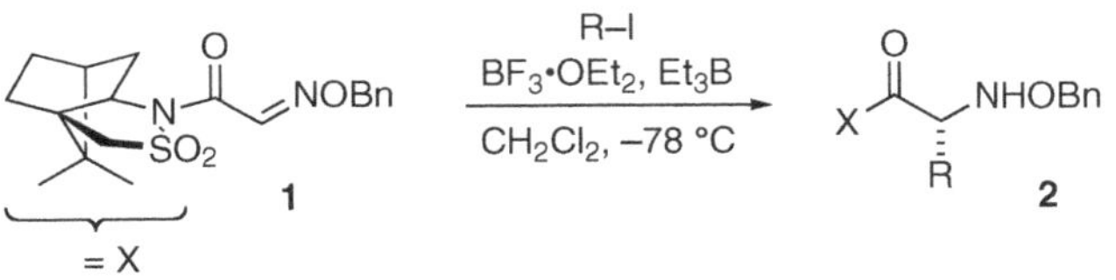

Entry	R^1	Yield[a] (%)	dr[b]
1	Et	80	95:5
2	*i*-Bu	83	97:3
3	*i*-Pr	80	96:4
4	*s*-Bu	69[c]	>98:2
5	*c*-C_6H_{11}	86	96:4
6	*t*-Bu	83	>98:2
7	$AcO(CH_2)_4$	41[d]	>98:2
8	$Cl(CH_2)_4$	15[d]	>98:2

[a]Isolated yield
[b]dr = diastereomer ratio
[c]Obtained as 1:1 mixture of diastereomers
[d]Major product was ethyl addition

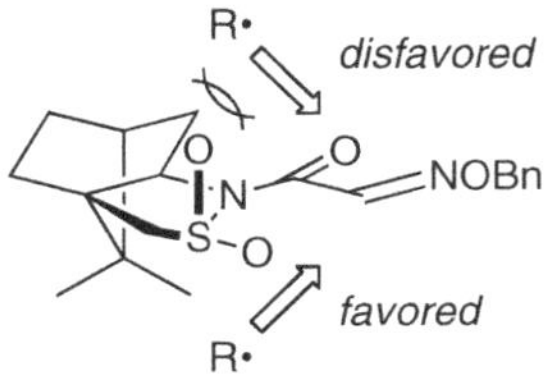

Fig. 3 Naito stereocontrol model proposed for additions to camphorsultam-functionalized glyoxylic oxime ethers

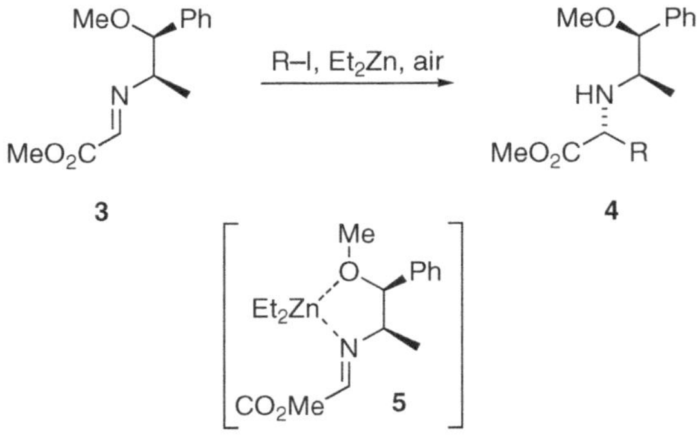

Scheme 1 Radical addition to glyoxylate imines

was useful to promote radical addition to chiral imines (Scheme 1) [57, 58]. Stereoselectivity was modest in additions to 1-phenylethylimines. However, two-point binding imines capable of chelating Zn(II), such as norephedrine-derived imine **3**, led to improved diastereomer ratios (Review: see [59]).

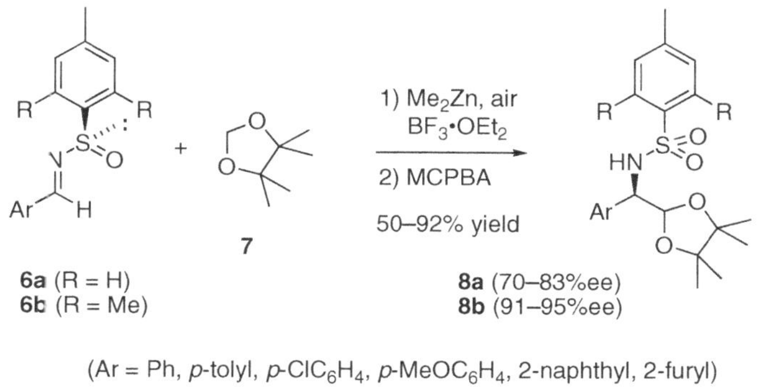

Scheme 2 Radical addition to *N*-sulfinylimines

Chiral *N*-sulfinylimines, pioneered by Davis and Ellman for stereocontrolled additions of carbanion reagents [60, 61], have also been employed successfully in radical additions. Tomioka and coworkers have used the dimethylzinc/air method to generate radicals by H-atom abstraction from ethers and acetals which then add to the C=N bond of the *N*-sulfinylimines **6** [62]. Although tetrahydrofuran and *tert*-butyl methyl ether gave modest stereocontrol, formaldehyde acetals were more promising, with the best results observed for addition of dioxolane **7** (Scheme 2) to substituted benzaldehyde imines (70–83% *ee*, measured after oxidation to sulfonamide). Subsequently, increasing the steric demand of the arenesulfinyl stereocontrol element led to improved enantioselectivities [63]. Recently the reaction has been extended to include triethylborane-mediated addition of an iodomethyl ester to *N*-alkoxycarbonylimines [64].

2.2 *Design of Chiral* N*-Acylhydrazones*

In the late 1990s we began our work in developing hydrazones as promising radical acceptors for asymmetric transformations. At that time, oxime ethers were the most commonly explored C = N radical acceptors (Review: see [65–67]). We considered that linking an amino or amido substituent to the C = N nitrogen atom could afford the same reactivity enhancement seen in oxime ethers, and would also provide opportunities for superior rotamer control needed for stereoselectivity. Both activation and stereocontrol would be achieved through one removable *N*-substituent on the imine, and the substituents on the carbon of the C=N bond would then be freed from those roles, offering potential for broader scope.

Our new type of chiral hydrazone, tailored for use in free radical addition reactions, incorporates Lewis acid activation and restriction of rotamer populations as key design elements. We hypothesized that the restricted rotation achieved through two-point binding would fix a substituent relative to the plane of the C=N bond to differentiate the enantiotopic approach trajectories (Fig. 4). The Lewis acid would

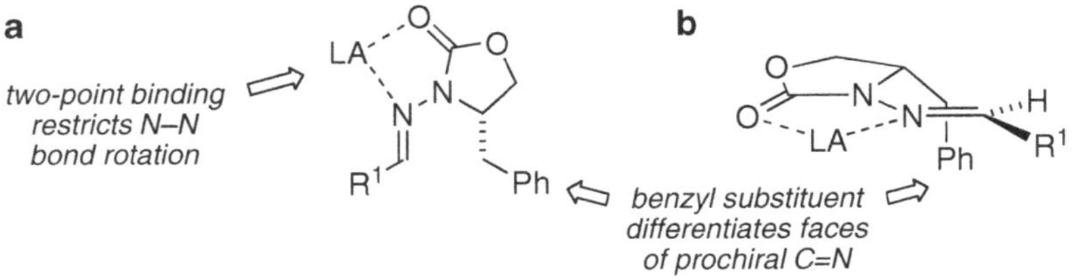

Fig. 4 (**a**) Lewis acid chelation induces rigidity and electron-deficiency into the *N*-acylhydrazone radical acceptor (LA = Lewis acid). (**b**) Benzyl substituent of 4-benzyl-2-oxazolidinone provides facial differentiation of the C=N bond

also lower the LUMO energy of the C=N π-bond, increasing its reactivity toward nucleophilic alkyl radicals [49, 68, 69]. This would ensure selective reactivity via the chelated structure, suppressing the non-selective background reaction. After radical addition, H-atom abstraction would occur to afford an *N*-acylhydrazine product; reductive cleavage of the N–N bond [70, 71] would provide an enantiomerically pure amine and release the chiral auxiliary for reuse. For the first test of our design, *N*-acylhydrazones derived from *N*-aminooxazolidinones [72] were chosen to satisfy all these design criteria (Fig. 4).

2.3 Preparation and Initial Reactivity Studies of Chiral N-Acylhydrazones

Although *N*-amino derivatives of oxazolidinones such as **9** (Scheme 3) were sporadically reported in the literature [73–75], the installation of the *N*-amino group was not well-developed. Therefore we began with a study of electrophilic amination of commercial oxazolidinones. A variety of electrophilic amination reagents were employed for this purpose, and the preferred reagents were *O*-(*p*-nitrobenzoyl) hydroxylamine ($NbzONH_2$), *O*-(diphenylphosphinyl)hydroxylamine ($Ph_2P(=O)ONH_2$), and NH_2Cl [76, 77]. The optimized procedure for *N*-amination with $NbzONH_2$ entailed deprotonation with NaH (or KH) in hot dioxane, followed by introduction of $NbzONH_2$ as a solid at ambient temperature [78]. With NH_2Cl the same procedure may be followed, but the stoichiometry of base and chloramine should be carefully controlled to nearly 1 equiv. in order to obtain reproducible yields in the amination. A survey of representative condensations of *N*-aminooxazolidinone **9** with aldehydes and ketones (Scheme 3) shows that the sequence is robust and general, affording a wide range of chiral *N*-acylhydrazones **10** [79, 80]. The reaction is readily scaled to multigram quantities, and the carbonyl component of these *N*-acylhydrazones may be exchanged with other carbonyl compounds [78].

Ketone-derived chiral *N*-acylhydrazones may also be prepared by direct condensation with *N*-aminooxazolidinones (Scheme 3) [77, 81]. Mixtures of *E*/*Z* isomers were usually obtained, although ketone *N*-acylhydrazones with highly branched

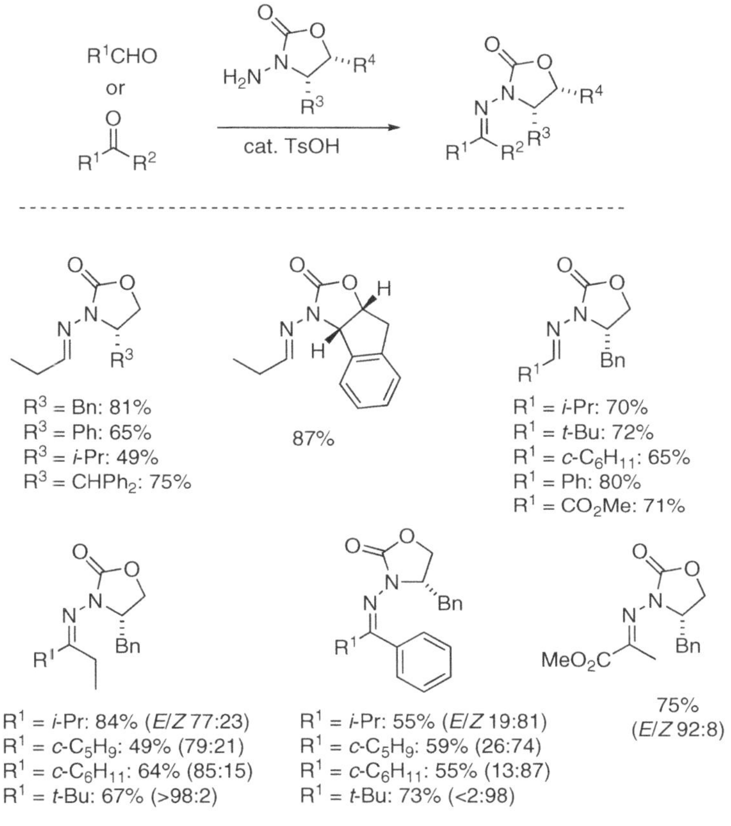

Scheme 3 Representative preparations of chiral *N*-acylhydrazones

tertiary butyl (*t*-Bu) substituents were formed as single isomers. A pyruvate-derived hydrazone was formed with high selectivity, and the major isomer was readily separated from its minor (*Z*)-isomer by flash chromatography [77]. Others have recently used these amination and condensation procedures to prepare very similar chiral *N*-acylhydrazones from ketones with excellent results [82].

2.3.1 Additions of Secondary and Tertiary Radicals

The first test of the chiral *N*-acylhydrazones was in tin-mediated radical addition of various secondary and tertiary iodides [79, 80]. Using the tin hydride method with triethylborane initiation [83, 84] (Bu_3SnH, Et_3B/O_2), with $ZnCl_2$ as a Lewis acid additive, addition of isopropyl iodide to *N*-acylhydrazone **10a** afforded adduct **11a** with a diastereomer ratio of 99:1 (Table 2). In contrast, **11a** was produced with poor selectivity (*dr* 2:1) in the absence of Lewis acid, which validated the two-point

Table 2 Tin-mediated radical addition to chiral *N*-acylhydrazone **10a** in the presence of $ZnCl_2$

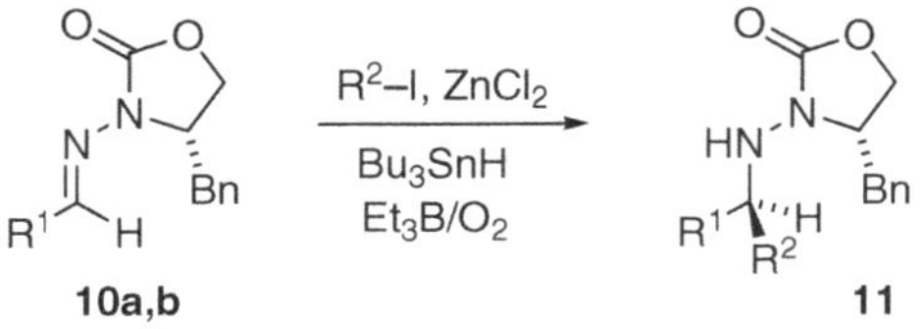

Entry	R^1	R^2	Product, Yield[a] (%) (*dr*)
1	Et (**10a**)	*i*-Pr	**11a**, 60 (99:1)
2	Et (**10a**)	c-C_5H_9	**11b**, 59 (96:4)
3	Et (**10a**)	c-C_6H_{11}	**11c**, 28 (97:5)
4	Et (**10a**)	*t*-Bu	**11d**, 54 (95:5)
11	Ph (**10b**)	*i*-Pr	**11e**, 42 (99:1)
12	Ph (**10b**)	c-C_5H_9	**11f**, 59 (96:4)
13	Ph (**10b**)	c-C_6H_{11}	**11 g**, 30 (99:1)
14	Ph (**10b**)	*t*-Bu	**11 h**, 83 (93:7)

Reaction conditions: Bu_3SnH (5 equiv.) and O_2 (7 mL/mmol) by syringe pump, *i*-PrI (10 equiv), Et_3B (10 equiv.), and Lewis acid (2 equiv.), 2:1 CH_2Cl_2/ether, –78°C → rt
[a]Isolated yield, %

Scheme 4 Role of oxazolidinone substituents on diastereoselectivity in isopropyl radical addition

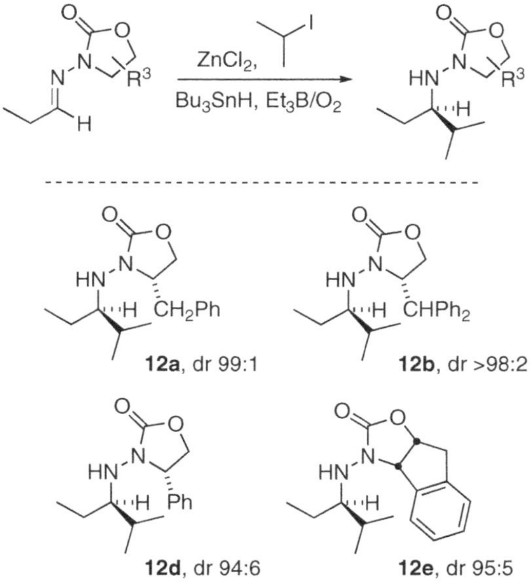

binding stereocontrol model (Fig. 3). Cyclopentyl, cyclohexyl, and *tert*-butyl iodides successfully added both to propionaldehyde *N*-acylhydrazone **10a** and to benzaldehyde *N*-acylhydrazone **10b** (Table 2). Ethyl radical generated from triethylborane can compete for the radical acceptor, and, as a result, the ethyl radical adduct was observed (<10% yield) in all cases.

In search of optimal stereocontrol, substituents on the oxazolidinone moiety were varied. Thus, isopropyl radical additions to several propionaldehyde *N*-acylhydrazones were compared for stereoselectivity (Scheme 4). High diastereoselectivities were observed in all adducts **12a–12e**, although a rigorous measurement was not obtained on **12c** ($R = i{-}\mathrm{Pr}$). All of the auxiliaries impart stereocontrol suitable for practical synthetic application [80].

2.3.2 Triethylborane-Mediated Radical Additions Without Tin

Triethylborane and diethylzinc may serve both as initiator and chain transfer agents in radical additions to C=N bonds [59, 65]. This raised the question of whether similar additions to chiral *N*-acylhydrazones may occur in the absence of tin hydride. Accordingly, we attempted tin-free additions of various halides to the propionaldehyde hydrazone **10a** in the presence of triethylborane, using $InCl_3$ as the Lewis acid [80]. These reactions were indeed successful with various secondary iodides (Table 3, Entries 2–4). Chloroiodomethane also gave successful addition, and the chloromethyl adduct bears functionality suitable for subsequent manipulations.

The use of photolysis also enabled a tin-free method devised by Fernández and Alonso for addition of the 1,3-dioxolan-2-yl radical to these chiral *N*-acylhydrazones [85]. Irradiation in the presence of 1 equiv. benzophenone led to H-atom abstraction from the solvent, 1,3-dioxolane, followed by intermolecular radical addition to chiral *N*-acylhydrazones (Scheme 5). The Lewis acid ($InCl_3$) facilitated excellent diastereoselectivity in addition of this formyl radical equivalent. The preferred diastereomer was that suggested by the Lewis acid chelate model, consistent with the model supported by our own observations (see Fig. 4). After N–N bond cleavage and oxidation at the formyl carbon, preparation of α-amino acids was achieved with high stereoselectivity. A one-pot protocol was also introduced for this reaction, preparing the *N*-acylhydrazone in situ for the radical addition; thus, for a series of aldehydes, the corresponding adducts **13a–13f** were obtained with yields ranging from 75 to 99%.

A limitation of the aforementioned methods is that they are unsuitable for the use of primary alkyl iodides. Under Et_3B/O_2 initiation conditions, the desired radical is generated by I-atom transfer from the alkyl iodide to ethyl radicals; this is not favorable in the case of primary iodides. Thus ethyl radical (from Et_3B) competes with the desired addition of a primary radical in these circumstances. Also, generating radicals by H-atom transfer from ethers or acetals has limited applicability because the radical precursor is generally also a solvent. Because of the expanded synthetic potential of primary alkyl iodides as radical precursors, there

Table 3 Tin-free radical additions to **10a** in the presence of $InCl_3$

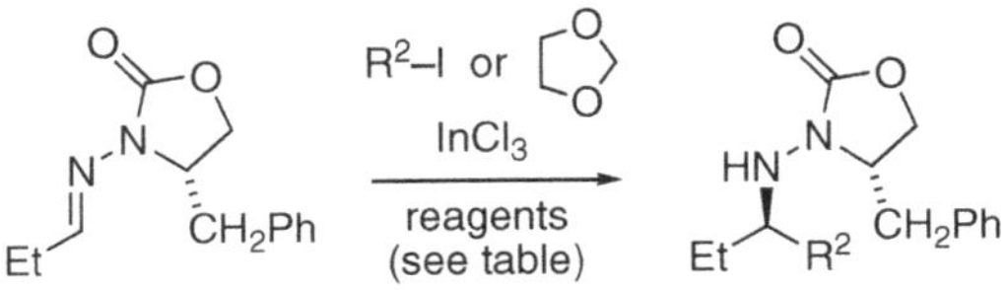

Entry	R^2	Method	Yield[a] (%)
1	Et	A	33
2	*i*-Pr	A	75[b]
3	*c*-C_5H_9	A	47
4	*c*-C_6H_{11}	A	56
5	$ClCH_2$	A	33
6	1,3-Dioxolan-2-yl	B	87[c]
7	1,3-Dioxolan-2-yl	C	96[d]

Reaction conditions: A: As in Table 2, minus Bu_3SnH. B: 1,3-dioxolane, Ph_2CO, $InCl_3$, hν, –78°C. C: Same as B, except one-pot; *N*-acylhydrazone prepared and used in situ
[a]Isolated yield
[b]*dr* >95:5 (^{1}H NMR)
[c]*dr* 91:9 (Alonso [85])
[d]*dr* 98:2 (Alonso [85])

Scheme 5 A one-pot condensation–radical addition of *N*-acylhydrazones

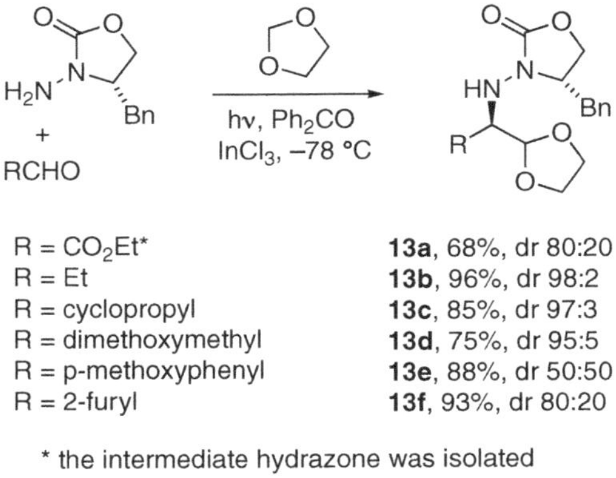

is great import in finding alternatives to accomplish I-atom transfers of broader utility for radical additions.

2.4 *Manganese-Mediated Radical Addition: Discovery and Method Development*

Our attention was drawn to photolytic radical generation with hexamethylditin because it showed promise in Kim's prior work with C = N radical acceptors, which included additions of primary radicals [86–92]. We noted, however, that manganese carbonyl [93–95] [$Mn_2(CO)_{10}$] (λ_{max} 340 nm, $\sigma_{Mn-Mn} \rightarrow \sigma^*_{Mn-Mn}$)

Table 4 Mn-mediated radical additions to *N*-acylhydrazones

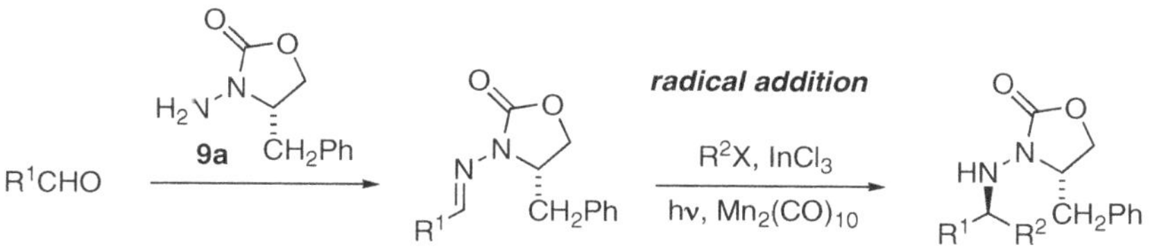

Entry	Aldehyde (or acetal)	Yield, hydrazone[a] (%)	Halide R^2X	Yield, radical addition[b] (%)	*dr*
1	CH_3CH_2CHO	81	CH_3CH_2I	85	–
2			CH_3I	48[c,d] (*S*)	95:5[e]
3			*n*-PrI	66 (*R*)	94:6[e]
4			*n*-BuI	78 (*R*)	95:5[e]
5			*n*-$C_5H_{11}I$	79 (*R*)	96:4[e]
6			*i*-BuI	54[c] (*R*)	95:5[f]
7			*i*-PrI	75 (*R*)	95:5[f]
8			$ClCH_2I$	63 (*R*)	93:7[e]
9			$Cl(CH_2)_3I$	52 (*R*)	96:4[f]
10			$Cl(CH_2)_4I$	55 (*R*)	96:4[e]
11			Cl_2CHBr	38[c,d] (*R*)	98:2[f]
12	CH_3CHO	66	CH_3CH_2I	66 (*R*)	95:5[e]
13	*n*-PrCHO	87		63 (*S*)	95:5[e]
14	*n*-BuCHO	89		72 (*S*)	97:3[e]
15	*n*-$C_5H_{11}CHO$	88		77 (*S*)	97:3[e]
16	*i*-BuCHO	85		65 (*S*)	95:5[f]
17	$ClCH_2CH(OMe)_2$	85		57 (*S*)	93:7[e]
18	$Cl(CH_2)_3CHO$	95		60 (*S*)	93:7[f]
19	$Cl(CH_2)_4CHO$	89		62 (*S*)	97:3[e]
20	$Cl_2CHCH(OEt)_2$	54		34[c] (*S*)	89:11[f]

Reaction conditions: (1) Aldehyde or acetal (5–10 equiv.), **9a**, *p*-toluenesulfonic acid, CH_2Cl_2, rt. (2) Hydrazone in deoxygenated CH_2Cl_2 (0.1 M), $InCl_3$ (2.2 equiv.), $Mn_2(CO)_{10}$ (1–2 equiv.), R^2X (10 equiv.), hν (300 nm, pyrex), 1–2 d, ca. 35°C

[a]Isolated yield

[b]Isolated yields of purified diastereomer mixtures. *R* or *S* denotes the configuration of the new stereogenic center. Addition of methyl iodide gives opposite configurations due to the lower priority of the methyl ligand

[c]20 equiv. of R^2X was used

[d]1,8-Diazabicyclo[5.4.0]undec-7-ene (DBU) was used in removal of Mn byproducts

[e]Ratio by HPLC (Chiralcel OD, 2-PrOH/hexane)

[f]Ratio by ^1H NMR

may be photolyzed without any sensitizer, leading to homolytic metal–metal bond cleavage. This produces two •$Mn(CO)_5$ radicals, which are well-known to abstract halogen atoms from alkyl halides [96]. Despite its longtime familiarity in organometallic chemistry, the first studies of the synthetic scope of this reactivity mode appeared in a series of papers by Parsons (For seminal examples see [97–100]).

Armed with these precedents, we applied these manganese-mediated photolytic conditions to the addition of ethyl iodide to *N*-acylhydrazone **10a** (Table 4) [101, 102]. Irradiation (300 nm) with $Mn_2(CO)_{10}$ using $InCl_3$ as a Lewis acid furnished the ethyl adduct in 85% yield (entry 1), a dramatic improvement over use of triethylborane or hexamethylditin. Control experiments revealed a requirement for both irradiation and $Mn_2(CO)_{10}$. Without $InCl_3$, the reaction proceeded sluggishly (21% yield, 2 days).

A variety of other halides, including methyl iodide and dihaloalkanes, were also effective (Table 4, entries 2–11). An exception was a low-yielding addition of 2-chloroethyl iodide, which presumably was compromised by fragmentation of the 2-chloroethyl radical. Similar versatility was displayed with respect to the hydrazone component; ethyl radical addition to various *N*-acylhydrazones occurred in good yields (entries 12–20). These adducts are epimeric to those derived from hydrazone **10a** with respect to the new stereogenic center, as a result of simply changing the roles of the aldehyde and iodide precursors. From the strategy standpoint, this combination of stereocontrol flexibility and functional group compatibility is advantageous for total synthesis applications.

2.5 *Hybrid Radical–Ionic Annulation*

2.5.1 Pyrrolidine Synthesis

Using radical addition reactions in the presence of electrophilic spectator functionalities was attractive, as such demonstrations of compatibility would illustrate the complementarity of radical conditions and strongly nucleophilic conditions. In this regard, it should be noted that dihalopropanes had been used as radical precursors (Table 4, entries 8–11), preserving halide functionality in most cases. However, in one case, addition of 3-chloro-1-iodopropane (1), characterization of the adduct indicated there was no chlorine present; this reaction had afforded pyrrolidine **14** [101, 102]. This outcome may be explained by sequential radical addition and S_N2-type cyclization in situ. Support for such a process also was found upon addition of ethyl iodide to the 3-chlorobutyraldehyde hydrazone (Table 4, entry 18), which gave the epimeric pyrrolidine (*epi*-**14**). These radical–polar crossover reactions (For selected recent examples of radical-polar crossover reactions see [103–112]), which may also be termed hybrid radical–ionic annulations, offer a novel way to achieve 1,2-bis-functionalization of the C=N bond.

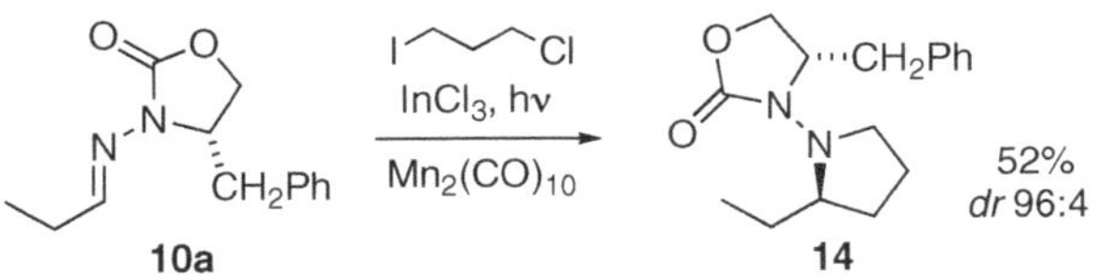

10a **14** (1)

Control of the steps, interrupting the pyrrolidine annulation after the radical addition step, has been observed upon slight modifications to the conditions using

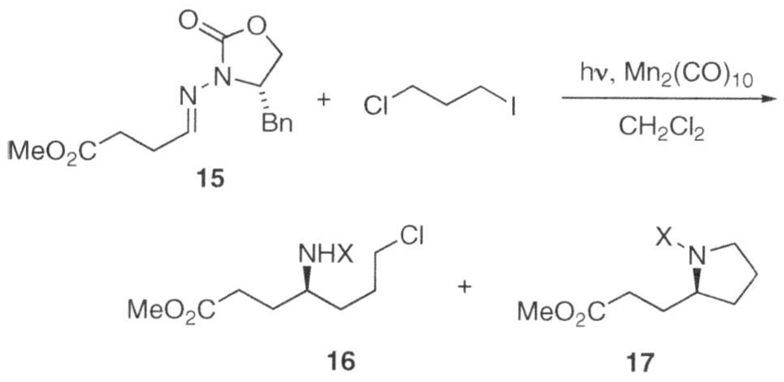

	16	17
15 h, with $InCl_3$:	45%	not observed
65 h, with $InCl_3$:	15%	48%
65 h, without $InCl_3$:	not observed	56%

Scheme 6 Control of stepwise annulation

hydrazone **15** (Scheme 6) [113]. After a 15-h reaction time, the initial radical adduct **16** was isolated in 45% yield, and, on extending the reaction time to 65 h, the radical adduct had mostly cyclized to the pyrrolidine **17**. When the reaction was carried out in the absence of Lewis acid, the only observed product was pyrrolidine **17**. Further studies of the scope of these pyrrolidine constructions are ongoing.

2.5.2 Stepwise Annulation in Piperidine Synthesis

The simple piperidine alkaloid coniine (For selected asymmetric syntheses of coniine see [13, 114–118]) (Scheme 7) offered a preliminary test case for hybrid radical–ionic annulation in alkaloid synthesis. We envisioned a radical addition followed by cyclization, and two alternative bond constructions can be considered. The exocyclic propyl group may be introduced as part of the radical acceptor (path A), or may originate in the radical precursor (path B).

In contrast to the pyrrolidine constructions, the addition of ethyl radical to a difunctional 5-chloropentanal hydrazone (Table 4, entry 19) did not result in cyclization in situ. Therefore, the chloride substituent was replaced with a tosylate to facilitate the polar cyclization. The Mn-mediated addition of 1-iodopropane to 5-tosyloxypentanal *N*-acylhydrazone **18** (Scheme 8) indeed provided the expected annulation product **19** in 59% yield. Unfortunately, unusually poor stereocontrol was observed (dr 3:1) [102]. The lack of high selectivity in the addition to hydrazone **18** is very unusual among all the examples of Mn-mediated radical addition observed to date. This, together with anomalously poor mass balance (no reactant hydrazone was recovered), provides evidence for a polar cyclization to form iminium ion **20** (Scheme 8) prior to radical addition; such a cyclization would be detrimental to stereoselectivity due to the loss of two-point binding of the Lewis acid.

Because of the problems with the premature ionic cyclization noted above, a stepwise approach was adopted for piperidine construction. From butyraldehyde hydrazone **21** (Scheme 9) and 4-chloro-iodobutane (**22**), Mn-mediated photolysis afforded the acyclic adduct **23** in 66% yield (dr 95:5); the cyclization did not

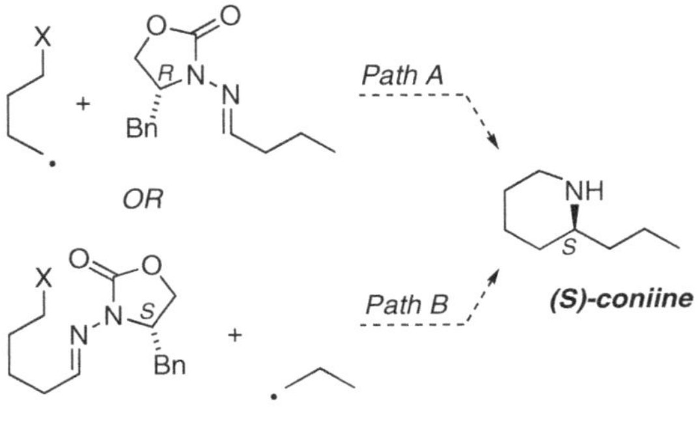

Scheme 7 Retrosynthetic disconnection of coniine

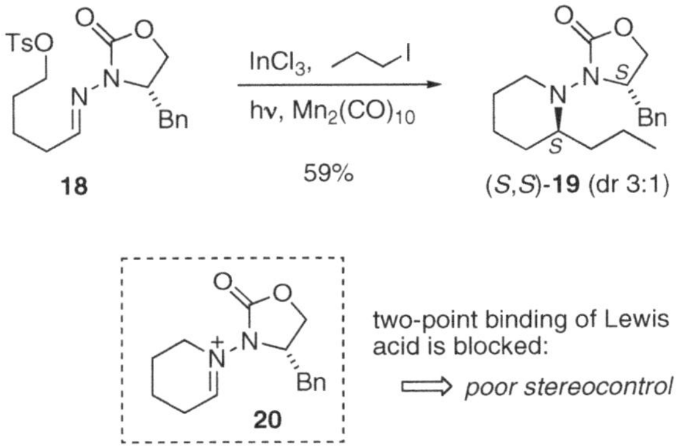

Scheme 8 Radical addition to 5-tosyloxypentanal hydrazone **18**

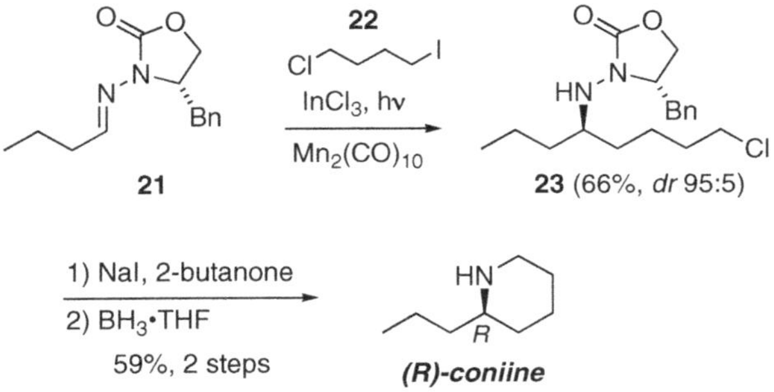

Scheme 9 Synthesis of (*R*)-coniine

occur in situ [101, 102]. Nevertheless, Finkelstein conditions afforded the piperidine, and reductive removal of the auxiliary afforded coniine in 34% overall yield for four steps. This reaction sequence offers a favorable comparison between radical- and carbanion-based syntheses using the same retrosynthetic disconnection [114, 118].

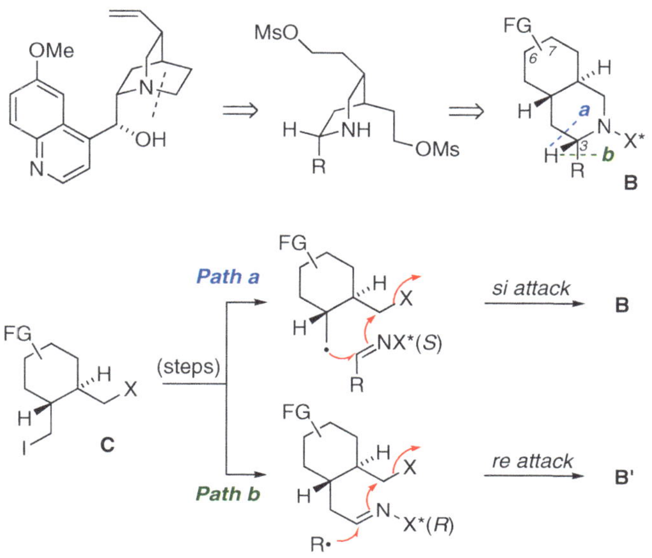

Scheme 10 Retrosynthetic analysis of quinine

2.5.3 Application to Formal Synthesis of Quinine

Our approach to the antimalarial alkaloid quinine focuses on strategic application of the Mn-mediated hybrid radical–ionic annulation [119, 120]. Retrosynthetic disconnection of the azabicyclic ring system to a piperidine **A** (Scheme 10) suggests isoquinolizidine **B** as a potential precursor, where functionality of C6 and C7 would eventually be exploited for ring cleavage. The radical–ionic annulation is then applied via disconnection of either of two C–C bonds in structure **B**, involving either ***Path a*** or ***Path b***. With enantiomeric chiral auxiliaries, these paths converge to structures **B** and **B′**, which differ only in the configuration of the removable chiral auxiliary. A pseudo-C2-symmetric precursor **C** would facilitate efficient access to the coupling components needed to test either of these strategies.

The ultimately successful radical–ionic annulation was carried out in the presence of protected 1,2-diol functionality at C6 and C7, specifically the coupling of **24** with **25** (Scheme 11). Stoichiometry was a concern, because iodide **25** had been prepared through several synthetic steps, making it impractical to rely on large excesses of radical precursor, as is commonly required for many intermolecular radical additions. Fortunately, the Mn-mediated coupling of **24** and **25** with only 1.25-fold excess of **25** proceeded in 93% yield in 1 mmol scale, giving **26** as a single diastereomer. Although completion of the hybrid radical–ionic annulation in situ during the Mn-mediated coupling has not yet been achieved, a stepwise process provided decahydroisoquinoline **27** in a quite satisfactory overall yield (85% for three steps) [119]. The low stoichiometric requirement in the coupling of the multifunctional iodide **25** to an

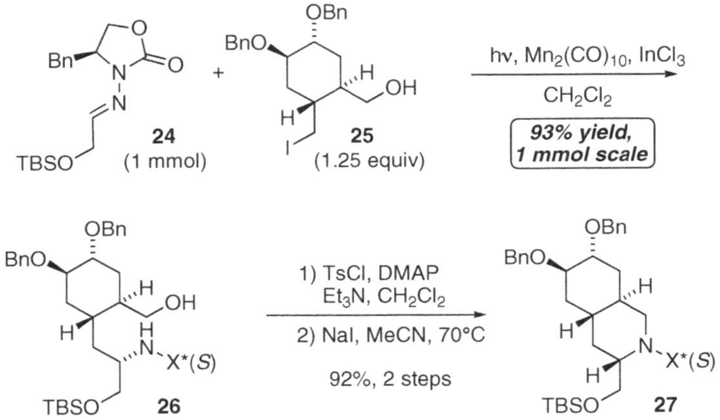

Scheme 11 Mn-mediated radical addition en route to quinine

imino compound is attractive and should enable broader applications of this Mn-mediated coupling process in complex target synthesis.

To complete a formal synthesis of quinine, quinolizidine **28** was converted in three steps to the piperidine **29** (Scheme 12). Unfortunately the cyclization to form an azabicyclic ring system was not regioselective; both hydroxyethyl groups cyclized, and the preferred product contained the azabicyclo[3.2.1]octane rather than the desired azabicyclo[2.2.2]octane. This necessitated a six-step sequence to differentiate the hydroxyethyl groups of **13**. This eventually furnished quincorine, which has previously been converted in two steps to quinine [120].

2.6 *Applications in Amino Acid Synthesis*

2.6.1 Synthesis of γ-Amino Acids

Although α- and β-amino acids have drawn more attention in synthetic chemistry, γ-amino acids such as **30** and **31** (Fig. 5) are also important targets from the perspective of bioorganic and medicinal chemistry (Reviews: see [121, 122; Examples: see [123–136]). Disconnections of C–C bonds as shown calls for iodides and hydrazones bearing oxygen-containing functional groups, an important challenge to the synthetic versatility of the Mn-mediated coupling reactions. With this in mind, we employed Mn-mediated radical addition for a novel synthesis of γ-amino acids **30** and **31** [137].

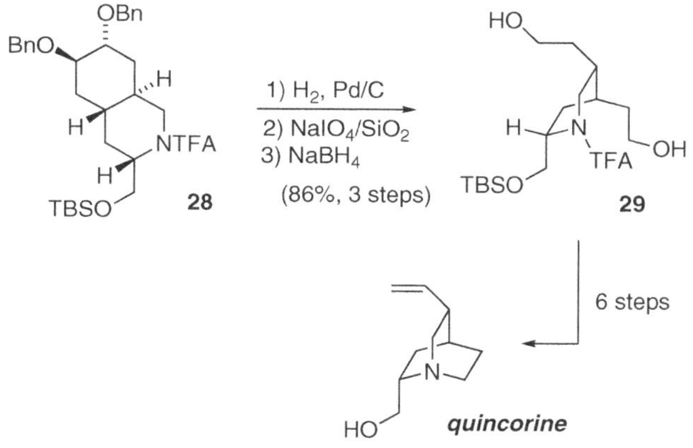

Scheme 12 Conversion of quinolizidine **28** to quincorine

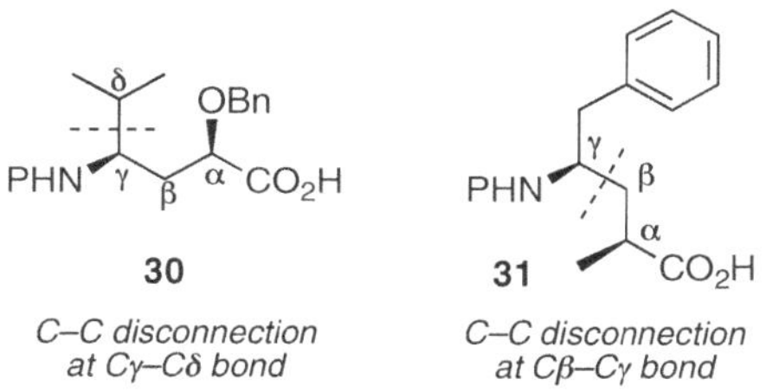

Fig. 5 Representative γ-amino acids with strategic bond disconnections at the γ–δ and β–γ carbons

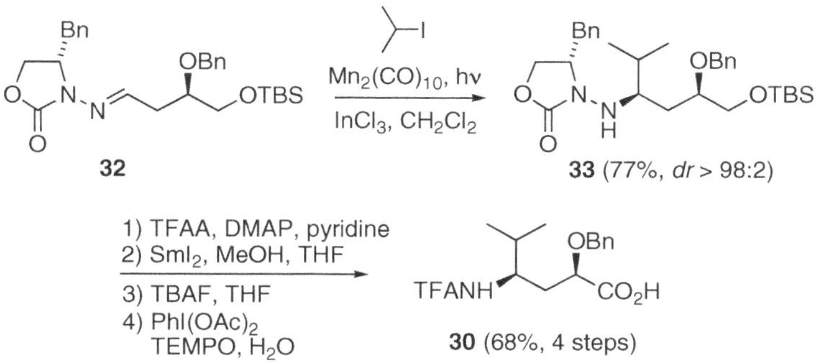

Scheme 13 Addition to a β-alkoxyhydrazone without β-elimination

Because disconnection of α-alkoxy-γ-amino acid **30** calls for a base-sensitive β-alkoxyhydrazone **32** (Scheme 13), there is a potential for β-elimination of the alkoxy group from the hydrazone precursor **32** which makes non-basic conditions critical. In fact, treatment of **32** with TBAF in THF led to just such a β-elimination (Marié J-C University of Iowa, unpublished results). However, the Mn-mediated

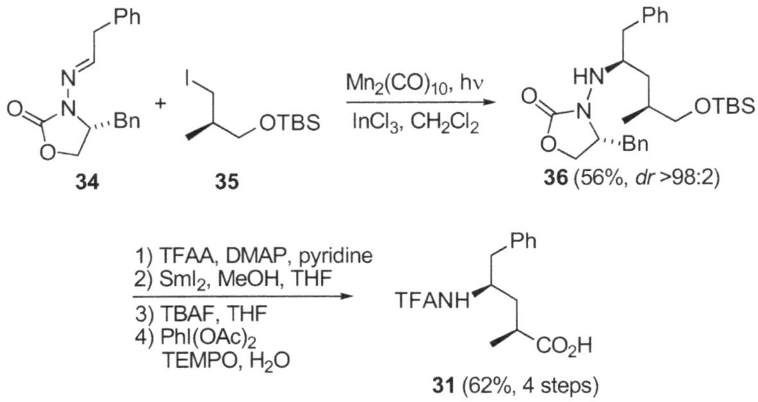

Scheme 14 Addition of a 3-silyloxyalkyl iodide to an *N*-acylhydrazone

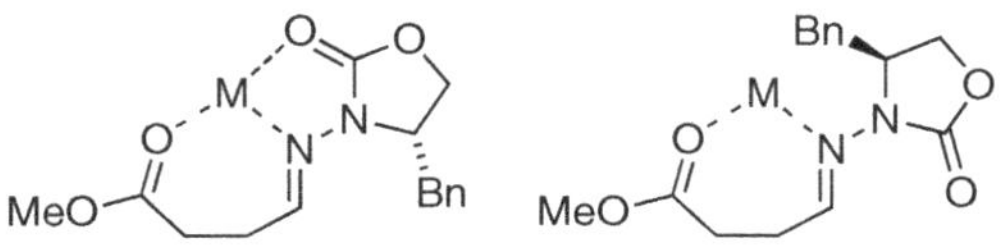

Fig. 6 Some hypothetical multipoint binding of a Lewis acid by *N*-acylhydrazones bearing additional ester functionality

radical addition of isopropyl iodide proceeded in 77% yield, without any evidence of β-elimination, to afford **33** as a single diastereomer. Reductive removal of the chiral auxiliary and oxidation to the carboxylic acid gave **30** in good overall yield [137].

Phenylacetaldehyde *N*-acylhydrazone **34** served as the radical acceptor for assembly of γ-amino acid **31** (Scheme 14), employing difunctional iodide **35** in the Mn-mediated radical addition (56% yield, single diastereomer) [137]. As with **33** (shown above), this radical adduct **36** was converted through the same four-step sequence to γ-amino acid **31**.

Although the aforementioned routes provided the desired γ-amino acids, it was desirable to develop a synthesis which incorporates the carboxylic acid oxidation state *prior to coupling*. We hypothesized that Mn-mediated radical addition would accomplish this objective, and therefore initiated a study of Mn-mediated coupling of alkyl iodides with γ-hydrazonoesters (Fig. 6) [113]. We had already shown that the Mn-mediated radical addition conditions offer excellent chemoselectivity, but it remained to be seen whether the stereocontrol model would be disrupted; would an additional Lewis basic ester function in the hydrazone interfere with the role of In (III) in two-point binding and rotamer control?

Table 5 Additions of alkyl iodides to γ-hydrazonoesters

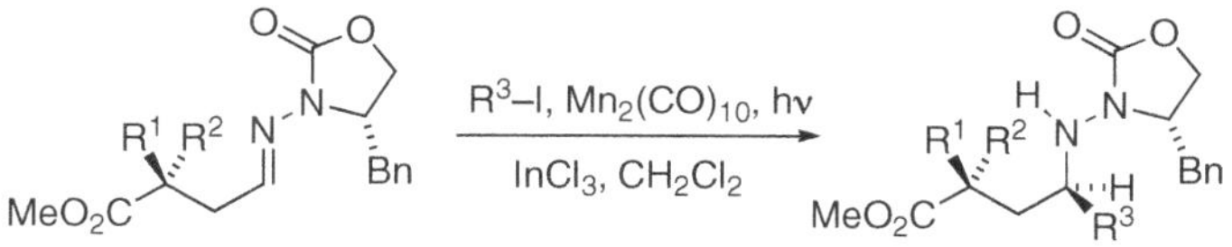

37a: R^1=R^2=H
37b: R^1=H, R^2=Me
37c: R^1=R^2=Me
37d: R^1=H, R^2=OBn

38–44

Entry	Hydrazone	Iodide R^{3}I	Product, yield (%)	*dr*
1	**37a**	*i*-PrI	**38a**, >99	94:6
2	**37b**		**38b**, 98	95:5
3	**37c**		**38c**, 96	99:1
4	**37d**		**38d**, 96	90:10
5	**37a**	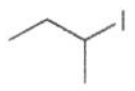	**39**, 82	–
6		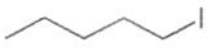	**40**, 66	94:6
7			**41**, 37	96:4
8			**42**, 61	97:3
9		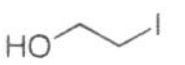	**43**, 33	96:4
10			**44**, 45	85:15

Prototypical radical additions were examined under Mn-mediated photolysis conditions with $InCl_3$ as the Lewis acid, coupling isopropyl iodide with a variety of γ-hydrazonoesters **37a–37d** (Table 5) bearing varied substitution at the position α to the ester. The α-methyl, α,α-dimethyl, and α-benzyloxy substituents appeared to have little effect on reaction efficiency and selectivity, as all provided the isopropyl adducts with consistently high diastereoselectivities and excellent yields (91–98%). Surprisingly, the selectivity was only slightly diminished in the absence of $InCl_3$ (entries 5 and 6); the yield in the absence of Lewis acid activation was modest but synthetically useful.

For the γ-hydrazonoesters, NMR experiments substantiated the typical two-point chelation model. A mixture of **37a** with $InCl_3$ in CD_2Cl_2 exhibited the H–C = N absorbance of the γ-hydrazonoester at 7.74 ppm, 0.30 ppm upfield from **37a** alone, consistent with precedent regarding Lewis acid coordination to imino compounds [138]. Also, the carbons of the oxazolidinone C = O and the hydrazone C = N were shifted downfield by 8 and 5 ppm on mixing with $InCl_3$. In contrast, the C = O carbon of the ester showed minimal change (<1 ppm). This suggests that the $InCl_3$ is chelated by the imino nitrogen and the oxazolidinone carbonyl in the usual way, without significant interference by the ester function.

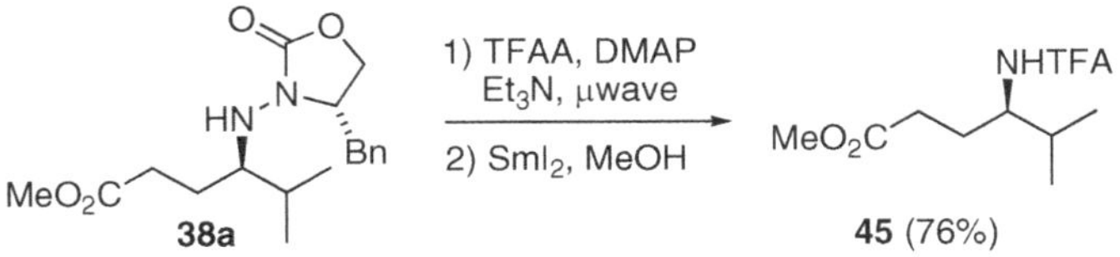

Scheme 15 Conversion of radical adduct to *N*-trifluoroacetamide

A range of iodides were next examined in reactions with and without $InCl_3$, starting with a comparison of secondary and primary iodides (Table 5). When secondary iodides were subjected to coupling with γ-hydrazonoester **37a** the yields were excellent (entries 1 and 5), while primary iodides gave the desired adducts in moderate yields (33–66%, entries 6–10). All of these reactions occurred with good-to-excellent diastereoselectivities, and it was worth noting that both silyl ether and alcohol functionality were compatible with the coupling.

To document the synthetic applicability of these reactions, N–N bond cleavage was needed. After trifluoroacetylation of **38a** (Scheme 15) under microwave irradiation [139–141], exposure to SmI_2 smoothly furnished known γ-aminoester **45** and offered proof of absolute configuration [142].

2.6.2 Synthesis of α,α-Disubstituted α-Amino Acids

Although radical additions to aldimine-type acceptors have now become well-established; intermolecular additions to ketimine radical acceptors are rare by comparison [143, 144]. We envisioned that the versatility of the Mn-mediated radical additions might offer potential access to a diverse range of *tert*-alkyl amines which are difficult to prepare by other means (Review: see [145]). In planning a study of such reactions we were aware of the importance of ensuring that the reaction takes place exclusively through a single C=N π-bond geometry. Aldehyde *N*-acylhydrazone derivatives are exclusively obtained in *E* geometry, making this issue of little relevance, but ketone hydrazones are generally formed as mixtures of *E* and *Z* isomers. Therefore we sought a ketone hydrazone which could be obtained predominantly as one isomer.

The *N*-amino-2-oxazolidinone **9a** was condensed with methyl pyruvate (**46**) to give hydrazone **47** (Scheme 16) as an *E*/*Z* mixture (dr 92:8), from which the minor (*Z*)-isomer was removed via flash chromatography to give pure (*E*)-**47** in 75% yield. Addition of ethyl iodide under Mn-mediated photolysis conditions in the presence of $InCl_3$ gave a moderate yield of **48a** (66% yield, dr 70:30), while the corresponding isopropyl adduct **48b** was very effectively produced (85% yield, dr 92:8). The N–N bond was cleaved upon conversion of isopropyl adduct **48b** to the benzoyl derivative and treatment with SmI_2/MeOH (Scheme 10) to afford known benzamide (*S*)-(+)-**49** [146] and confirm the assigned configuration.

In the additions to **47** it was noted, through variations of the stoichiometric loading of Lewis acid, that amounts of $InCl_3$ less than 2 equiv. resulted in lower

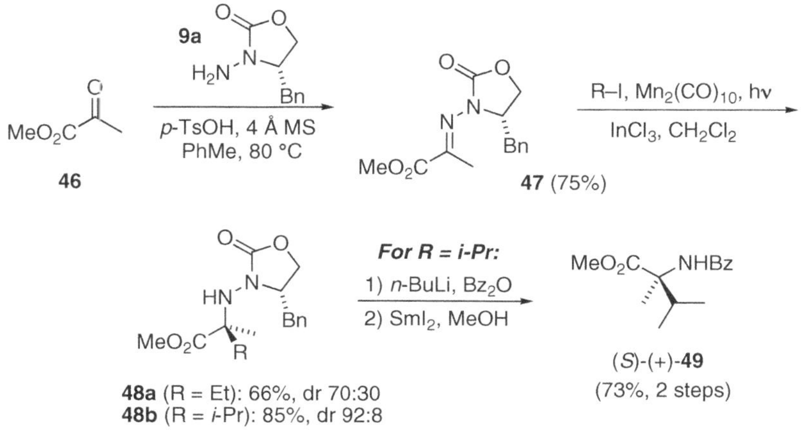

Scheme 16 Radical addition to ketimine **47**

diastereoselectivity. From this it may be inferred that the Lewis acid, aside from its usual chelation by the *N*-acylhydrazone, may interact with another Lewis basic site (e.g., the ester).

2.7 *Considerations for Synthesis Design Using Mn-Mediated Radical Addition*

2.7.1 Functional Group Compatibility

In the forgoing sections there are numerous examples illustrating the use of Mn-mediated radical additions to couple compounds containing more than one functional group. Although there are still combinations left to be explored, the examples published to date already illustrate that various useful functionalities may be tolerated within either of the precursors.

In the radical precursor, the alkyl iodide may be accompanied by alkyl chloride, alcohol, benzylic ether, or silyl ether functionalities. The alkyl chlorides have certain limitations on the location relative to the radical; 2-chloroethyl radical may eliminate chloride prior to radical addition, and the adduct from 3-chloropropyl radical may cyclize after radical addition.

In the *N*-acylhydrazone radical acceptor, the functionalities tolerated include alkyl chloride, benzylic ether, silyl ether, and ester. An alkoxy leaving group may be accommodated at the β-carbon of an *N*-acylhydrazone without β-elimination, which complements the functional group tolerance of basic organometallic reagents. Depending on the chain length, a subsequent cyclization may occur in radical adducts containing alkyl chloride.

What types of functionalities are not compatible? Applications to total synthesis are rigorous proving grounds, often revealing that potentially useful methodologies are in fact limited to simple monofunctional precursors. Despite the tolerances for functional groups noted above, application of Mn-mediated radical additions to total synthesis objectives uncovered some cases of incompatibility which deserve further mention.

In the early stages of developing a synthetic route to quinine, the $Mn_2(CO)_{10}$-mediated coupling of iodide **50** and *N*-acylhydrazone **51** was attempted in 10:1 PhH/MeCN (2). Unfortunately no coupling product could be found, so each of the spectator functionalities had to be examined in control experiments in order to find the structural features which might be interfering with the reaction. Mn-mediated radical additions of iodides containing the silyl ether moiety had previously been successful in various contexts [101, 102, 113, 119, 120, 137], so interference by the silyl ether was ruled out. However, it was noted that a precipitate formed on mixing hydrazone **51** with $InCl_3$ in 10:1 PhH/MeCN, and the precipitate remained insoluble even at higher ratios of CH_3CN. This suggested closer examination of the *N*-acylhydrazone to determine whether complexation of the *N*-acylhydrazone with $InCl_3$, normally required to facilitate radical addition, had been disrupted by the basic methoxyquinoline. A second concern was that the compatibility of an electron-rich methoxyaryl group with the Mn-mediated radical additions had not previously been established.

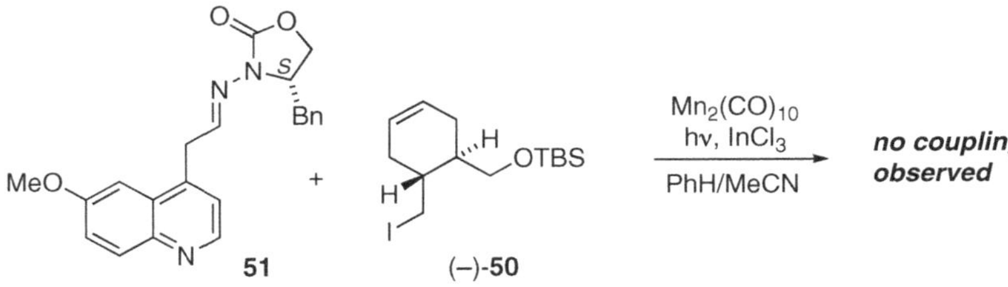

(2)

To address this question, a series of control experiments was carried out with aromatic hydrazones **52a**, **52b**, and **52c**, using Mn-mediated radical addition of iodide **53**. From hydrazones **52a** and **52b**, the expected products **54a** and **54b** were obtained in moderate yield (ca. 40%), confirming that the electron-rich methoxyaryl substituent was compatible with the coupling. Pyridine-containing hydrazone **52c**, however, gave none of the desired coupling product **54c**, indicating that heteroaromatic nitrogen may have interfered with the Mn-mediated coupling reaction (Scheme 17).

When attempting the coupling of iodide **50** (2) with simplified hydrazones en route to quinine, another issue of compatibility arose. Although OH and OTBS groups were already known to be well-tolerated, the couplings of **50** were inefficient. This raised suspicion about the potential side reactions of an alkene moiety in the radical intermediate. Despite changing the identity of the spectator functional groups, or the roles of the two precursors, so that the alkene was in the radical acceptor (e.g., **55a–55d**), the reaction remained inefficient. Yields of the desired product remained

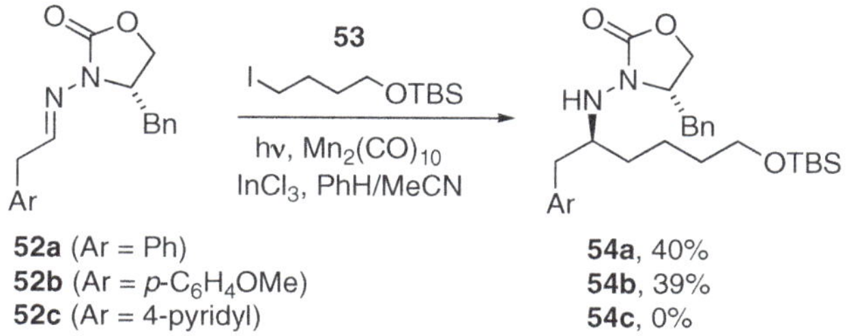

Scheme 17 Control experiments with electron-rich aromatics

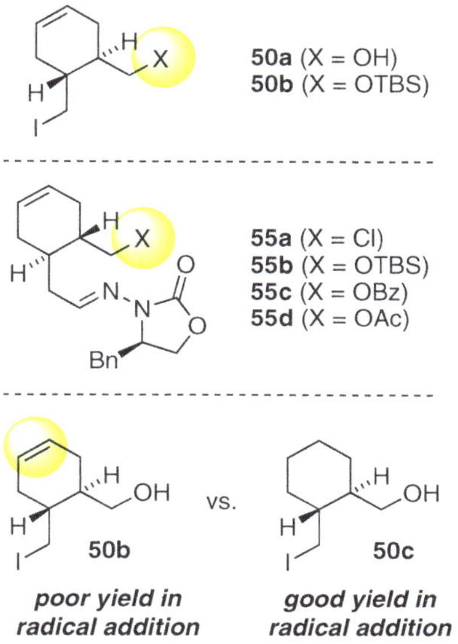

Scheme 18 Functionality variations in Mn-mediated coupling attempts for quinine synthesis

generally less than 30% despite extensive efforts toward improvement. Finally, a control experiment with saturated iodide **50c** revealed that the alkene indeed was detrimental to the success of this coupling. It is unclear whether this incompatibility is a general one or a peculiarity of the examples shown in Scheme 18.

In summary, there are some useful functional groups bearing either acidic protons (alcohol) or electrophilic centers (alkyl chloride, ester) which are tolerated as spectators in the Mn-mediated radical addition reaction. These compatibilities complement those of other chiral amine synthesis methods involving strongly nucleophilic (and basic) organometallic reagents, thereby offering some useful options for synthetic design. Ethers and silyl ethers are also compatible. Selected examples involving pyridines and alkenes as spectator groups revealed some complications, the scope of which remains unclear, so for synthetic design purposes these types of moieties may need to be introduced later in the route.

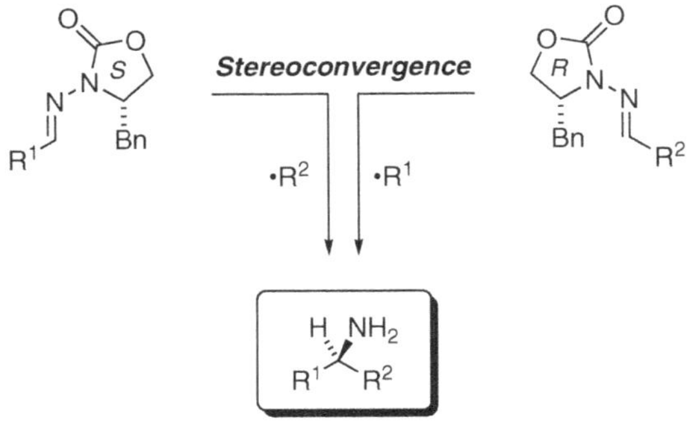

Scheme 19 Stereoconvergent routes to chiral amines

2.7.2 Stereoconvergence for Flexibility in Synthetic Application

Considering the examples discussed above, it is clear that the Mn-mediated radical additions offer useful functional group compatibilities in both the radical precursor and *N*-acylhydrazone acceptor. In addition, the epimeric configuration can be selected by either (1) employing the enantiomeric auxiliary, or (2) interchanging the roles of R^1 and R^2 in the alkyl halide and aldehyde precursors of Table 4 (For other applications of this strategy see [147, 148]). Thus, stereoconvergent construction of alternative C–C bonds at the chiral amine stereocenter (Scheme 19) can be readily conceived, so that the roles of these precursors can be chosen on the basis of synthetic strategy rather than rather than on the basis of functional group limitations of the methodology. Such strategic flexibility contributes to the synthetic potential of these radical addition reactions.

3 Asymmetric Catalysis of Radical Addition

Although the forgoing sections have illustrated the viability of stereocontrolled radical addition to C=N bonds as a route to chiral amines, imparting the stereocontrol through asymmetric catalysis remains a challenge. To date, most efforts toward this goal have required very high catalyst loading (usually stoichiometric) or have limited scope. In this section, a summary of these studies illustrates some of the highlights and limitations.

Naito reported the first asymmetric radical additions to C=N bonds with a non-covalent mode of stereocontrol. Building upon earlier work exploiting chiral glyoxylate imines as radical acceptors in the presence of Lewis acids, Naito employed Lewis acids bearing a chiral bisoxazoline ligand for addition to achiral glyoxylate oxime ether **56** (Scheme 20). With $MgBr_2$ and ligand **58**, enantioselectivity up to 52% *ee* was obtained at stoichiometric loading [54]. Jorgensen published an alternative approach, attempting catalysis with a combination of Cu(I) and tolBINAP (**59**) [149].

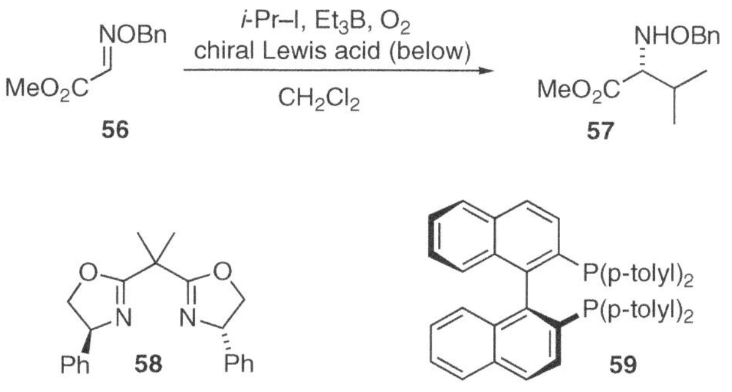

58•$MgBr_2$ (1 equiv)
97% yield, 52% ee
(Naito)

59•$CuPF_6$ (20 mol%)
5% yield, 33% ee
(Jorgensen)

Scheme 20 Asymmetric addition of isopropyl iodide to glyoxylate oxime ether

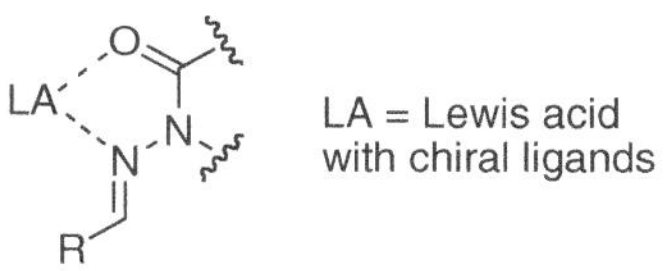

Fig. 7 Two-point binding of a Lewis acid by a generalized *N*-acylhydrazone structure

Table 6 Comparison of Lewis acids in isopropyl addition to **60a**[a]

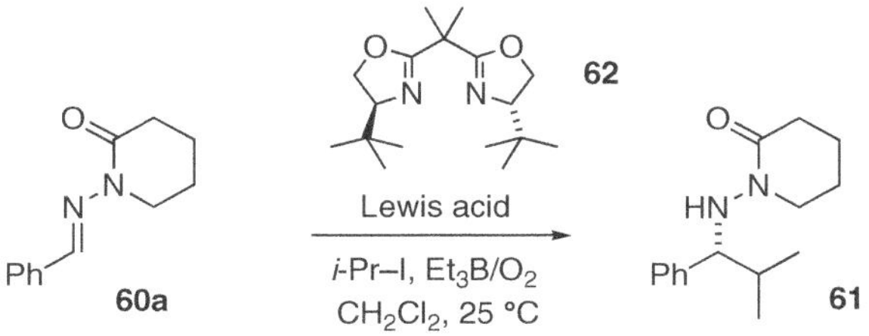

Entry	Lewis acid	Solvent, additive	Yield[b] (%)	*ee*[c]
1	$InCl_3$	CH_2Cl_2	33	57
2	$Mg(ClO_4)_2$	CH_2Cl_2	41	66
3	$Cu(OTf)_2$	CH_2Cl_2	41	59
4[d,e]	$Cu(OTf)_2$	PhH/CH_2Cl_2 (2:1), 4 Å MS	66	**95**
5[d,e,f]	$Cu(OTf)_2$	PhH/CH_2Cl_2 (2:1), 4 Å MS	**94**	86

[a]Reaction conditions: Lewis acid (1 equiv.), *tert*-butylbisoxazoline ligand (1 equiv.), 2-iodopropane (6 equiv.), Et_3B/O_2 (6 equiv.)
[b]Isolated yield, %
[c]Enantiomeric excess, determined by HPLC
[d]Et_3N was added during workup
[e]Preformed aquo complex $Cu(tBu\text{-}Box)(H_2O)_2(OTf)_2$ was used
[f]10 equiv. of *i*-PrI and Et_3B were used

Table 7 Varying the radical precursors and acceptors in Cu(II) catalyzed addition to *N*-acylhydrazones

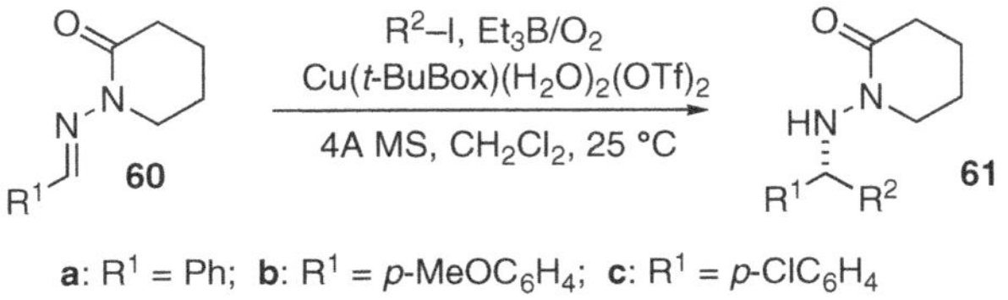

Entry[a]	Halide	Hydrazone	Catalyst load (equiv.)	% Yield[b]	% *ee*[c]
1	*i*PrI	**60a**	1	66	95
2	*i*-PrI	**60b**	1	46	90
3	*i*-PrI	**60c**	1	53	81
4	EtI[d]	**60a**	1	88	83
5	*c*-C_5H_9I[d]		1	86	84
6	*c*-$C_6H_{11}I$[d]		1	84	89
7	$ClCH_2I$[d]		1	44[e]	95
8	*i*PrI		0.5	71	81
9	*i*PrI		0.2	83	58
10	*i*PrI		0.1	74	46

[a]Reaction conditions: see Table 6
[b]Isolated yield, %
[c]Enantiomeric excess, % (hexane:2-propanol, Chiralcel OD or AD)
[d]10 equiv. of alkyl halide was used
[e]56% recovery of unreacted hydrazone

Unfortunately this gave very low enantioselectivity, with a yield lower than the catalyst loading.

We sought to demonstrate the first example of catalytic asymmetric induction in radical addition to C=N bonds. Our effort toward this goal was built upon the two-point binding motif of *N*-acylhydrazones (Fig. 7), which could potentially transmit stereocontrol from chiral ligands accompanying a Lewis acid.

The first successes exploited additions of isopropyl iodide, with radical initiation by triethylborane and oxygen. Using valerolactam-derived achiral *N*-acylhydrazone acceptor **60a** (Table 6), highly enantioselective isopropyl radical additions were promoted by 1 equiv. each of Lewis acid and bisoxazoline ligand **62**. With $InCl_3$, $Mg(ClO_4)_2$, and $Cu(OTf)_2$ as the Lewis acids only modest yields of **61** were obtained (entries 1–3), but selectivities in the range of 57–66% *ee* set a new standard for selectivity in radical additions to C=N bonds [150]. Using benzene/CH_2Cl_2 as the solvent (entry 4), the selectivity increased further to 95% *ee* (66% yield). The less polar solvent system presumably facilitated the assembly of a ternary complex from the ligand, Lewis acid, and substrate. The yield improved to 94% with larger amounts of 2-iodopropane and Et_3B (entry 5), but this came at the expense of some selectivity.

A series of radical precursors and acceptors were employed, showing some versatility accompanied by high enantioselectivity (Table 7) [150]. Using stoichiometric amounts of the preformed aquo complex Cu(*t*Bu-Box)$(H_2O)_2(OTf)_2$, isopropyl

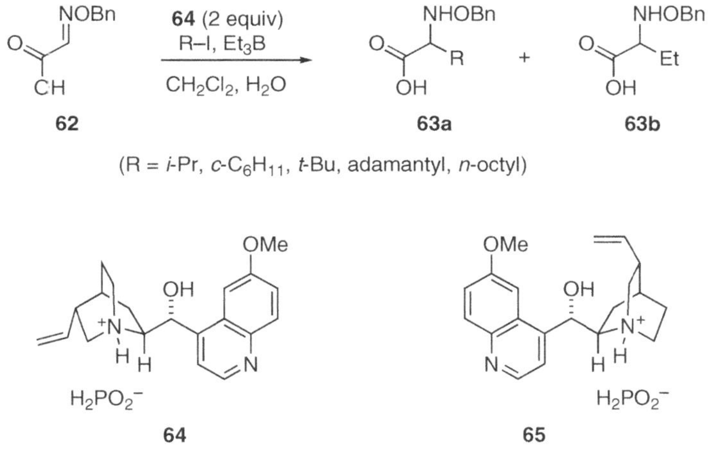

Scheme 21 Radical addition in the presence of cinchona alkaloid salts of hypophosphorous acid

additions to *N*-acylhydrazones prepared from benzaldehyde, *p*-chlorobenzaldehyde, and *p*-methoxybenzaldehyde were highly enantioselective (entries 1 and 2). Additions of various radicals, including chloromethyl, to **60a** were also successful (entries 3–6). Turnover of the chiral catalyst was examined by lowering the catalyst loading (Table 7, entries 8–10). The yield remained high, while enantioselection decreased. However, the yield and enantioselection (74% yield, 46% *ee*) at 10 mol% catalyst loading showed evidence of catalyst turnover—the first example of asymmetric catalysis in radical addition to C=N bonds.

The glyoxylic oxime ether **62** was subjected by Jang et al. to radical addition of several simple alkyl iodides in with triethylborane/oxygen initiation in the presence of cinchona alkaloid salts of hypophosphorous acid (Scheme 21) [151]. Although an excess of the cinchona alkaloid was required, transmission of stereochemical information was confirmed. A stereocontrol model was proposed with a combination of hydrogen bonding and π-stacking to activate the radical acceptor and constrain its structure.

More recently, the Jang group has reported that *N*-benzoylhydrazones are also suitable substrates for stereocontrolled radical addition in the presence of cinchona alkaloids [152]. With hydrazones prepared from a series of substituted benzaldehydes (Scheme 22), very high enantioselectivities (98–99% *ee*) were observed using the *O*-benzyl alkaloid derivative **68** (40 mol%). The high selectivity observed with aromatic hydrazones was not retained during addition to the *N*-benzoylhydrazone prepared from octanal (80% *ee*, not shown). Cyclohexyl, *tert*-butyl, and adamantyl radical additions were effective, but the use of primary radical (from *n*-octyl iodide) was detrimental to both efficiency and selectivity (55% yield, 81% *ee*).

Binaphthol-derived chiral Bronsted acids have been exploited to catalyze addition to *N*-aryl imines (Scheme 23) [153]. In these reactions, the use of 30 mol% of catalyst **71** was sufficient to promote addition of isopropyl and *tert*-butyl iodides to imine **69** with enantiomeric excesses in the 73–84% range, although the isolated

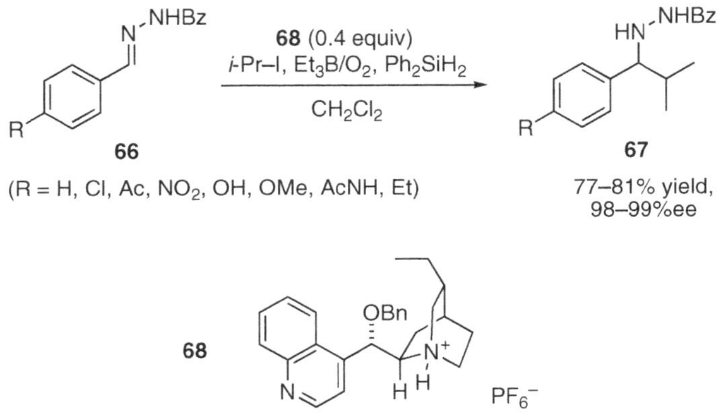

Scheme 22 Catalytic asymmetric radical addition to *N*-benzoylhydrazones in the presence of a cinchona alkaloid salt

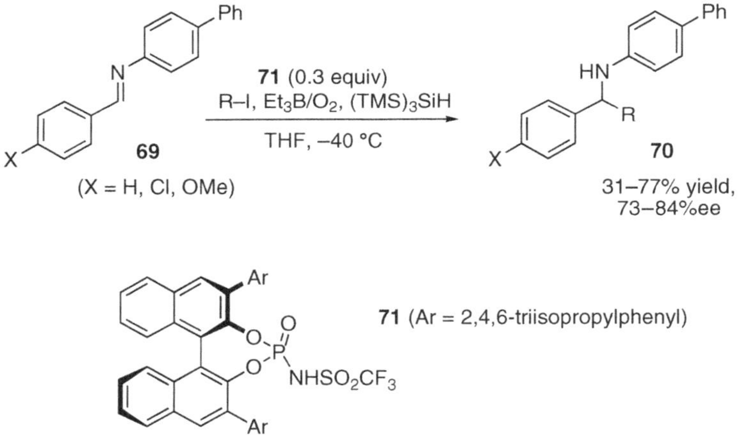

Scheme 23 Binaphthol-derived Bronsted acid catalyst for asymmetric radical addition to *N*-arylimines

yields were moderate. As in most reactions using the triethylborane–oxygen initiation system, the products were accompanied by significant amounts of the ethyl adduct (**76**, R = Et, from triethylborane).

4 Summary

Radical addition to imino compounds has emerged as a general approach with broad versatility that complements non-radical methodology. Our work has discovered and elaborated this concept into synthetically useful methodology. First, highly stereoselective intermolecular additions of alkyl iodides to chiral hydrazones in the presence of $Mn_2(CO)_{10}$ accommodate a broad range of functionality, including esters and unprotected hydroxyl groups which may not be compatible with carbanion chemistry. The viability of this strategy for stereocontrol has now been established in applications to target-directed synthesis. Second, we have developed a means of catalytic asymmetric induction; excellent enantioselectivity is obtained at stoichiometric loading, and moderate enantioselectivity is observed in conditions which demonstrate catalyst turnover. Although these are promising results, a method for asymmetric catalysis in radical addition to C=N bonds which offers high turnover numbers and broad scope is a challenge still to be met.

Acknowledgment The generous support by NSF (CHE-0096803 and CHE-0749850) and NIH (R01-GM67187) of our radical addition chemistry method development and applications to natural product synthesis is acknowledged with gratitude. Portions of this chapter are excerpted from a previous Topics in Current Chemistry volume: Gregory K. Friestad "Asymmetric Radical Addition to Chiral Hydrazones." In *Topics in Current Chemistry: Radicals in Synthesis III*. Gansauer A, Heinrich M (eds). Springer-Verlag: Berlin, 2012, vol. 320, pp. 61–92.

References

1. Friestad GK (2009) Addition of carbanions to azomethines. In: Enders D, Shaumann E (eds) Science of synthesis Vol 40a: Compounds with one saturated carbon-heteroatom bond: amines and ammonium salts. Thieme, Stuttgart
2. Yamada K-I, Tomioka K (2008) Chem Rev 108:2874–2886
3. Friestad GK, Mathies AK (2007) Tetrahedron 63:2541–2569
4. Ding H, Friestad GK (2005) Synthesis 2815–2829
5. Alvaro G, Savoia D (2002) Synlett 651–673
6. Kobayashi S, Ishitani H (1999) Chem Rev 99:1069–1094
7. Bloch R (1998) Chem Rev 98:1407–1438
8. Davis FA, Zhou P, Chen B-C (1998) Chem Soc Rev 27:13–18
9. Enders D, Reinhold U (1997) Tetrahedron Asymmetr 8:1895–1946
10. Denmark SE, Nicaise OJ-C (1996) J Chem Soc Chem Commun 999–1004
11. Stork G, Dowd SR (1963) J Am Chem Soc 85:2178–2180
12. Wittig G, Frommeld HD, Suchanek P (1963) Angew Chem Int Ed Engl 2:683
13. Guerrier L, Royer J, Grierson DS, Husson H-P (1983) J Am Chem Soc 105:7754–7755
14. Enders D, Diez E, Fernandez R, Martin-Zamora E, Munoz JM, Pappalardo RR, Lassaletta JM (1999) J Org Chem 64:6329–6336
15. Ischay MA, Yoon TP (2012) Eur J Org Chem 3359–3372
16. Rowlands GJ (2009) Tetrahedron 65:8603–8655
17. Renaud P, Sibi M (eds) (2001) Radicals in organic synthesis. Wiley-VCH, New York
18. Curran DP, Porter NA, Giese B (1995) Stereochemistry of radical reactions: concepts guidelines and synthetic applications. Wiley-VCH, New York
19. Jasperse CP, Curran DP, Fevig TL (1991) Chem Rev 91:1237–1286
20. Giese B (1986) Radicals in organic synthesis: formation of carbon–carbon bonds. Pergamon, New York
21. Hart DJ (1984) Science 223:883–887
22. Friestad GK (2012) In: Gansauer A, Heinrich M (eds) Topics in current chemistry: radicals in synthesis III, vol 320. Springer, Berlin, pp 61–92

23. Friestad GK (2010) In: Nugent T (ed) Chiral amine synthesis. Methods, developments and applications. Wiley-VCH, Weinheim, pp 51–74
24. Miyabe H, Yoshioka E, Kohtani S (2010) Curr Org Chem 14:1254–1264
25. Friestad GK (2001) Tetrahedron 57:5461–5496
26. Friestad GK (2005) Eur J Org Chem 3157–3172
27. Fallis AG, Brinza IM (1997) Tetrahedron 53:17543–17594
28. Nishiyama H, Kitajima T, Matsumoto M, Itoh K (1984) J Org Chem 49:2298–2300
29. Stork G, Kahn M (1985) J Am Chem Soc 107:500–501
30. Cusak A (2012) Chem Eur J 18:5800–5824
31. Gauthier DR Jr, Zandi KS, Shea KJ (1998) Tetrahedron 54:2289–2338
32. Fensterbank L, Malacria M, Sieburth SM (1997) Synthesis 813–854
33. Fleming I, Barbero A, Walter D (1997) Chem Rev 97:2063–2192
34. Bols M, Skrydstrup T (1995) Chem Rev 95:1253–1277
35. Friestad GK (1999) Org Lett 1:1499–1501
36. Friestad GK, Massari SE (2000) Org Lett 2:4237–4240
37. Friestad GK, Jiang T, Fioroni GM (2003) Tetrahedron Asymmetr 14:2853–2856
38. Friestad GK, Massari SE (2004) J Org Chem 69:863–875
39. Friestad GK, Fioroni GM (2005) Org Lett 7:2393–2396
40. Friestad GK, Jiang T, Mathies AK (2007) Tetrahedron 63:3964–3972
41. Friestad GK, Mathies AK (2007) Tetrahedron 63:9373–9381 (Corrigendum: Friestad GK, Mathies AK (2007) Tetrahedron 63:13039)
42. Friestad GK, Jiang T, Mathies AK (2007) Org Lett 9:777–780
43. Friestad GK, Jiang T, Fioroni GM (2008) Tetrahedron 64:11549–11557
44. Friestad GK (2005) Eur J Org Chem 3157–3172
45. Hart DJ, Seely FL (1988) J Am Chem Soc 110:1631–1633
46. Hart DJ, Krishnamurthy R, Pook LM, Seely FL (1993) Tetrahedron Lett 34:7819–7822
47. Bhat B, Swayze EE, Wheeler P, Dimock S, Perbost M, Sanghvi YS (1996) J Org Chem 61:8186–8199
48. Hanamoto T, Inanaga J (1991) Tetrahedron Lett 32:3555–3556
49. Russell GA, Yao C-F, Rajaratnam R, Kim BH (1991) J Am Chem Soc 113:373–375
50. Russell GA, Wang L, Rajaratnam R (1996) J Org Chem 61:8988–8991
51. Miyabe H, Shibata R, Ushiro C, Naito T (1998) Tetrahedron Lett 39:631–634
52. Miyabe H, Shibata R, Sangawa M, Ushiro C, Naito T (1998) Tetrahedron 54:11431–11444
53. Miyabe H, Ushiro C, Naito T (1997) J Chem Soc Chem Commun 1789–1790
54. Miyabe H, Ushiro C, Ueda M, Yamakawa K, Naito T (2000) J Org Chem 65:176–185
55. McNabb SB, Ueda M, Naito T (2004) Org Lett 6:1911–1914
56. Bertrand MP, Feray L, Nouguier R, Stella L (1998) Synlett 780–782
57. Bertrand MP, Feray L, Nouguier R, Perfetti P (1999) Synlett 1148–1150
58. Bertrand MP, Coantic S, Feray L, Nouguier R, Perfetti P (2000) Tetrahedron 56:3951–3961
59. Bertrand M, Feray L, Gastaldi S (2002) Comptes Rend Acad Sci Paris Chimie 5:623–638
60. Zhou P, Chen B, Davis FA (2004) Tetrahedron 60:8003–8030
61. Ellman JA, Owens TD, Tang TP (2002) Acc Chem Res 35:984–995
62. Akindele T, Yamamoto Y, Maekawa M, Umeki H, Yamada K, Tomioka K (2006) Org Lett 8:5729–5732
63. Kindele T, Yamada K, Sejima T, Maekawa M, Yamamoto Y, Nakano M, Tomioka K (2010) Chem Pharm Bull 58:265–269
64. Yamada K, Konishi T, Nakano M, Fujii S, Cadou R, Yamamoto Y, Tomioka K (2012) J Org Chem 77:1547–1553
65. Miyabe H, Ueda M, Naito T (2004) Synlett 1140–1157
66. Booth SE, Jenkins PR, Swain CJ, Sweeney JB (1994) J Chem Soc Perkin Trans 1:3499–3508
67. Booth SE, Jenkins PR, Swain CJ (1998) J Braz Chem Soc 9:389–395 [Chemical Abstracts 130:38274]
68. Renaud P, Gerster M (1998) Angew Chem Int Ed 37:2563–2579

69. Guerin B, Ogilvie WW, Guindon Y (2001) In: Renaud P, Sibi M (eds) Radicals in organic synthesis. Wiley-VCH, New York
70. Burk MJ, Feaster JE (1992) J Am Chem Soc 114:6266–6267
71. Sturino CF, Fallis AG (1994) J Am Chem Soc 116:7447–7448
72. Evans DA, Kim AS (1995) In: Paquette LA (ed) Encyclopedia of reagents for organic synthesis, vol 1. Wiley, New York
73. Kim M, White JD (1977) J Am Chem Soc 99:1172–1180
74. Ciufolini MA, Shimizu T, Swaminathan S, Xi N (1997) Tetrahedron Lett 38:4947–4950
75. Evans DA, Johnson DS (1999) Org Lett 1:595–598
76. Hynes J Jr, Doubleday WW, Dyckman AJ, Godfrey JD Jr, Grosso JA, Kiau S, Leftheris K (2004) J Org Chem 69:1368–1371
77. Friestad GK, Ji A (2008) Org Lett 10:2311–2313
78. Shen Y, Friestad GK (2002) J Org Chem 67:6236–6239
79. Friestad GK, Qin J (2000) J Am Chem Soc 122:8329–8330
80. Friestad GK, Draghici C, Soukri M, Qin J (2005) J Org Chem 70:6330–6338
81. Qin J, Friestad GK (2003) Tetrahedron 59:6393–6402
82. Lim D, Coltart DM (2008) Angew Chem Int Ed 47:5207–5210
83. Nozaki K, Oshima K, Utimoto K (1991) Bull Chem Soc Jpn 64:403–409
84. Brown HC, Midland MM (1972) Angew Chem Int Ed Engl 11:692–700
85. Fernández M, Alonso R (2003) Org Lett 5:2461–2464
86. Kim S, Lee IY, Yoon J-Y, Oh DH (1996) J Am Chem Soc 118:5138–5139
87. Kim S, Yoon J-Y (1997) J Am Chem Soc 119:5982–5983
88. Ryu I, Kuriyama H, Minakata S, Komatsu M, Yoon J-Y, Kim S (1999) J Am Chem Soc 121:12190–12191
89. Jeon G-H, Yoon J-Y, Kim S, Kim SS (2000) Synlett 128–130
90. Kim S, Kim N, Yoon J-Y, Oh DH (2000) Synlett 1148–1150
91. Kim S, Song H-J, Choi T-L, Yoon J-Y (2001) Angew Chem Int Ed 40:2524–2526
92. Kim S, Kavali R (2002) Tetrahedron Lett 43:7189–7191
93. Pauson PL (1995) In: Paquette LA (ed) Encyclopedia of reagents for organic synthesis, vol 2. Wiley, New York
94. Meyer TJ, Caspar JV (1985) Chem Rev 85:187
95. Gilbert BC, Parsons AF (2002) J Chem Soc Perkin Trans 2:367–387
96. Herrick RS, Herrinton TR, Walker HW, Brown TL (1985) Organometallics 4:42–45
97. Giese B, Thoma G (1991) Helv Chim Acta 74:1135–1142
98. Gilbert BC, Kalz W, Lindsay CI, McGrail PT, Parsons AF, Whittaker DTE (1999) Tetrahedron Lett 40:6095–6098
99. Gilbert BC, Lindsay CI, McGrail PT, Parsons AF, Whittaker DTE (1999) Synth Commun 29:2711–2718
100. Gilbert BC, Kalz W, Lindsay CI, McGrail PT, Parsons AF, Whittaker DTE (2000) J Chem Soc Perkin Trans 1:1187–1194
101. Friestad GK, Qin J (2001) J Am Chem Soc 123:9922–9923
102. Friestad GK, Qin J, Suh Y, Marié J-C (2006) J Org Chem 71:7016–7027
103. Callaghan O, Lampard C, Kennedy AR, Murphy JA (1999) J Chem Soc Perkin Trans 1:995–1001
104. Jahn U, Muller M, Aussieker S (2000) J Am Chem Soc 122:5212–5213
105. Harrowven DC, Lucas MC, Howes PD (2001) Tetrahedron 57:791–804
106. Rivkin A, Nagashima T, Curran DP (2003) Org Lett 5:419–422
107. Denes F, Chemla F, Normant JF (2003) Angew Chem Int Ed 42:4043–4046
108. Bazin S, Feray L, Vanthuyne N, Bertrand MP (2005) Tetrahedron 61:4261–4274
109. Ueda M, Miyabe H, Sugino H, Miyata O, Naito T (2005) Angew Chem Int Ed 44:6190–6193
110. Denes F, Cutri S, Perez-Luna A, Chemla F (2006) Chem Eur J 12:6506–6513
111. Maruyama T, Mizuno Y, Shimizu I, Suga S (2007) J Am Chem Soc 129:1902–1903
112. Miyata O, Takahashi S, Tamura A, Ueda M, Naito T (2008) Tetrahedron 64:1270–1284

113. Friestad GK, Banerjee K (2009) Org Lett 11:1095–1098
114. Enders D, Tiebes J (1993) Liebigs Ann Chem 173–177
115. Yamazaki N, Kibayashi C (1997) Tetrahedron Lett 38:4623–4626
116. Reding MT, Buchwald SL (1998) J Org Chem 63:6344–6347
117. Wilkinson TJ, Stehle NW, Beak P (2000) Org Lett 2:155–158
118. Kim YH, Choi JY (1996) Tetrahedron Lett 37:5543–5546
119. Korapala CS, Qin J, Friestad GK (2007) Org Lett 9:4246–4249
120. Friestad GK, Ji A, Baltrusaitis J, Korapala CS, Qin J (2012) J Org Chem 77:3159–3180
121. Ordonez M, Cativiela C (2007) Tetrahedron Asymmetr 18:3–99
122. Trabocchi A, Guarna F, Guarna A (2005) Curr Org Chem 9:1127–1153
123. Matthew S, Schupp PJ, Leusch H (2008) J Nat Prod 71:1113–1116
124. Kunze B, Bohlendorf B, Reichenbach H, Hofle G (2008) J Antibiot 61:18–26
125. Oh D-C, Strangman WK, Kauffman CA, Jensen PR, Fenical W (2007) Org Lett 9:1525–1528
126. Milanowski DJ, Gustafson KR, Rashid MA, Pannell LK, McMahon JB, Boyd MR (2004) J Org Chem 69:3036–3042
127. Williams PG, Luesch H, Yoshida WY, Moore RE, Paul VJ (2003) J Nat Prod 66:595–598
128. Horgen FD, Kazmierski EB, Westenburg HE, Yoshida WY, Scheuer PJ (2002) J Nat Prod 65:487–491
129. Dado GP, Gellman SH (1994) J Am Chem Soc 116:1054–1062
130. Hanessian S, Luo X, Schaum R, Michnick S (1998) J Am Chem Soc 120:8569–8570
131. Sanjayan GJ, Stewart A, Hachisu S, Gonzalez R, Watterson MP, Fleet GWJ (2003) Tetrahedron Lett 44:5847–5851
132. Seebach D, Schaeffer L, Brenner M, Hoyer D (2003) Angew Chem Int Ed 42:776–778
133. Farrera-Sinfreu J, Zaccaro L, Vidal D, Salvatella X, Giralt E, Pons M, Albericio F, Royo M (2004) J Am Chem Soc 126:6048–6057
134. Vasudev PG, Ananda K, Chatterjee S, Aravinda S, Shamala N, Balaram P (2007) J Am Chem Soc 129:4039–4048
135. Sasse F, Steinmetz H, Heil J, Höfle G, Reichenbach H (2000) J Antibiot 53:879–885
136. Höfle G, Glaser N, Leibold T, Karama U, Sasse F, Steinmetz H (2003) Pure Appl Chem 75:167–178
137. Friestad GK, Deveau AM, Marié J-C (2004) Org Lett 6:3249–3252
138. Lin CH (1993) Synth React Inorg Met Org Chem 23:1097–1106
139. Salazar J, Lopez SE, Rebollo O (2003) J Fluorine Chem 124:111–113
140. Iranpoor N, Zeynizadeh B (1999) J Chem Res (S) 124–125
141. Prashad M, Hu B, Har D, Repic O, Blacklock TJ (2000) Tetrahedron Lett 41:9957–9961
142. Allylation of the α-carbon of **15** established correlation with a known derivative. Schaum R (1998) PhD Thesis, Université de Montreal, Montreal, Canada
143. Torrente S, Alonso R (2001) Org Lett 3:1985–1987
144. Miyabe H, Yamaoka Y, Takemoto Y (2005) J Org Chem 70:3324–3327
145. Ramon DJ, Yus M (2004) Curr Org Chem 8:149–183
146. Obrecht D, Bohdal U, Broger C, Bur D, Lehmann C, Ruffieux R, Schönholzer P, Spiegler C, Müller K (1995) Helv Chim Acta 78:563–580
147. Enders D (1984) In: Morrison JD (ed) Asymmetric synthesis. Academic, New York
148. Husson H-P, Royer J (1999) Chem Soc Rev 28:383–394
149. Halland N, Jørgensen KA (2001) J Chem Soc Perkin Trans 1:1290–1295
150. Friestad GK, Shen Y, Ruggles EL (2003) Angew Chem Int Ed 42:5061–5063
151. Cho DH, Jang DO (2006) Chem Commun 5045–5046
152. Jang DO, Kim SY (2008) J Am Chem Soc 130:16152–16153
153. Lee S, Kim S (2009) Tetrahedron Lett 50:3345–3348

Top Curr Chem (2014) 343: 33–74
DOI: 10.1007/128_2013_492

Published online: 14 November 2013

Stereoselective Formation of Amines by Nucleophilic Addition to Azomethine Derivatives

André B. Charette and Vincent Lindsay

Abstract This chapter describes state-of-the-art methods to prepare α-chiral amines by the addition of nonstabilized nucleophiles to imine derivatives. The first part of the chapter illustrates the most effective diastereoselective addition reaction (substrate controlled and chiral auxiliary based methods) whereas the second part focuses on catalytic asymmetric methods.

Keywords Azomethine · Chiral amines · Chiral auxiliaries · Chiral catalysts · Chiral ligands · Nucleophilic addition

Contents

Abbreviations

DuPHOS	[2,5-Dimethylphospholano]benzene
EWG	Electron-withdrawing group
PG	Protecting group

A.B. Charette (✉) and V. Lindsay
Université de Montréal, Faculty of Arts and Sciences, Department of Chemistry,
PO Box 6128, Station Downtown, Montreal, QC H3C 3J7, Canada
e-mail: andre.charette@umontreal.ca

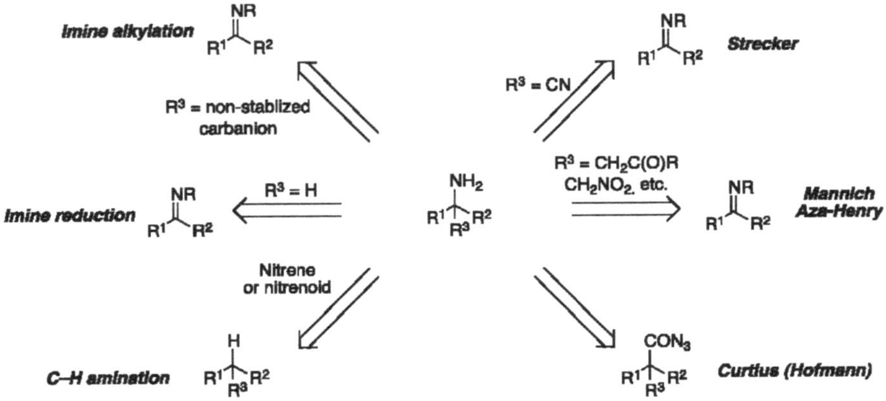

Fig. 1 Most common approaches to α-chiral amines

1 Introduction

The ubiquity of α-chiral amines in bioactive molecules and natural products has been a constant source of inspiration for the development of new and more effective methods for their formation. Various approaches to α-chiral amines are shown in Fig. 1. Common approaches to generate α-chiral amines include imine reduction [1] and Strecker [2] and Mannich type (aza-Henry) reactions [3, 4].

It is also possible to generate α-chiral amines starting from α-chiral carboxylic acids by a Curtius rearrangement. The process occurs with retention of configuration (1) [5]:

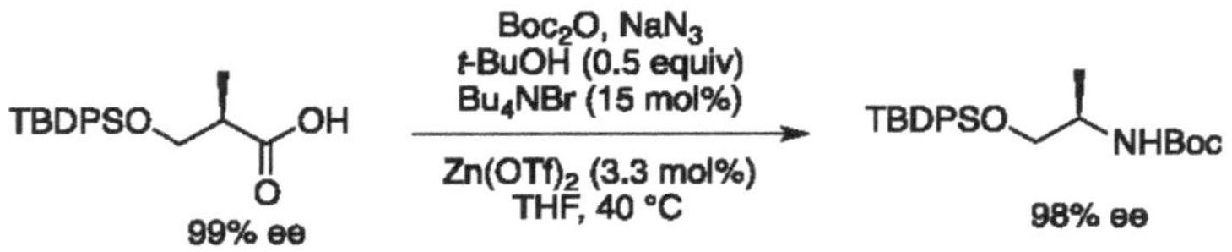

(1)

This chapter will cover the recent advances in the area of stereoselective azomethine alkylation only [6]. The sections are divided according to the type of electrophiles and nucleophiles used in the reactions and whether the processes required stoichiometric or catalytic amounts of a chiral inductor.

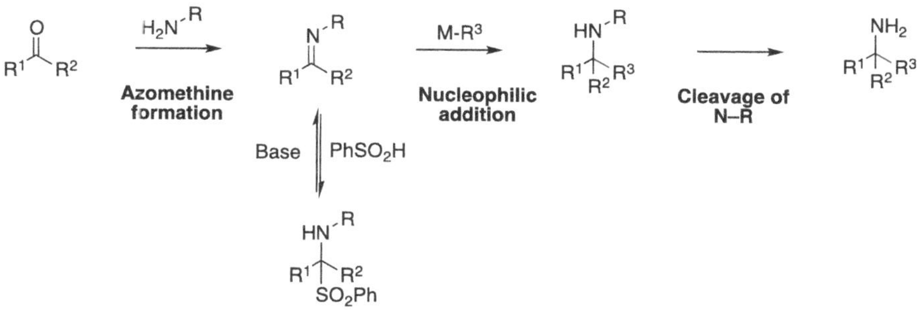

Scheme 1 Preparation of α-chiral amines from carbonyl derivatives

2 1,2-Addition of Unstabilized Carbanions to Chiral Azomethine Derivatives

Recent advances in the synthesis of various azomethine derivatives from aldehydes have greatly contributed to increasing the scope of accessible substrates for addition reactions.

These compounds are typically prepared by condensation of protected amines with aldehydes or ketones, leading respectively to aldimines and ketoimines with elimination of water (Scheme 1). In cases where imines are not very stable (some alkyl-substituted imines with enolizable protons), it is possible to isolate them in the form of their sulfinic acid adducts. These adducts break down to imines in the presence of base (or excess nucleophile). Following the addition of the nucleophile, the amino protective group is removed to provide access to a free amine.

The overall efficiency of these processes is often regarded as a combination of these three reactions. Ideally, the imine used or its equivalent must be accessible in high yield from an aldehyde or a ketone, possessing the appropriate electrophilicity for a nucleophilic addition under mild conditions, and the *N*-protective group should be easily removed afterwards.

2.1 *Electrophilicity of Azomethine Derivatives*

The electrophilicity of azomethines is highly dependent upon the nature of the *N*-substituent of the substrate. Each type of *N*-substituted imine has its own reactivity profile and is not usually interchangeable for a given methodology. These azomethines can be subdivided into six main classes depending of the type of *N*-substituent, each of which have significantly different electrophilicities, Lewis basicities, and modes of activation: (1) *N*-alkyl groups, (2) *N*-aryl groups, (3) *N*-heteroatoms, (4) *N*-electron-withdrawing groups, (5) iminium salts, and (6) *N*-ylides (Fig. 2).

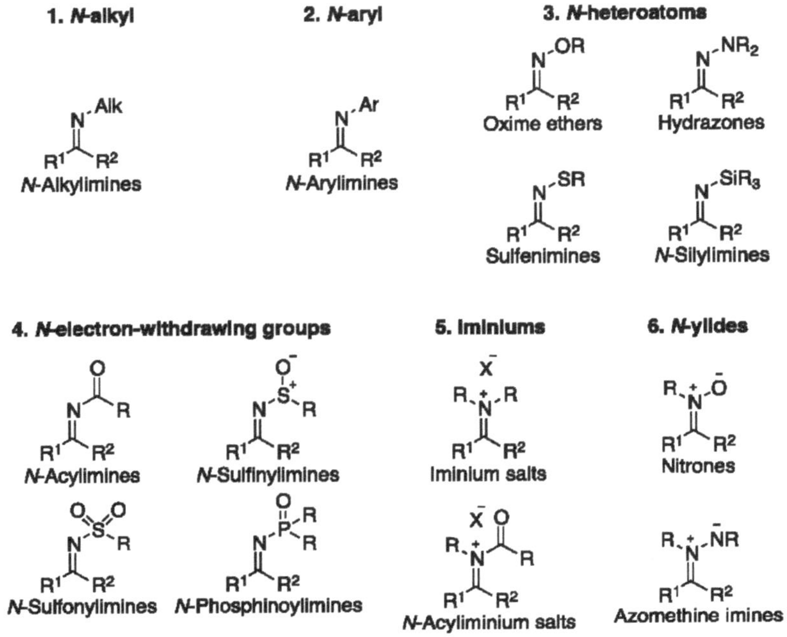

Fig. 2 Classes of azomethine derivatives in nucleophilic addition reactions

Compared to carbonyl derivatives, simple *N*-alkyl or *N*-aryl imines are usually more basic and less electrophilic, which means that they are sometimes more readily activated with a Lewis acid or a transition metal. In order to increase the electrophilicity of the C=N bond, various strategies have been employed, including the use of an electron-withdrawing substituent as a substituent directly linked to the imine carbon atom such as in glyoxal derivatives. Since the presence of an α-electron-withdrawing substituent is not always desired in the final product, a more general approach consists of using *N*-EWG imines as electrophiles, where the EWG, acting as a protective and an activating group, can be cleaved after the addition to liberate a free amine. In order to quantify the effect of such *N*-substitution on the imine electrophilicity, various data can be used (Fig. 3) [7, 8]. *N*-Acylimines and *N*-tosylimines are the most electrophilic of the *N*-EWG imines, while *N*-alkyl and *N*-aryl substituted imines are significantly less reactive than aldehydes. Thus, strong nucleophiles such as Grignard reagents are often needed for the addition to occur to simple *N*-alkylimines. On the other hand, the use of *N*-EWG imines as electrophiles is particularly common in catalytic asymmetric addition reactions using various transition metals such as rhodium, copper, iridium, or palladium. It should be noted that, due to the purely electronic nature of these calculations, the values depicted here are independent of the different steric effects at the approach of the nucleophile and of the mode of activation employed for the addition.

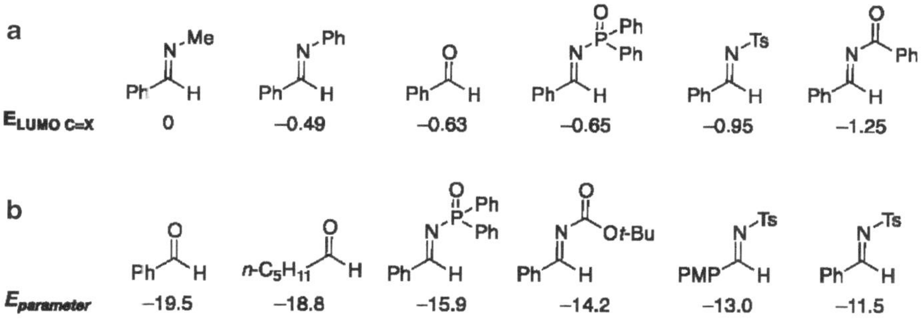

Fig. 3 Relative electrophilicity of C=X functionalities. (**a**) LUMO energies in eV (geometry optimized at B3LYP/6-31G(d)) [7]. (**b**) Mayr's *E* parameters [8]

2.2 *Diastereoselective Addition to Chiral Imines*

Two classes of diastereoselective nucleophilic addition reactions to imines will be discussed separately. The first involves nucleophilic addition reactions to chiral imines in which the chirality resides on the C-terminus (Fig. 4, type 1). One of the most reliable ways to achieve the preparation of enantiomerically enriched α-chiral amines is the use covalently-bound chiral auxiliaries, which can be located on the imine *N*-substituent (Fig. 4, type 2). These methodologies are typically designed to allow cleavage of the N–R* bond to liberate a free amine.

2.2.1 Chiral Imines and Derivatives Obtained from Chiral Carbonyl Derivatives

When the imine electrophile is derived from a chiral aldehyde, several competing predictive models are available, where the sense of induction greatly depends on the nature of the chiral group and the nucleophile employed. In analogy to the extensively studied addition of nucleophiles to carbonyl substrates, the predictive Cram, Felkin–Ahn, and Cornforth models are also applicable to imine chemistry. Thus in principle a non-chelating chiral group at the α-position of the acyclic imine will afford the Felkin–Ahn product issued from minimization of torsional strain at the transition state and subsequent attack on the most accessible face of the imine (Fig. 5a). Conversely, the presence of a chelating functionality at this position leads to the product issued from the addition on the least hindered face of the chelate intermediate according to Cram's cyclic model, under the appropriate reaction conditions favoring the formation of such chelate (Fig. 5b). It is noteworthy that Cram's cyclic model is usually effective only with *N*-alkyl or *N*-arylimine derivatives where the most Lewis basic site of the substrate is the nitrogen atom. As in carbonyl addition chemistry, it is sometimes possible to access either diastereomer when appropriate reaction conditions are used.

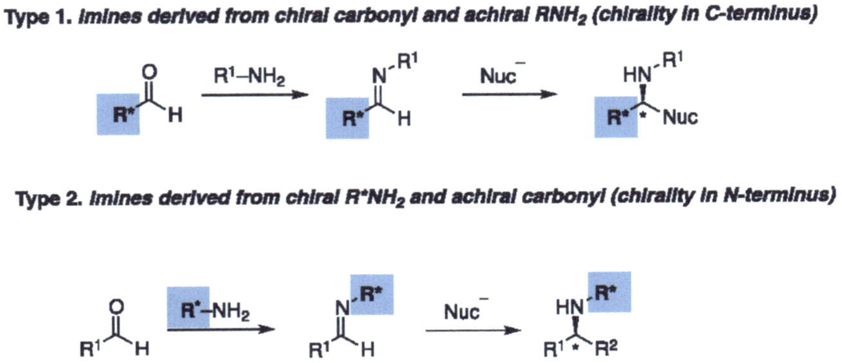

Fig. 4 Diastereoselective addition to chiral imines

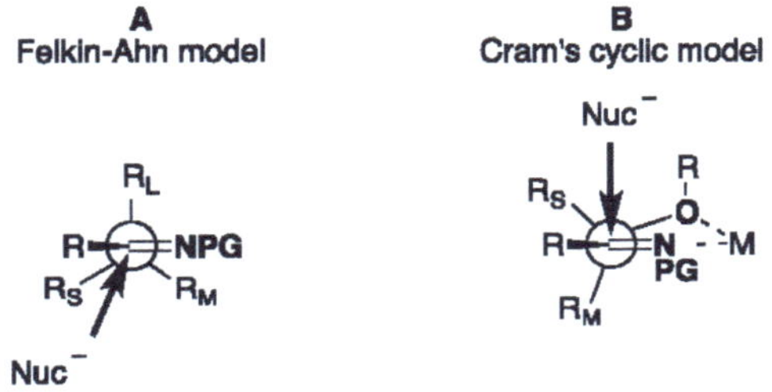

Fig. 5 Diastereoselective addition to chiral imines

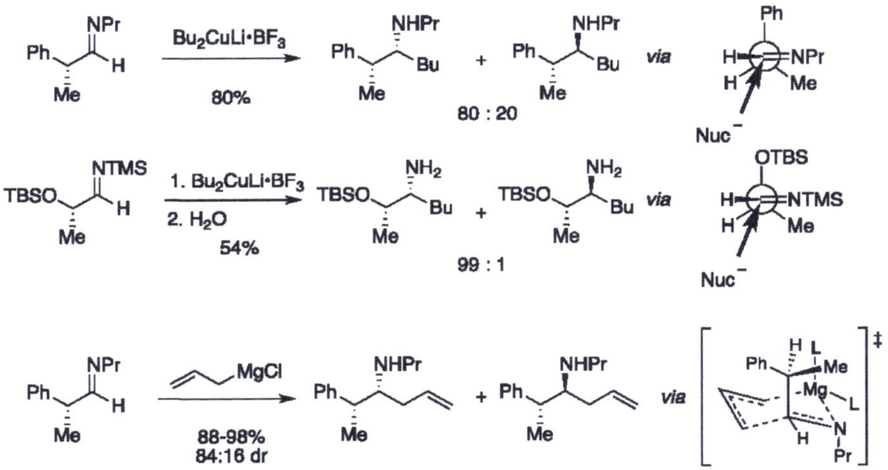

Scheme 2 Representative examples of diastereoselective addition to chiral imines

While successful examples of the Felkin–Ahn model with simple alkylmetals and imines remain scarce, alkylcopper or alkylcuprate reagents complexed with BF_3 can usually afford a good diastereoselectivity for the Felkin adduct (Scheme 2)

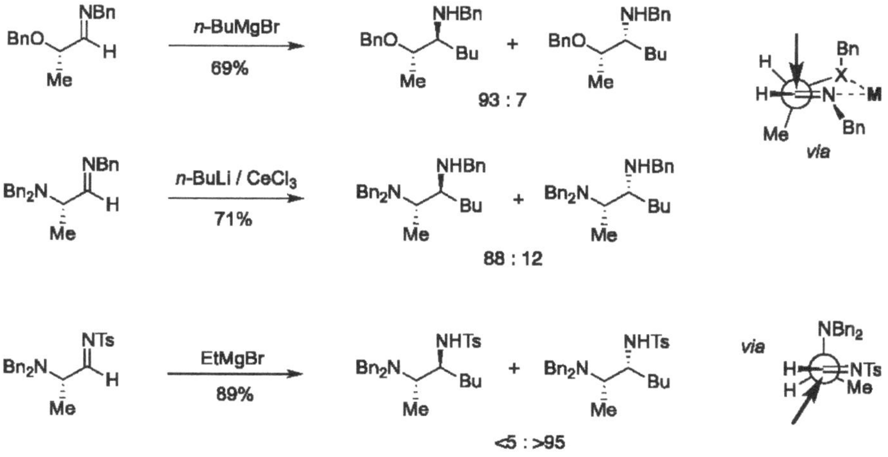

Scheme 3 Diastereoselective addition reactions to *N*-benzyl and *N*-tosylimines

[9, 10]. The use of allyl Grignard reagents is more common and can provide a reasonable level of diastereoselection [9].

The nucleophilic addition reaction under chelation control conditions (see Cram's cyclic model, Fig. 5) is a much more general and widespread means to access α-chiral amines from chiral imines. Various types of α-chiral chelating groups can be used in such processes, including α-chiral benzyl ethers or α-chiral *N*,*N*′-dibenzylamines (Scheme 3) [11, 12]. When the *N*-benzylimine was replaced with *N*-tosylimine the opposite diastereoselectivity issued from a non-chelation control (Felkin–Ahn) was observed for the corresponding 1,2-diamine product [12]. Diastereoselective reactions can also be obtained via chelation control with hydrazones [13–21], oxime ethers [22–24], or nitrones [25–37] derived from chiral aldehydes through an analogous stereoinduction mechanism, where the size of the chelate varies depending on the nature of the imine derivative used.

Diastereoselective addition reactions to imines derived from the chiral pool have provided access to unnatural α-amino acids (Scheme 4) [38, 39].

Alternatively, a chiral acetal, aminal, or thioacetal located at the α-position could be used as removable chiral auxiliaries to provide the corresponding aldehydes after hydrolysis (Fig. 6) [19, 40–42].

The stereocontrol of this strategy relies on a selective chelate formation on one of the Lewis basic sites located on the C-terminus. Both approaches have been shown to lead to very high levels of diastereocontrol for the addition step, with alkyl-, aryl-, or alkynyl-metal nucleophiles.

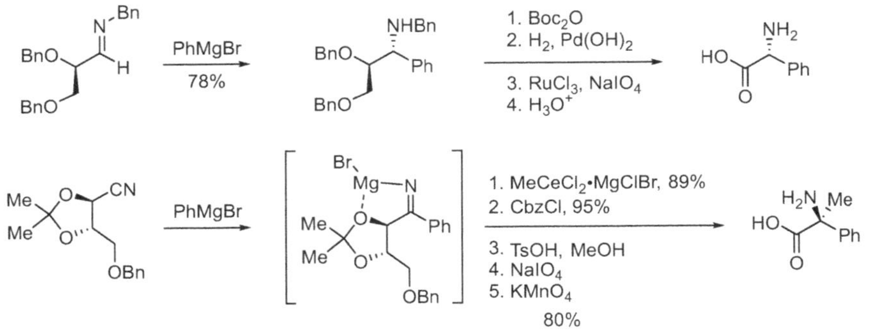

Scheme 4 Access to α-amino acids

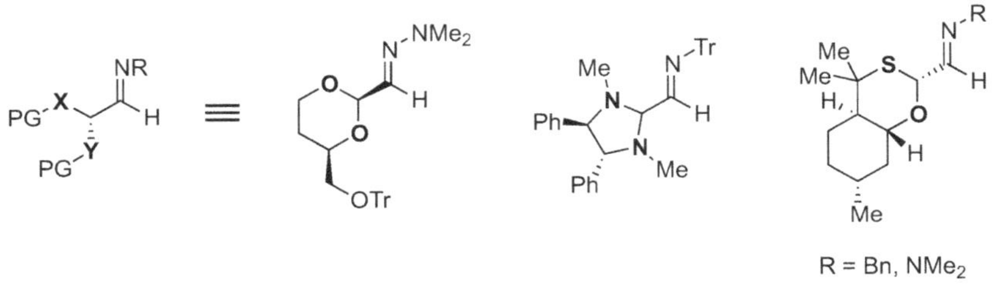

Fig. 6 Representative examples of C-terminus chiral auxiliaries in C=N addition reactions

2.2.2 Chiral Imines and Derivatives Obtained from Chiral Amines

The use of chiral auxiliaries located at the *N*-terminus of the imine substrate is one of the most efficient mean to access α-chiral amines via the addition of organometallic reagents to C=N bonds. Chiral imines derived from *N*-alkylamines, hydrazines, hydroxyamine (oximes and nitrones), and sulfinylamines have been the most widely utilized auxiliaries.

In analogy to the Felkin–Ahn model for C-terminus auxiliaries, a stereochemical model has been proposed to account for the observed diastereoselectivities with N-terminus auxiliaries relying on non-chelation control (Fig. 7a) [9, 43]. Successful examples using these auxiliaries are quite rare [44–47].

The use of α-methyl benzylamine as auxiliary with cuprates/BF_3 mixtures can lead to good diastereoselectivities in certain cases (Scheme 5) [48, 49]. Hydrogenolysis of the benzylic amine leads to the destructive chiral auxiliary cleavage.

Chiral auxiliaries relying on chelation control at the N-terminus are much more widespread and structurally diverse. When the electrophile is an imine, the diastereochemical outcome of the addition can be predicted via a chelation model (Fig. 7b) [50–56].

Several examples of chiral auxiliaries, often derived from amino acids, are illustrated in Fig. 8. While auxiliaries of types I and II are compatible with a variety

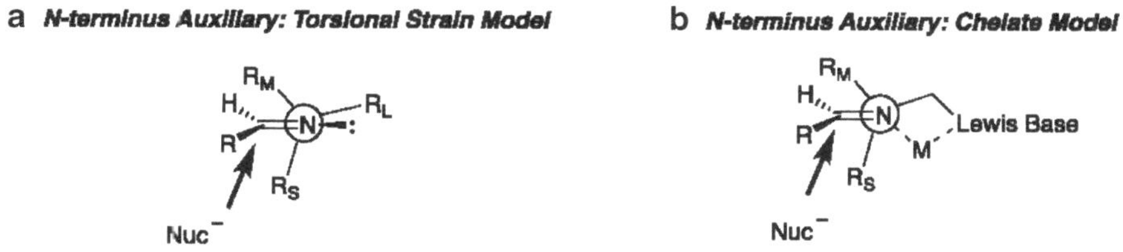

Fig. 7 Torsional vs chelate model in nucleophilic addition reactions

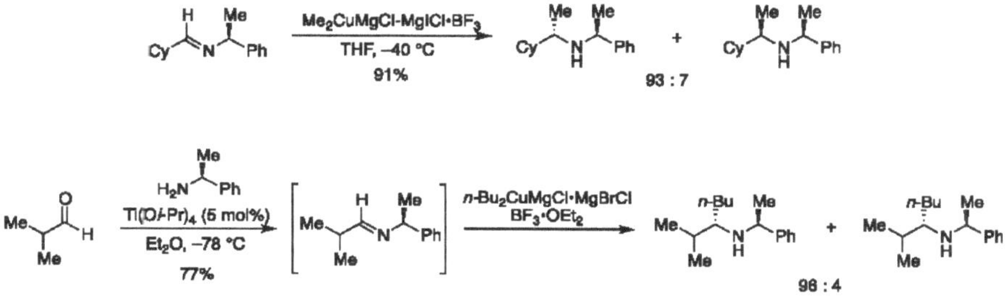

Scheme 5 Example of α-methyl benzylamine as chiral auxiliary

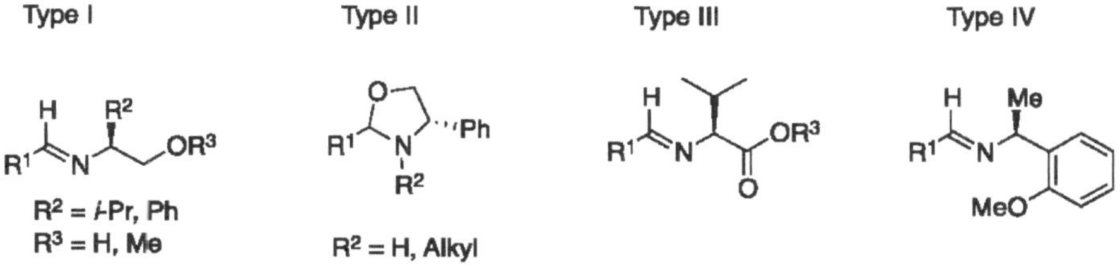

Fig. 8 Imine-derived chiral auxiliaries based on the chelate model

of reagents such as Grignards, organolithium, organozinc, organocuprates, or organocerium nucleophiles [50–59], type III auxiliaries have been limited to allylation reactions under Barbier-type conditions due to ester compatibility issues with more nucleophilic carbanion equivalents [60–68]. The addition of a methoxy group at the ortho position of α-methyl benzylamine leads to high diastereoinduction with alkyl- and allyllithium reagents [69–72]. Recent examples of the use of these auxiliaries are shown in Scheme 6 [70, 73–75].

Chiral hydrazones where the chirality is located on the hydrazine moiety of the precursor have found widespread use for the synthesis of α-chiral amines via the diastereoselective nucleophilic addition of various organometallic reagents. As in the previous case, chelate formation activates the hydrazone towards the nucleophilic addition, which can occur either through an internal or an external delivery,

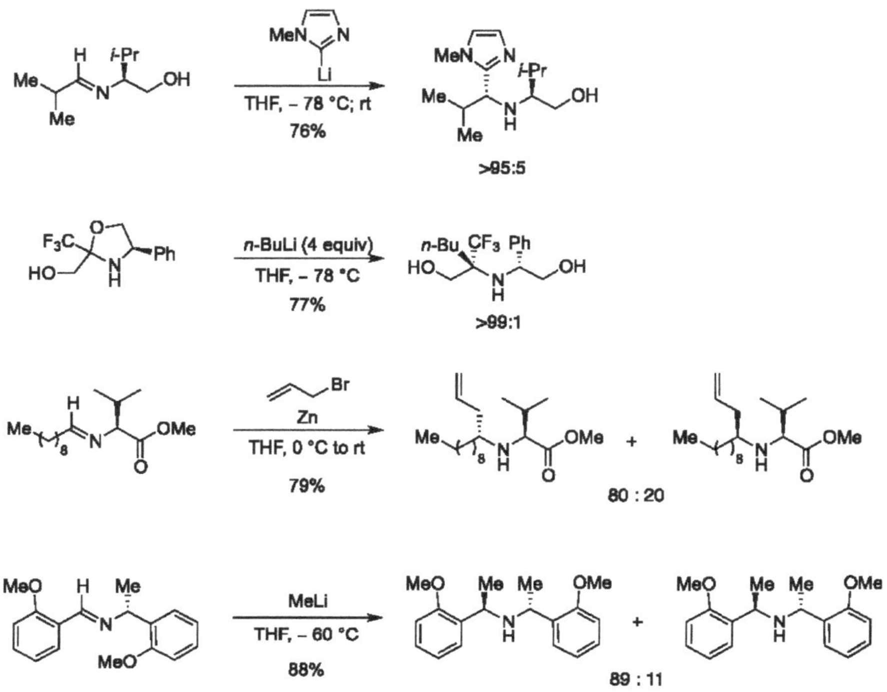

Scheme 6 Representative examples of N-terminus chiral auxiliaries derived from alkylamines

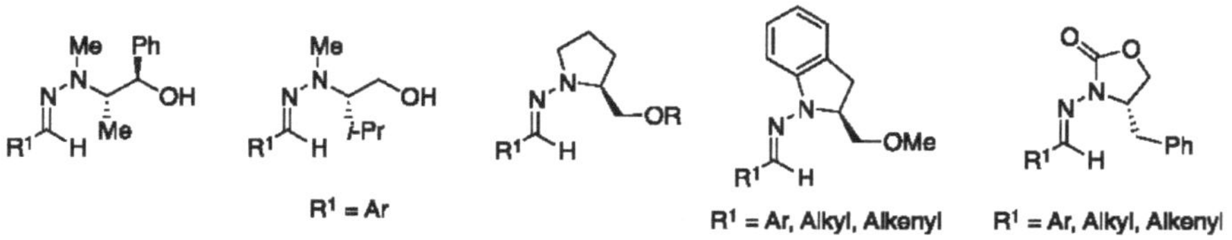

Fig. 9 Representative examples of hydrazone-based chiral auxiliaries

depending on the conditions used. For instance, chiral auxiliaries bearing Lewis basic alkoxides, ethers, and oxazolidinones that are able to form five- or six-membered chelates have been developed (Fig. 9). Chiral hydrazones derived from *N*-aminopyrrolidinol also called SAMP (or RAMP for the *R*-isomer) are general electrophiles for the stereoselective addition of a variety of organometallic reagents including organolithium, Grignards, and organocerium species [76–84]. Several specific transformations are shown in Scheme 7 [84–86]. Hydrazones derived from aliphatic and aromatic aldehydes and protected

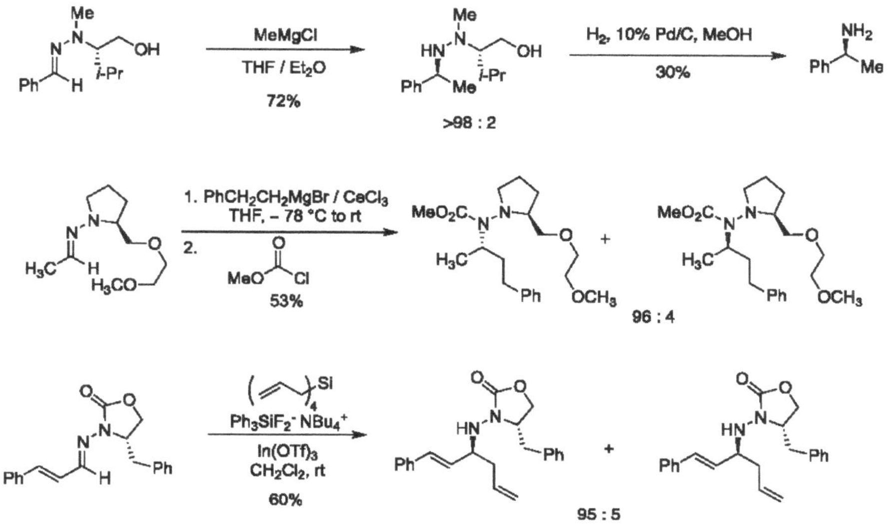

Scheme 7 Representative examples of diastereoselective addition using hydrazone based chiral auxiliaries

hydrazinoalcohols are usually tolerated and compatible with both sp^3 and sp^2 nucleophiles. The use of a chiral *N*-acylhydrazone equivalent as auxiliary has been shown to be effective for the indium-mediated addition of allylsilanes to both aromatic and aliphatic electrophiles [86–89]. In addition, nucleophiles for these reactions also includes carbon radicals generated from alkyl iodides [90–101]. In all cases, reductive cleavage of the N–N bond leads to the free α-chiral amine.

Chiral oximes derived from *O*-alkyl hydroxylamines have been used successfully as electrophiles for diastereoselective addition reactions [102–107]. Although these substrates are usually considered to be less general than their hydrazone counterparts, a number of successful examples of their use have been reported [102, 108].

One of the most practical and reliable types of *N*-bound auxiliary for the diastereoselective addition of Grignard reagents to the C=N bond consists of a chiral sulfinyl group directly bound to nitrogen. The most efficient auxiliaries of this type reported to date are definitely **A** [109–114] and **B** [115–129] (Fig. 10), while the parent camphor-derived analogue **C** is not as general and has not been extensively used [130, 131].

The *N*-toluenesulfinylimines (Fig. 10, **A**) are only compatible with allyl- and benzyl-Grignard reagents due to the competitive attack on the sulfinyl group with other nucleophiles such as alkyl-, vinyl-, and arylmetals. Conversely, the use of a much hindered *N*-*tert*-butanesulfinyl group (Fig. 10, **B**) allows for chemoselective addition reactions to the imine functionality with these nucleophiles, generally with very high diastereocontrol. A chair-like six-membered transition-state where the

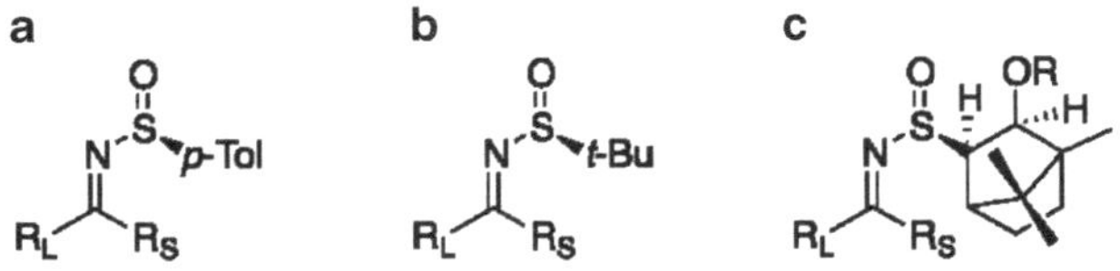

Fig. 10 Most efficient sulfinylimines based chiral auxiliaries

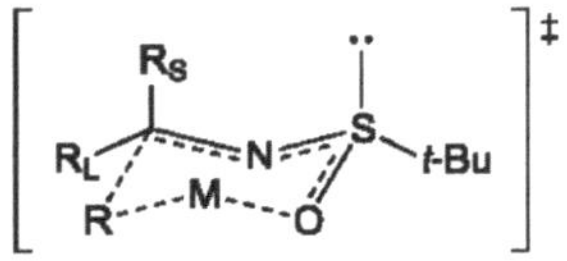

Fig. 11 Chelation model to explain the selectivity with *N-tert*-butylsulfinylimines

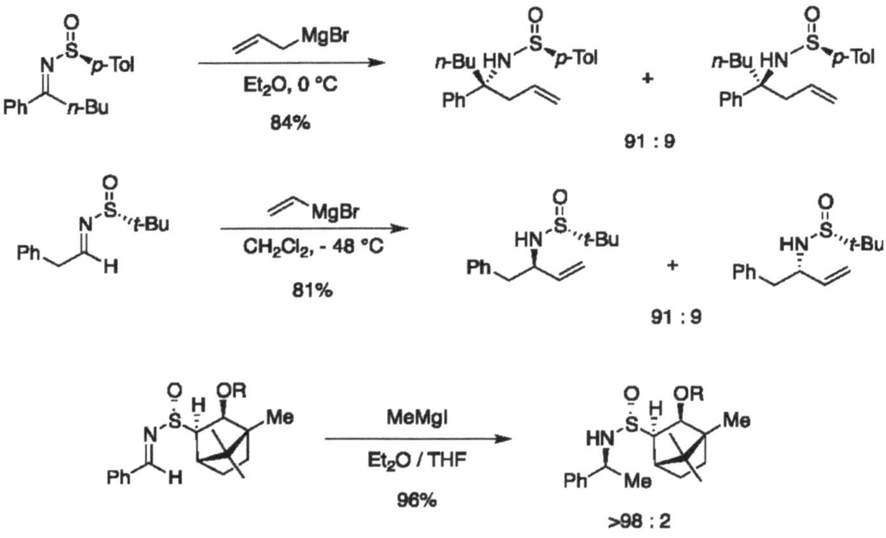

Scheme 8 Representative examples of diastereoselective addition to *N*-sulfinylimines

sulfur substituent and the large group of the imine are placed in equatorial positions accounts for the observed relative configuration of the products (Fig. 11). Some additional examples of the use of these auxiliaries are shown in Scheme 8 and in Table 1 [109, 116, 130] (for a review see [132]).

Vinyl-substituted nucleophiles can be generated in situ through carbo- or hydrometallation of alkynes, using organocuprates (2) or Rh-based catalysts (3), respectively, and subsequently react with *N*-sulfinylimines, leading to polysubstituted allylamines in high regio- and stereoselectivity [113, 127]. It should be noted that esters are compatible in the rhodium-catalyzed addition.

Table 1 *N-tert*-Butanesulfinyl group as chiral auxiliary in the stereoselective addition reactions [115–129]

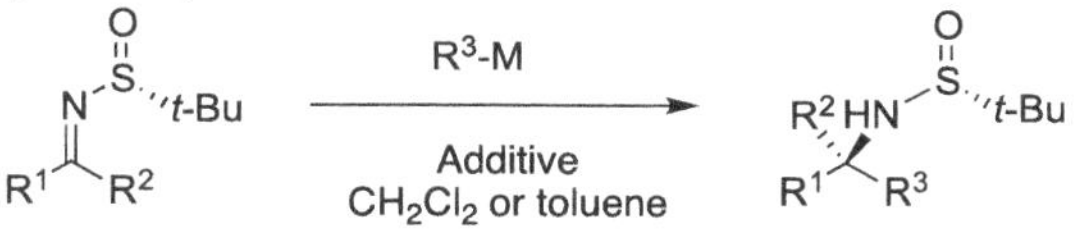

Entry	R^1	R^2	R^3-M	Yield (%)	dr	Additive
1	Et	H	PhMgBr	>99	96:4	–
2	*i*-Pr	H	EtMgBr	>99	97:3	–
3	Ph	H	EtMgBr	98	92:8	–
4	*i*-Pr	Me	PhLi	93	97:3	Me_3Al
5	Ph	Me	*n*-BuLi	86	98:2	Me_3Al
6	Ph	*n*-Bu	MeLi	>99	99:1	Me_3Al
7	2-Naphthyl	Me	PhLi	62	99:1	Me_3Al

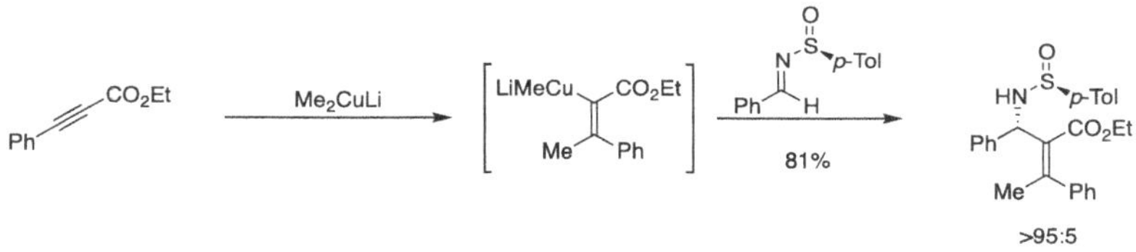

(2)

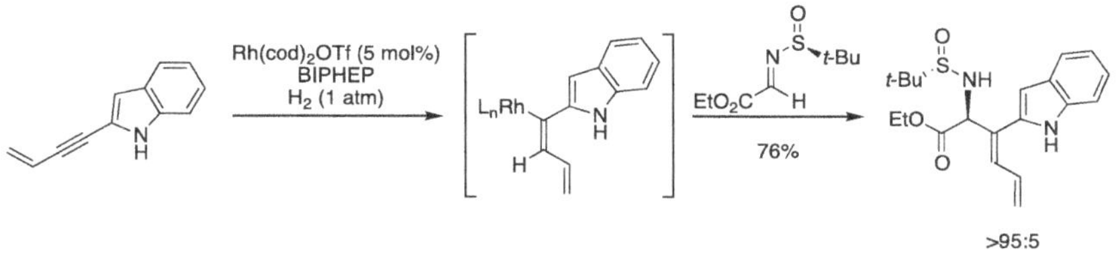

(3)

It is also possible to achieve the diastereoselective addition of arylboronic acids through the use of Rh(I) catalysts, affording a mild and highly chemoselective alternative to aryl-Grignard reagents as stoichiometric nucleophiles (Scheme 9a) [125, 126]. In this type of process, only a catalytic amount of the active arylrhodium nucleophile is generated at once through transmetallation and the reaction tolerates a wide range of imines.

The versatile *N-tert*-butylsulfinamine is readily cleaved to liberate the free amine under a variety of relatively mild acidic conditions, the most widely used being HCl in MeOH.

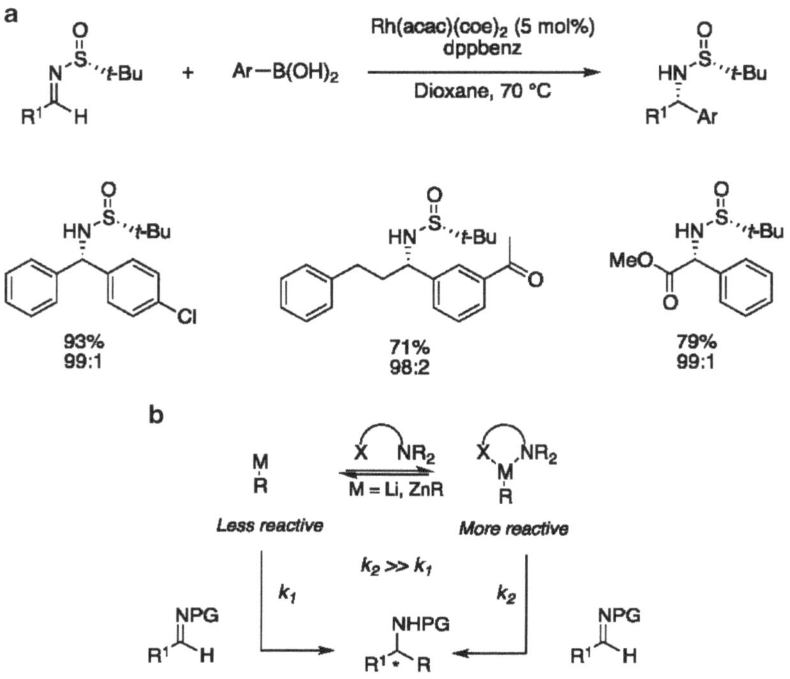

Scheme 9 (**a**) Representative examples of Rh-catalyzed arylboronic acids addition reactions. (**b**) Ligand accelerated nucleophilic addition

2.3 *Stoichiometric Amounts of Chiral Reagents*

The addition of stoichiometric chiral nucleophiles to achiral imines is not much used in synthesis given the efficiency of chiral auxiliary based and catalytic asymmetric methods. Most of the successful methods reported to date using stoichiometric amounts of chiral nucleophiles are limited to allylation or crotylation reactions. An obvious advantage of such a strategy is the direct formation of enantiopure α-chiral homoallylic amines, without the need for the cleavage of a chiral auxiliary. Several examples of these reactions are provided in (4–6) [133–142]:

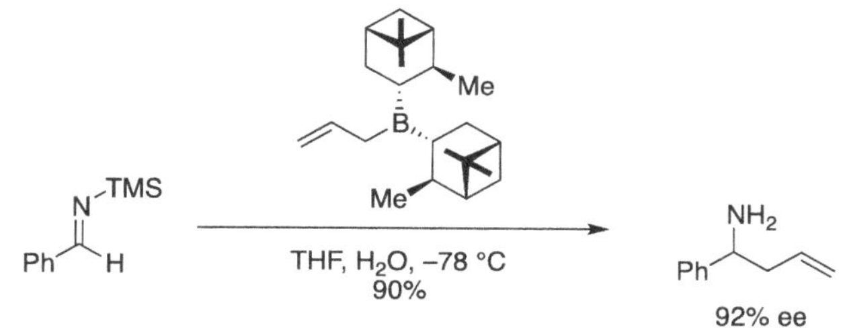

(4)

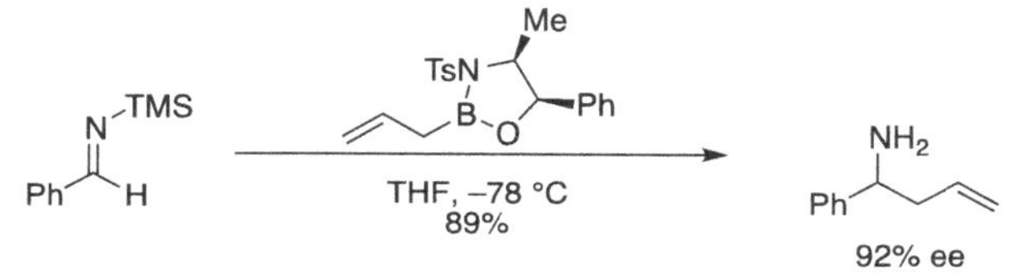

(5)

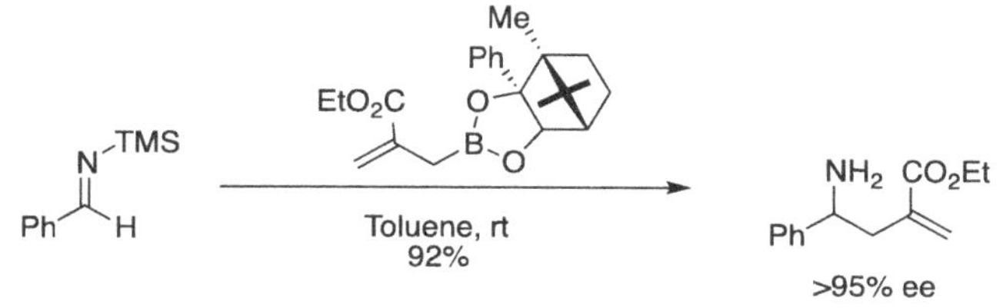

(6)

The addition of a chiral crotylboron reagent bearing diisopinocamphenyl units allows for the stereospecific formation of β-chiral homoallylic amines in good dia- and enantioselectivities (7) [139]:

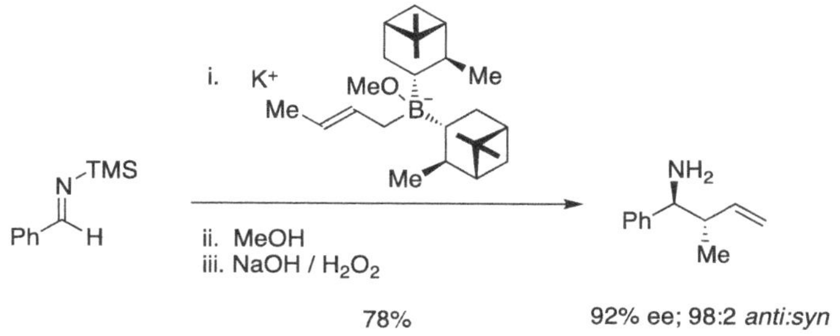

(7)

The allylboration of cyclic *N*-alkylimines was reported to be highly enantioselective when a BINOL-derived allylboron nucleophile was used (8) [140]:

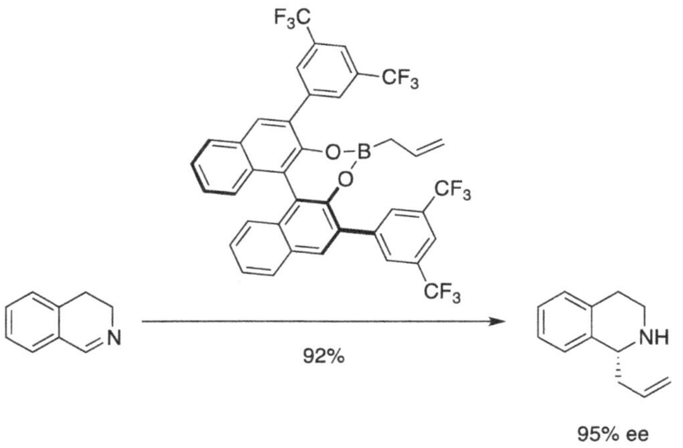

(8)

The enantioselective allylation of ketoimines is also possible with enantiopure *B*-allyl-10-phenyl-9-borabicyclo[3.3.2]-decane, although generally with lower yields than with the parent aldimines (9) [141]:

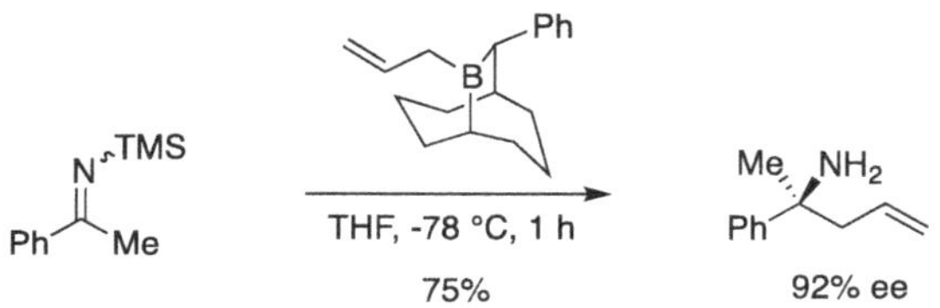

(9)

More recently, several very effective and practical chiral allylsilanes have been reported to add to *N*-acylhydrazones with excellent stereocontrol (Table 2) [143–146]. Good yield and stereoselectivity are typically observed for hydrazones derived from both aldehydes and ketones (entries 1–4). The reaction with crotylsilane reagents also proceeds with high stereocontrol providing either the syn or the anti depending on the configuration of the starting reagent (entries 5, 6).

Chiral allyl- or crotylzinc reagents bound to bis(oxazoline) ligands are also competent nucleophiles in this type of process, and the stereoselectivity observed is typically high with a cyclic *N*-alkylimine or glyoxal-derived oximes [147, 148].

N-Acylhydrazones react with a chiral allylindium species formed in situ in a mixture of In(0), allyl iodide, and stoichiometric amounts of a chiral BINOL derivative to afford *N*-acyl homoallylhydrazines (10) [149]. The structure of the BINOL ligand was optimized to allow a highly efficient catalytic asymmetric process, in which only 10 mol% of this chiral ligand is necessary [150].

Table 2 Addition of allylsilanes to *N*-acylhydrazones [145, 146]

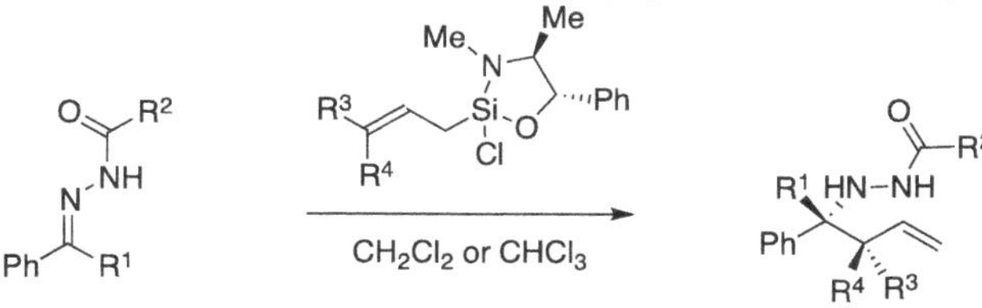

Entry	R^1	R^2	R^3	R^4	Yield (%)	dr	*ee* (%)
1	H	Me	H	H	86	–	88
2	Me	Ph	H	H	86	–	90
3	*i*-Pr	Ph	H	H	80	–	97
4	CO_2Me	Ph	H	H	76	–	93
5	H	Ph	Me	H	89	95:5	97
6	H	Ph	H	Me	81	96:4	95

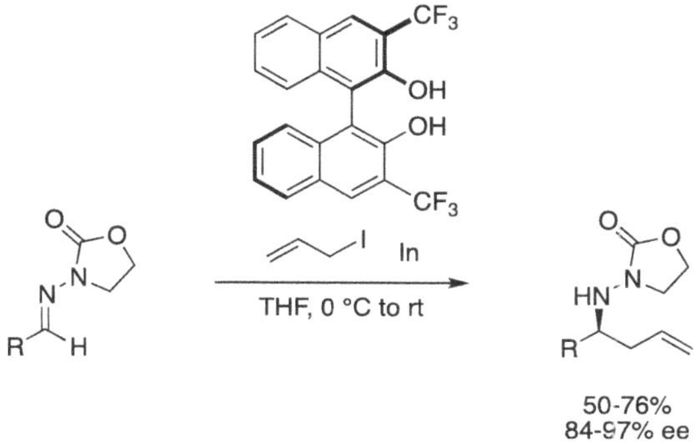

(10)

2.4 *Catalytic Asymmetric Nucleophilic Addition*

The importance of asymmetric catalysis in the last 40 years and the progress made in a vast array of transformations has stimulated organic chemists to develop new strategies for the synthesis of enantiomerically enriched α-chiral amines via addition reactions. Although the new methodologies have not been published at the same rate as carbonyl addition reactions, a large number of very effective catalytic asymmetric addition reactions to imines are now available. The most efficient and practical methods are summarized in the next sections. Since most methodologies are effective with only one class of carbon nucleophiles, each subsection below will be divided by the type of nucleophile that is used.

Table 3 Early examples of nucleophilic addition reactions to imines employing substoichiometric amounts of a chiral Lewis base

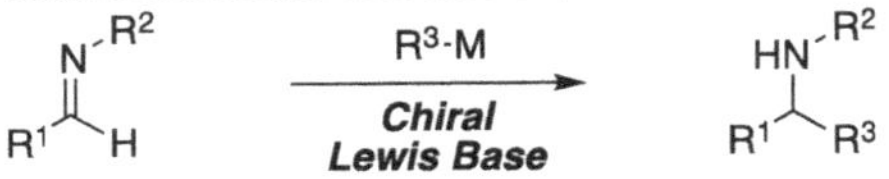

Entry	Chiral Lewis base (mol%)	R^1	R^2	R^3-M	Yield (%)	*ee* (%)	References
1	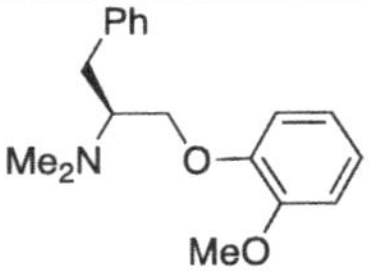 (30 mol%)	Ph	PMP	BuLi	99	60	[152]
2	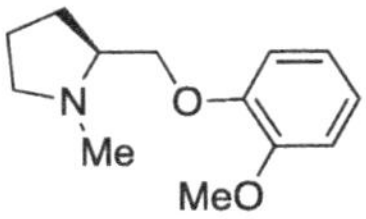 (25 mol%)	Ph	Ph	BuLi	44	6	[156]
3	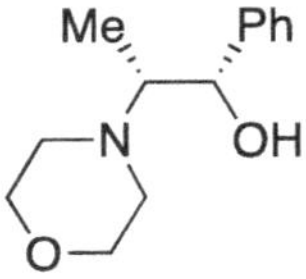 (50 mol%)	Ph	$P(O)Ph_2$	Et_2Zn	69	85	[158]
4	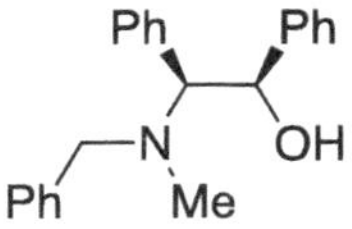 (50 mol%)	Ph	$P(O)Ph_2$	Et_2Zn	67	90	[159]
5	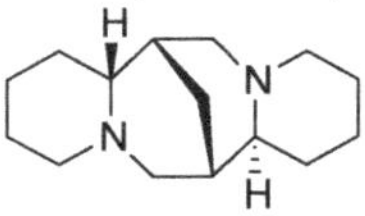 (20 mol%)	$Ph(CH_2)_2$	PMP	BuLi	91	79	[160–163]
6	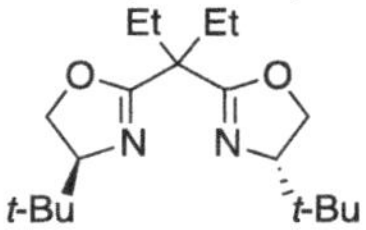 (10 mol%)	Ph	PMP	MeLi	97	69	[164, 165]
7	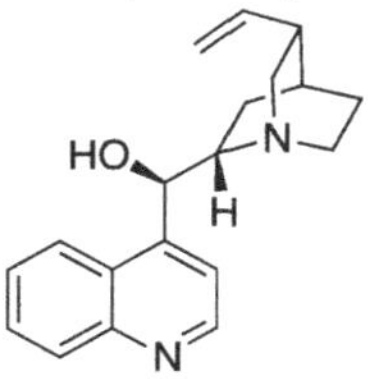 (50 mol%)	Ph	$P(O)Ph_2$	Et_2Zn	77	87	[166]

2.4.1 Background: Lewis Base Activation of the Nucleophile vs Transition Metal Catalysis

One of the first strategies explored for the synthesis of α-chiral amines through the addition of nonstabilized nucleophiles to imine derivatives consists of using stoichiometric and substoichiometric amounts of a chiral bidentate Lewis base capable of activating the Lewis acidic organometallic reagent via complexation of the metal center (Scheme 9b). The key for the success of this approach mainly resides in the fact that the reagent complexed with the chiral Lewis base is far more nucleophilic than that of the uncomplexed species (R–M). If the rate of the background reaction (k_1) is significantly lower than that of the complexed species (k_2), then one may hope to observe good enantioselectivities if a suitable chiral Lewis base is used. One problem that researchers had faced for many years is the fact that very reactive organolithium or Grignard reagents were always used early on. These reagents react very rapidly with imines even in the absence of a Lewis base. The key issue early on was to discover nucleophiles or reaction conditions in which the rate of the background reaction was suppressed.

A summary of the early results using organolithium and diorganozinc reagents is shown in Table 3 [151–166]. Most of these methods require a considerable amount of the chiral ligand (10–50 mol%) to be effective and each process is not very general.

One of the rare successful example was reported, using *N*-formylarylimines and chiral [2.2]paracyclophane-based *N*,*O*-ligand as catalyst (Table 4) [167]. Due to the instability of these imines, they had to be formed in situ from their corresponding sulfinate adduct by the addition of 1 equiv. extra of Et_2Zn acting as a base. A related methodology was reported for the addition of Ph_2Zn to these substrates (for other examples of enantioselective addition of diorganozinc reagents to these imines, see [168–170]). The generality of this methodology with regards to nucleophiles that can be used still remains to be established. The formamide is easily hydrolyzed under mild acidic conditions (HCl, MeOH, 50°C).

A second and most successful approach involving nonstabilized nucleophiles and imine derivatives employs a chiral catalyst capable of transmetalation with a non-reactive nucleophilic partner added in stoichiometric amounts (Scheme 10). This active chiral nucleophile formed catalytically is then capable of stereoselective addition reactions to the C=N bond to form enantioenriched α-chiral amine derivatives.

This area of research has generated considerable interest from the synthetic organic community, leading to the development of the most efficient methodologies for the synthesis of α-chiral amine derivatives through addition reactions. These methods will be reviewed in the following sections.

Table 4 Catalytic asymmetric addition to *N*-formylimines.

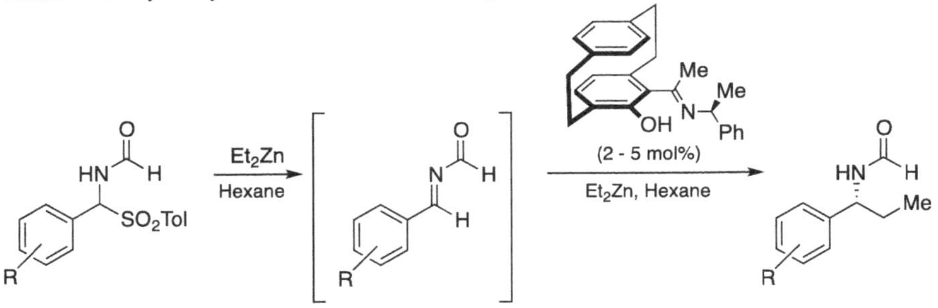

Entry	R	Yield (%)	*ee* (%)
1	H	>99	95
2	4-Cl	97	90
3	4-MeO	97	95
4	4-MeO_2C	90	94
5	2,6-Cl_2	98	95
6	3-Cl	99	93

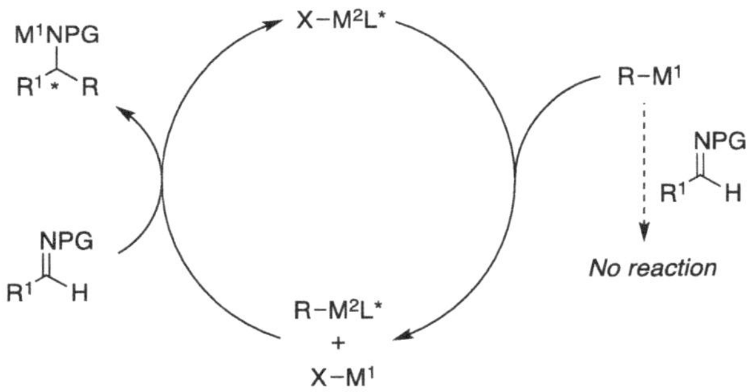

Scheme 10 Nucleophilic addition reactions involving a transmetalation process

2.4.2 Addition of Alkylmetals (sp^3 Carbon)

One of the first very efficient systems studied is the Cu-catalyzed addition of diorganozinc reagents. In this type of process, imine derivatives bearing either electron-withdrawing groups or arenes are usually required. The practicality of the method often relies on the availability of the chiral ligand, on the generality and accessibility of the nucleophiles used, and on the mildness of the reaction conditions to liberate the free amine following the stereoselective addition. A general mechanism for this reaction is shown in Scheme 11. A diorganozinc species added in stoichiometric amounts is capable of transmetalation with the Cu salt used as catalyst (typically CuOTf or $Cu(OTf)_2$ that is in situ reduced into a Cu(I) species), forming an active Cu-alkyl nucleophile, which adds to the imine derivative. The resulting product can then transmetalate with one of the diorganozinc species,

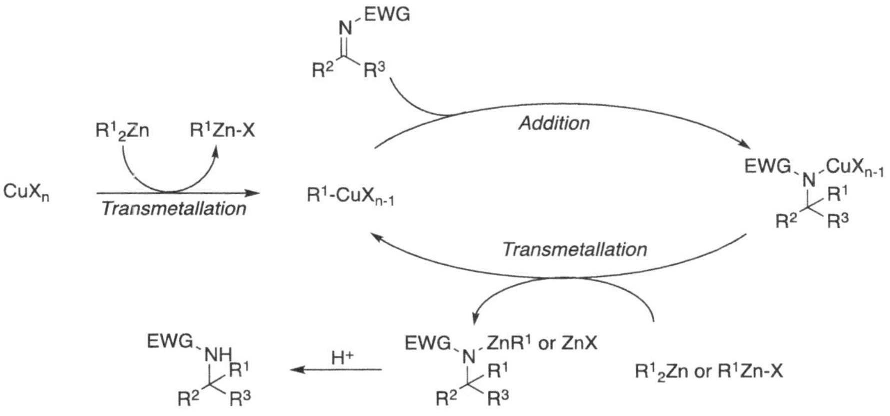

Scheme 11 Postulated mechanism for the Cu-catalyzed addition of diorganozinc reagents to imine derivatives

regenerating the active nucleophilic catalyst and the zinc salt of the amine product. If a chiral ligand is complexed to the Cu center throughout the process, then a catalytic enantioselective reaction can be envisioned via appropriate control of the chiral environment around the Cu center.

The first successful method of this type employed chiral amido-phosphines ligands for the $Cu(OTf)_2$-based catalyst with *N*-sulfonylimines as electrophiles (Table 5) [171–173]. Further optimization of the chiral phosphine ligand led to three amido-phosphine ligands, which afford a highly enantioselective method for aromatic as well as aliphatic imine derivatives with Et_2Zn, while the addition of Me_2Zn or $(i\text{-Pr})_2Zn$ furnishes slightly lower enantiomeric excesses (for use of a ferrocene-based related ligand see [174]). The *N*-tosyl group can be removed using samarium diiodide in a mixture of THF and HMPA.

N-Phosphinoylimines are also reactive electrophiles in this type of reaction, and the use of a chiral bis(phosphine) monoxide derived from Me-DuPHOS leads to excellent enantioselectivities for a number of *N*-phosphinoylimine derivatives and diorganozinc reagents (Table 6) [175–184]. Notably, it was found that the phosphine oxide unit does not need to be chiral for high enantioinduction as long as the phospholane unit is chiral [182].

Alkyl-substituted *N*-phosphinoylimines are compatible electrophiles but it is preferable to use their air-stable *p*-toluenesulfinic acid adducts, which get converted to the corresponding imine under the reaction conditions (Scheme 12) [177]. This reaction is also possible with trifluoromethylketoimines as electrophiles, leading to good enantioselectivity for the corresponding *N*-phosphinoylamines (Scheme 12) [183, 184]. In all cases, the *N*-phosphinoyl group is cleaved under mild acidic conditions (HCl, MeOH).

A chiral binaphthylthiophosphoramide ligand also induces high enantioselectivity for this reaction with Et_2Zn and aromatic substrates (11) [185, 186] (for use of related ligands for similar reactions see [187, 188]):

Table 5 Cu-catalyzed enantioselective addition of diorganozinc reagents to *N*-sulfonylimines using chiral amido-phosphine ligands

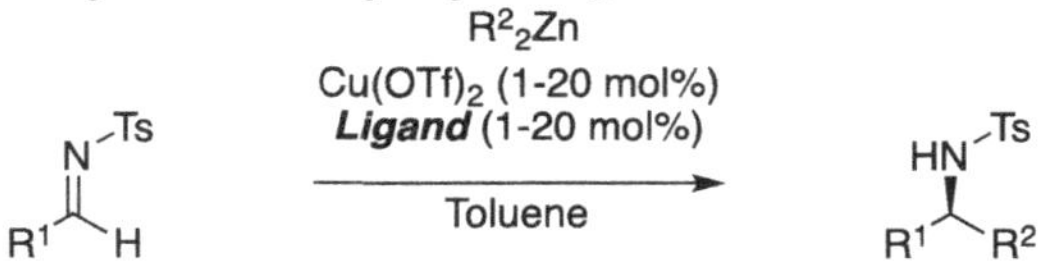

Entry	Ligand	R^1	R^2	Yield (%)	*ee* (%)
1	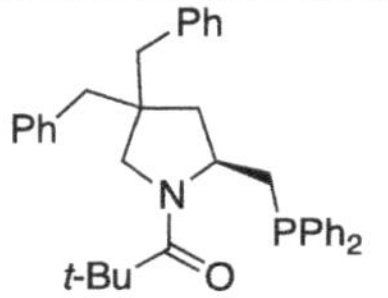	Ph	Et	98	93
2	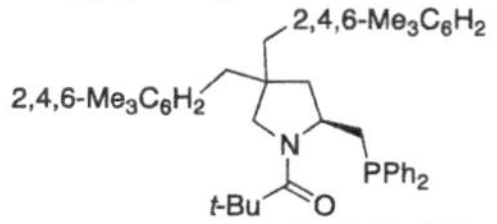	Ph	Et	97	96
3	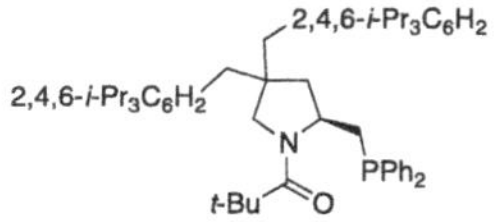	Ph	Et	96	88
4	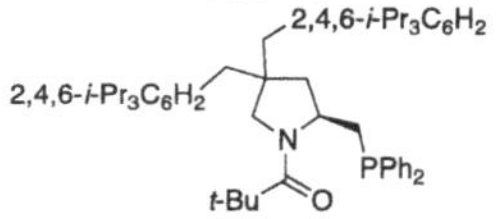	Ph	Me	97	87
5	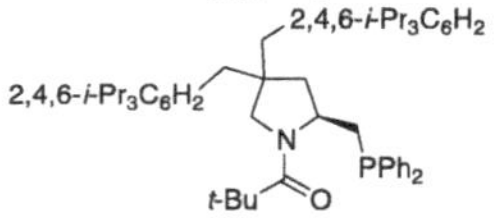	Ph	*i*-Pr	92	78
6	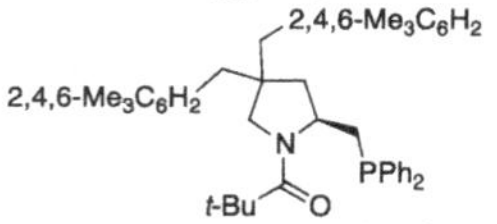	*c*-C_6H_{11}	Et	84	96
7	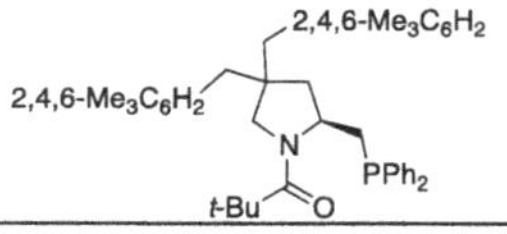	$Ph(CH_2)_2$	Et	69	93

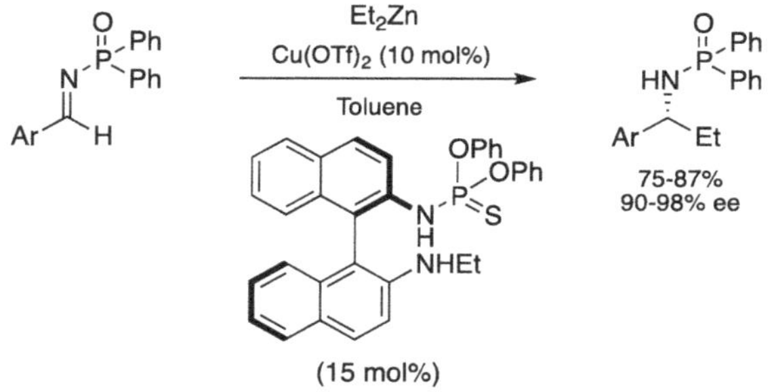

(11)

Table 6 Cu-catalyzed enantioselective addition of diorganozinc reagents to *N*-phosphinoylimines using Me-DuPHOS monoxide

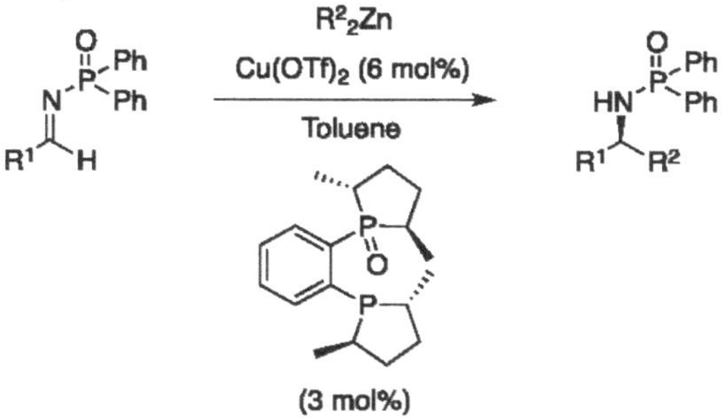

Entry	R^1	R^2	Yield (%)	*ee* (%)
1	Ph	Et	96	98
2	1-Naphthyl	Et	96	97
3	2-Furyl	Et	97	96
4	*c*-C_3H_5	Et	95	94
5	Ph	Me	87	97
6	Ph	*i*-Pr	84	95
7	Ph	*n*-$C_{10}H_{21}$	73	97
8	Ph	TBSO-$(CH_2)_6$-	52	90

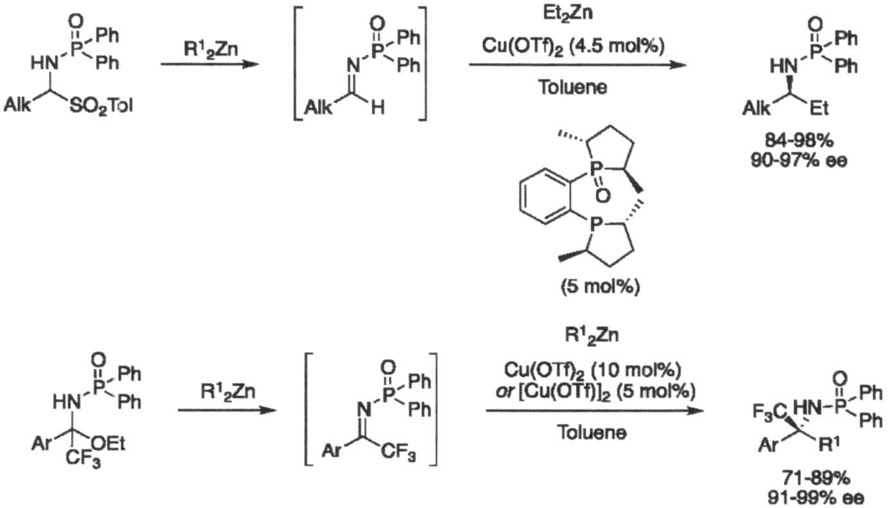

Scheme 12 Enantioselective addition reactions to *N*-phosphinoylimines

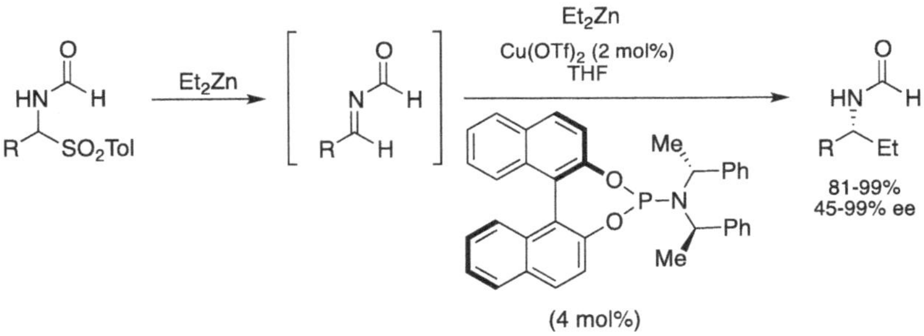

Scheme 13 Copper phosphoramide catalyzed dialkylzinc addition

N-Formylimines formed in situ from their corresponding sulfinic acid adducts are also good substrates for these reactions. A monodentate chiral phosphoramidite copper complex can catalyze the addition reaction, producing good enantiomeric excesses with aryl-substituted imines (Scheme 13) [189]. While all imine derivatives tested afforded high yield in this reaction, alkyl-substituted imines afforded only poor enantioselectivity (45–70% *ee*).

Early transition metal complexes derived from zirconium and hafnium are excellent catalysts for the addition of diorganozinc reagents to *N*-*o*-anisidylimines when peptide-based ligands are used [190–194]. Combinatorial screening of peptides led to the discovery that several chiral ligands in combination with Zr(O*i*-Pr)$_4$ were effective at catalyzing this transformation (12):

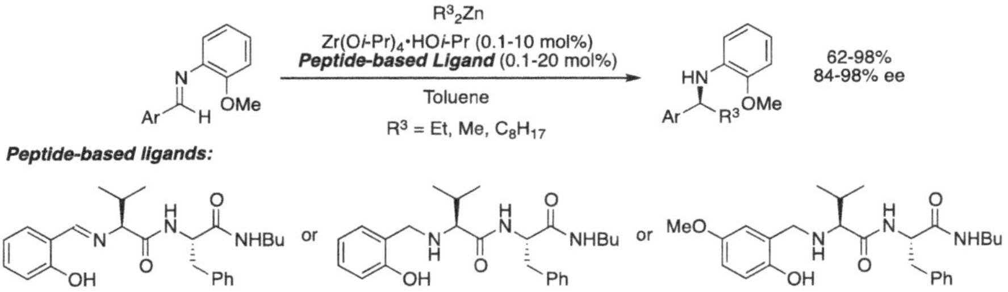

(12)

For alkyl-substituted imines, a three-component procedure was developed in which an in situ generation of the imine from the corresponding amine and aldehyde was necessary to get high yields and selectivities (Table 7).

The use of an analogous hafnium-based catalyst instead generally improves the yield while still affording high enantioselectivity (Scheme 14) [193].

These reactions are also compatible with a variety of electron-poor ketoimine derivatives (13) [194].

The *N*-aryl group can be removed under oxidative conditions such as upon treatment with iodosobenzene diacetate or $AgNO_3/(NH_4)_2S_2O_8$.

Table 7 Zr-catalyzed enantioselective addition of diorganozinc reagents to *N*-*o*-anisidylimines using peptide-based ligands

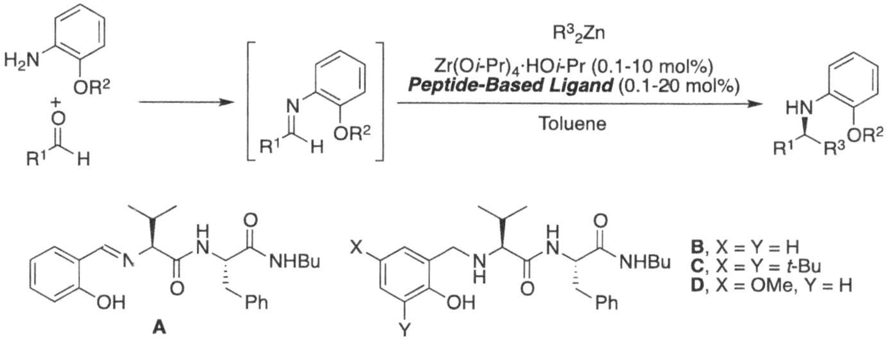

Entry	Ligand	R^1	R^2	R^3	Yield (%)	*ee* (%)
1	**A**	Ph	Me	Et	82	93
2	**A**	2-Furyl	Me	Et	87	97
3	**B**	Ph	Me	Me	79	88
4	**B**	Ph	Me	n-C_8H_{17}	63	98
5	**A**	3-Pyridyl	Me	Et	>98	85
6	**D**	n-C_4H_9	Me	Et	69	97
7	**D**	c-C_3H_5	Me	Et	83	98
8	**D**	Br-$(CH_2)_5$-	Me	Et	57	95
9	**C**	$TBSOCH_2CC$	Ph	Et	70	>98

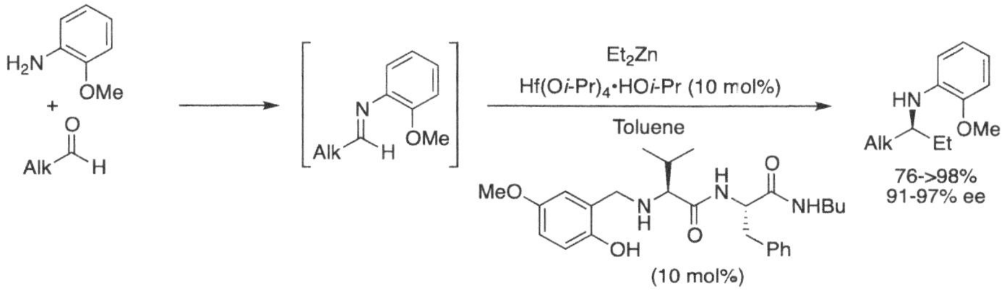

Scheme 14 Addition to *N*-arylimines

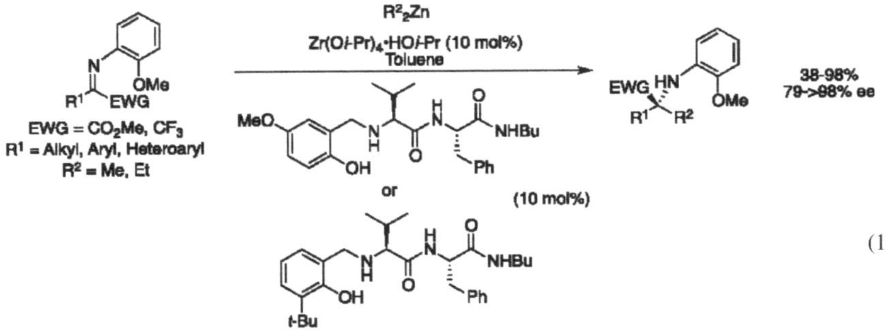

(13)

A rhodium-catalyzed dimethylzinc addition to *N*-tosylimine derivatives provides high yields and selectivities of the addition product (14) [195]:

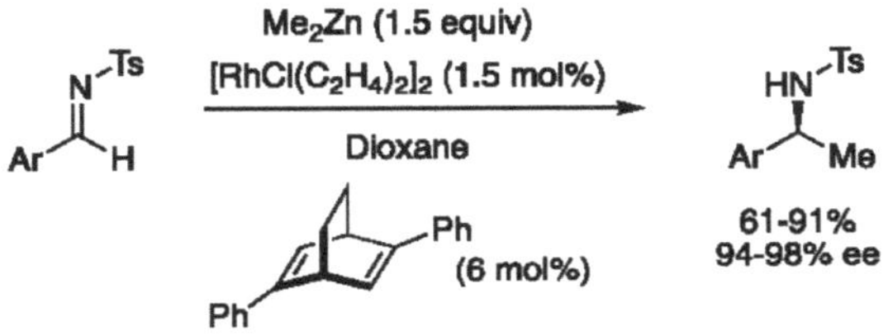

(14)

2.4.3 Addition of Alkenyl, Aryl, and Heteroarylmetals (sp^2 Carbon)

The generation of enantiomerically enriched allylic amines by the catalytic asymmetric addition of alkenylmetal to imine derivatives is an extremely important transformation. Since sp^2 hybridized carbon nucleophiles are less reactive than their sp^3 counterparts, one challenge associated with this process is the preparation of not only stereodefined vinylmetals but also those that will be reactive enough to undergo addition reactions with imines. A chiral Ni(0) catalyst can catalyze both, the carbometalation of an alkyne followed by its oxidative coupling with the imine leading to an enantioenriched allylic amine (15) [196, 197]. The process occurs with moderate to good enantioselectivity when a chiral phosphine is added to the reaction mixture:

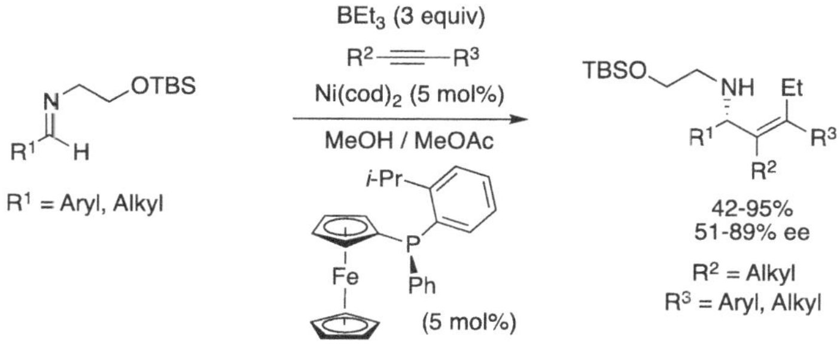

(15)

Alternatively, a chiral Ir(I) catalyst was also found to be competent in effecting an enantioselective oxidative coupling reaction, and the presence of H_2 terminates the catalytic cycle to lead to the synthesis of enantioenriched allylic amines (16) [198]:

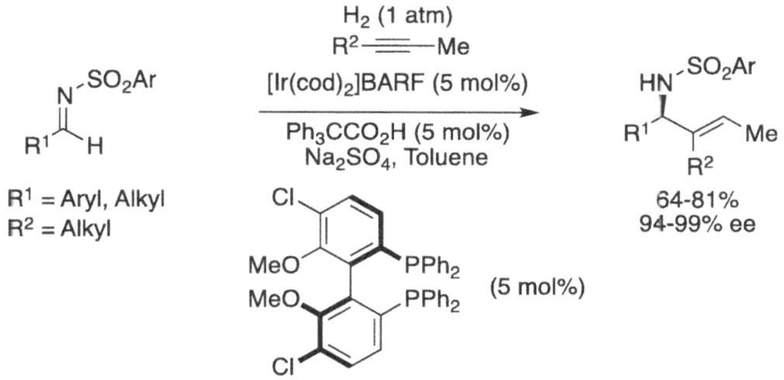

(16)

An analogous Rh(I) catalyst allows the use of acetylene to generate a chiral *cis*-dienylmetal equivalent that is capable of undergoing a highly enantioselective alkenylation of *N*-sulfonylimines (17) [199]:

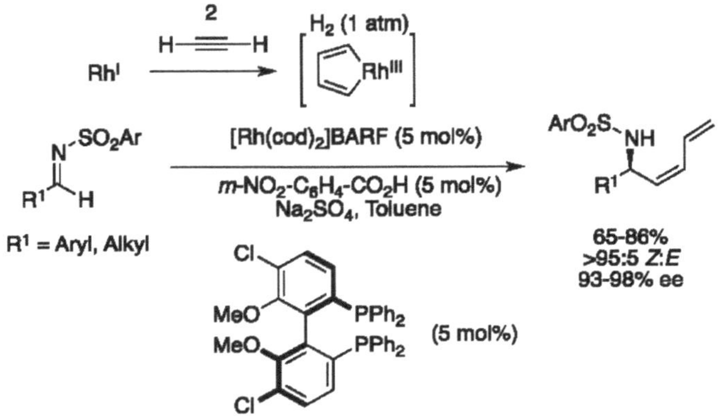

(17)

Vinylboronates can add to α-iminoester intermediates formed in situ through enantioselective organocatalysis with a chiral diol. A wide range of substituents on the amine or the alkenylboronate nucleophile are tolerated (Scheme 15) [200].

The catalytic asymmetric arylation of imines has emerged as one of the most powerful tools for the synthesis of the biologically important biarylmethylamine derivatives.

The Rh(I)-catalyzed enantioselective addition of arylmetal equivalents to *N*-sulfonylimines has become one the most powerful ways to effect this process. A wide variety of chiral phosphorus-based ligands have been tested for this reaction, and the most efficient methods are summarized in Table 8 [201–209]. Aryltitaniums, arylstannanes, arylboroxines, or arylboronic acid can be used as arylmetal equivalents added in stoichiometric amounts, usually with high yields and enantioselectivities.

The Rh(I)-catalyzed enantioselective arylation of alkenylimines is also possible under similar conditions (18), and the enantioselective arylation of aliphatic *N*-tosylimines with arylboronic acids is highly efficient using a chiral bis(phosphine) (19). *N*-Boc imines can also be used effectively as electrophiles only if the sulfinate adduct is employed as substrate (20):

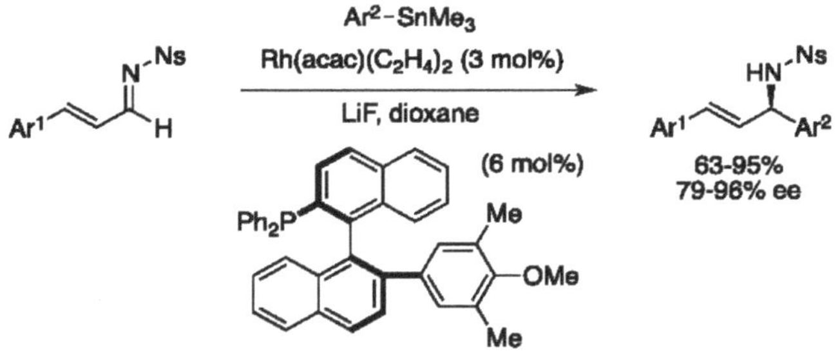

(18)

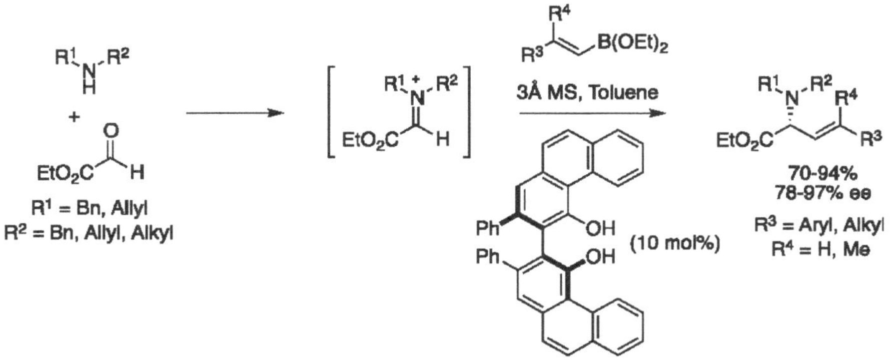

Scheme 15 Organocatalyzed vinylboronate addition reactions

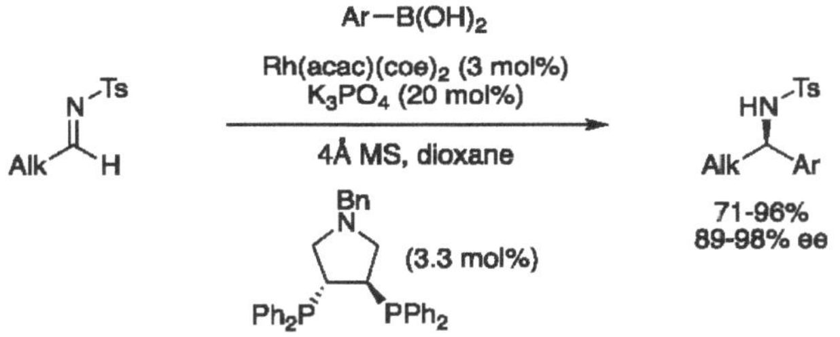

(19)

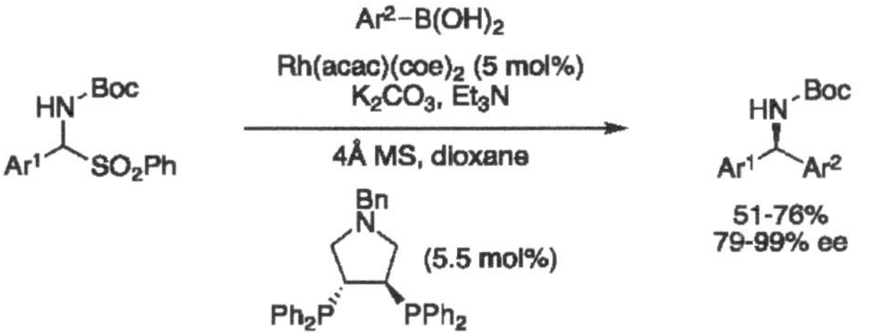

(20)

Chiral dienes have been found to be the most powerful ligands for the addition of boroxines to *N*-sulfonylimines, affording yields, enantiomeric excesses, and substrate generality superior to those of most of the other chiral ligands presented above. Selected examples are shown in Table 9 [210, 211].

Table 8 Rh(I)-catalyzed enantioselective arylation of *N*-sulfonylimines using chiral phosphine ligands

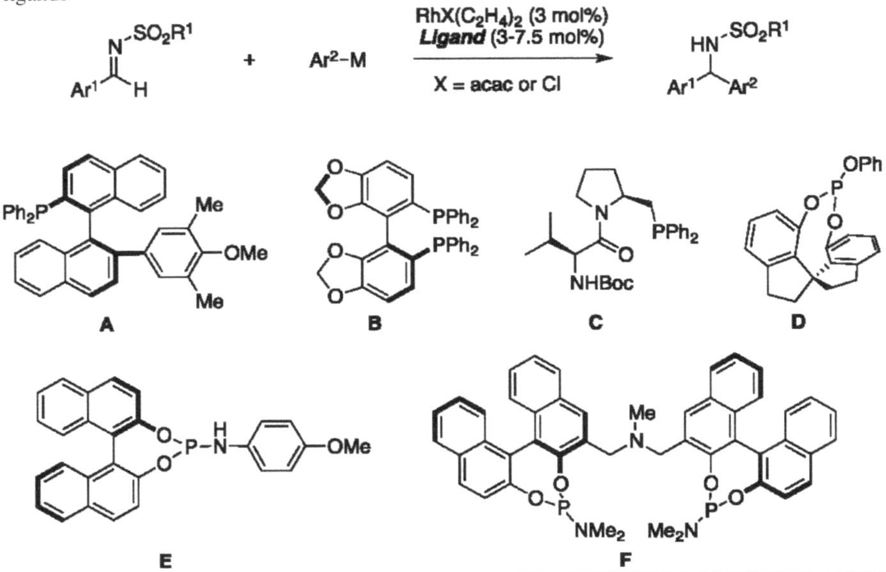

Entry	Ligand	Ar^1	Ar^2-M	R^1	Yield (%)	*ee* (%)	References
1	**A**	4-Cl-C_6H_4	Ph-$SnMe_3$	4-$NO_2C_6H_4$	83	92	[202]
2	**B**	4-Cl-C_6H_4	Ph-Ti(O*i*-Pr)$_3$	2,4,6-(*i*-Pr)$_3C_6H_4$	95	94	[203]
3	**C**	4-Me-C_6H_4	(4-$PhC_6H_4BO)_3$	4-MeC_6H_4	86	72	[204]
4	**D**	4-Cl-C_6H_4	Ph-B$(OH)_2$	4-MeC_6H_4	85	93	[205]
5	**E**	4-Cl-C_6H_4	Ph-B$(OH)_2$	NMe_2	95	95	[206]
6	**F**	4-Cl-C_6H_4	Ph-B$(OH)_2$	4-MeC_6H_4	98	95	[209]

2.4.4 Addition of Alkynylmetals (sp Carbon)

Early success in the enantioselective addition of alkynylmetal equivalents to imine derivatives to form these products in enantioenriched form was only possible using stoichiometric amounts of chiral reagents [212–215]. The synthesis of enantioenriched propargylamines by a catalytic asymmetric alkynylcopper addition reaction can be achieved by a three-component coupling, involving the reaction of an amine, an aldehyde, and a terminal alkyne with a chiral Cu catalyst (Scheme 16) [216–218] (for related methodologies see [219–225]). It is noteworthy that this transformation can be carried out either in toluene or in water with similar efficiency, and the reaction is also possible with preformed α-iminoesters as electrophiles.

Secondary amines, leading to iminium intermediates instead of active electrophiles, can also be used with copper catalysts derived from CuBr and several chiral

Table 9 Rh(I)-catalyzed enantioselective arylation of *N*-sulfonylimines using chiral diene ligands

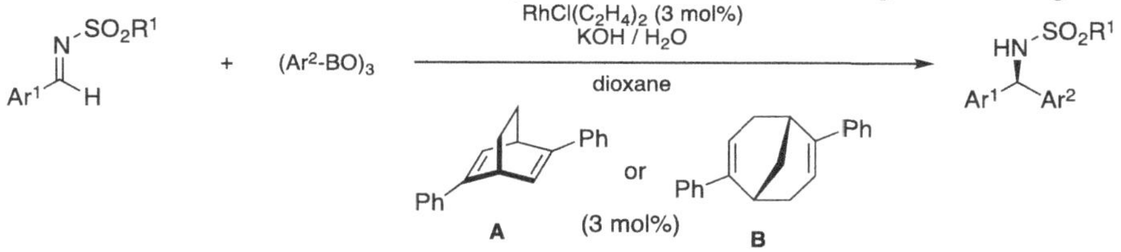

Entry	Ligand	Ar[1]	Ar[2]	R[1]	Yield (%)	*ee* (%)
1	**A**	4-Cl-C_6H_4	Ph	4-Me-C_6H_4	96	98
2	**B**	4-Cl-C_6H_4	Ph	4-NO_2-C_6H_4	96	98
3	**A**	4-CF_3-C_6H_4	Ph	4-Me-C_6H_4	97	95
4	**B**	4-CO_2Me-C_6H_4	Ph	4-NO_2-C_6H_4	95	95
5	**A**	4-MeO-C_6H_4	Ph	4-Me-C_6H_4	96	99
6	**B**	4-MeO-C_6H_4	Ph	4-NO_2-C_6H_4	98	99
7	**A**	1-Naphthyl	Ph	4-Me-C_6H_4	95	98
8	**B**	1-Naphthyl	Ph	4-NO_2-C_6H_4	94	96
9	**A**	Ph	4-Cl-C_6H_4	4-Me-C_6H_4	99	99
10	**B**	Ph	4-Cl-C_6H_4	4-NO_2-C_6H_4	96	99
11	**A**	Ph	4-MeO-C_6H_4	4-Me-C_6H_4	97	96
12	**B**	Ph	4-MeO-C_6H_4	4-NO_2-C_6H_4	94	98

ligands, affording the propargylic amine in excellent enantioselectivity (Scheme 17) [226–234].

Chiral Zr-catalysts based on peptides can be used with mixed alkynylzinc reagents and *N*-anisidylimines to produce the aryl-protected propargyl amines in good yields and enantioselectivities (21) [235]:

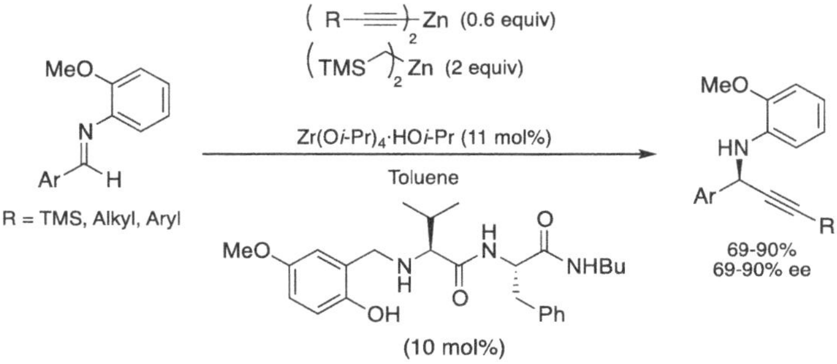

(21)

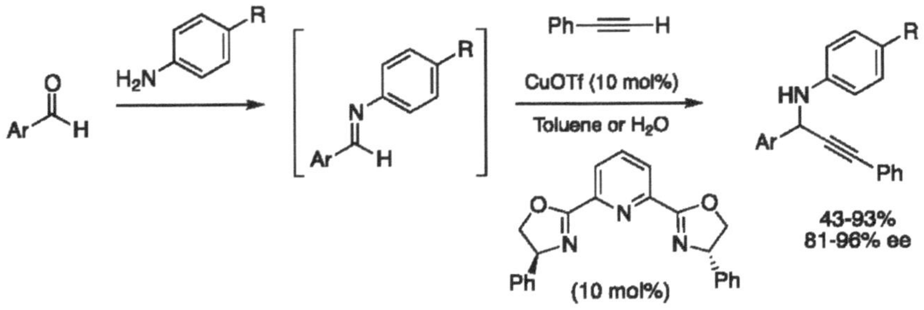

Scheme 16 Copper•pybox catalyzed alkynylation reaction

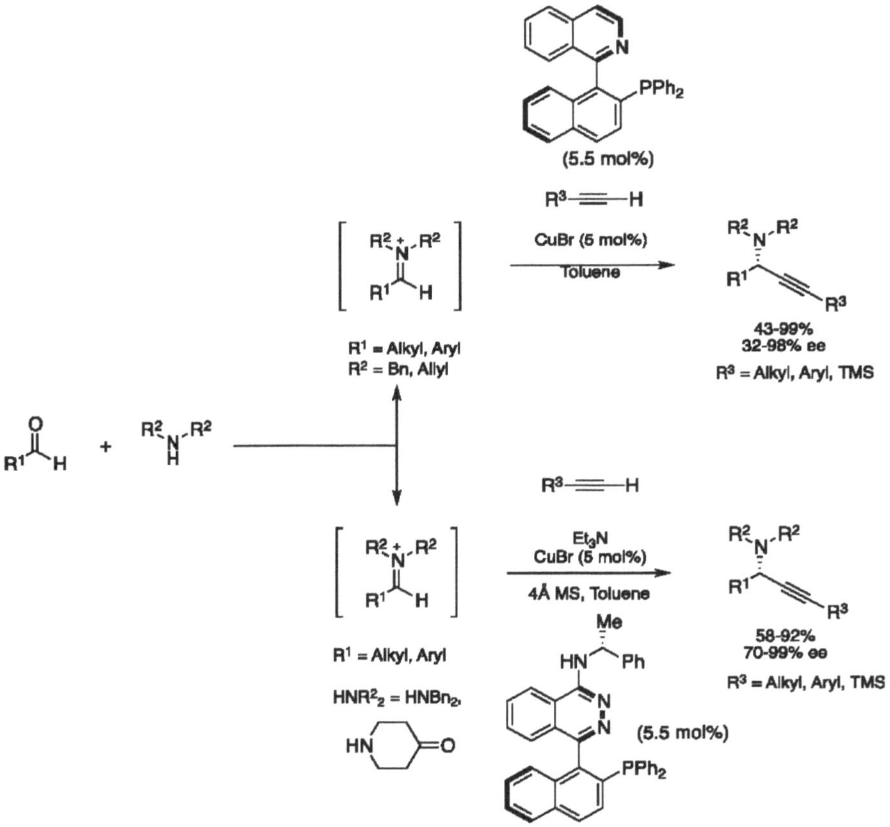

Scheme 17 Enantioselective copper-catalyzed alkynylation of iminium salts

2.4.5 Addition of Allylmetal

Allylmetal nucleophiles display a unique reactivity compared to other types of nucleophiles. If the metal bearing the allyl unit is Lewis acidic, the reaction can occur via an initial Lewis acid/Lewis base interaction followed by an allyl group transfer via a six-membered ring. Chiral Pd- and Zr-based chiral complexes are found to be highly efficient catalysts in the addition of various achiral allylstannanes to *N*-benzyl- and *N*-*o*-anisidylimines, respectively, in good enantioselectivity for aryl-substituted imines (22, 23) [236–241]. The allylstannane can be replaced with an allylsilane reagent with similar efficiency [237, 239], and the analogous carboethoxyallylation is also possible [240].

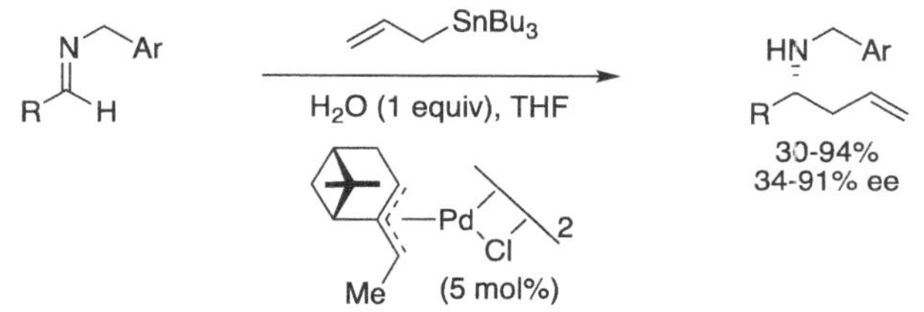

(22)

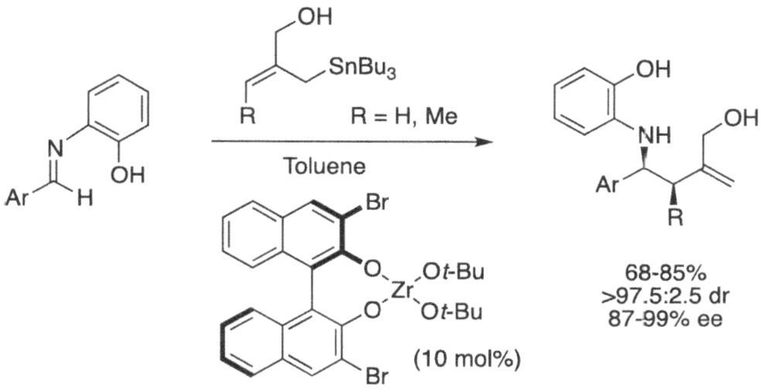

(23)

In addition to these methods, a few other enantioselective methods involving Zn- or Cu-based catalysts were developed for the addition of allylstannane and allylsilane nucleophiles to α-iminoesters, although the enantioselectivities of these processes are modest [242–244].

Allylboronates are found to be good nucleophiles in the enantioselective CuDuPHOS catalyzed allylation of *N*-benzylketoimine electrophiles (24) [245]. It is proposed that the fluoride counterion accelerates the transmetallation of the allyl group from the boronate species to produce the chiral allylcopper complex that later reacts with the imine.

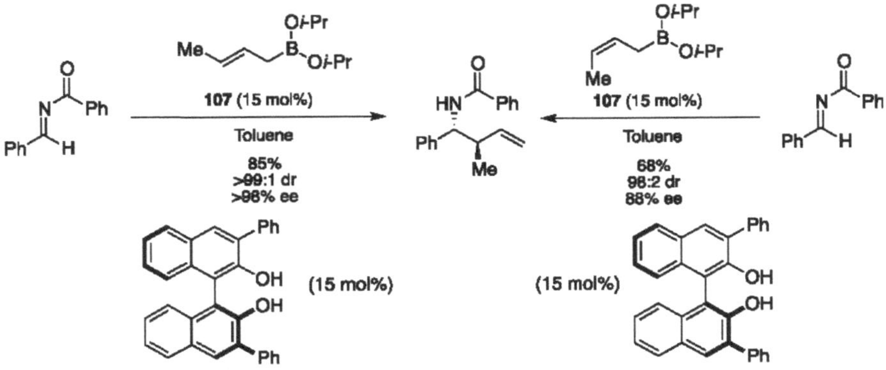

Scheme 18 Crotylation reactions of *N*-acylimines

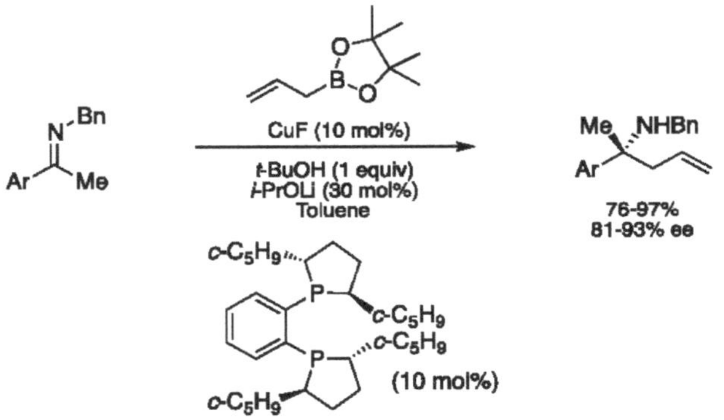

(24)

The organocatalyzed enantioselective allylboration of *N*-acylimines is possible with BINOL derivatives (25, Scheme 18) [246]. This reaction is believed to proceed via ligand exchange between an isopropoxy group of the allylboronate and BINOL, followed by allylboration of the imine activated through hydrogen bonding with the remaining free hydroxyl group of the organocatalyst:

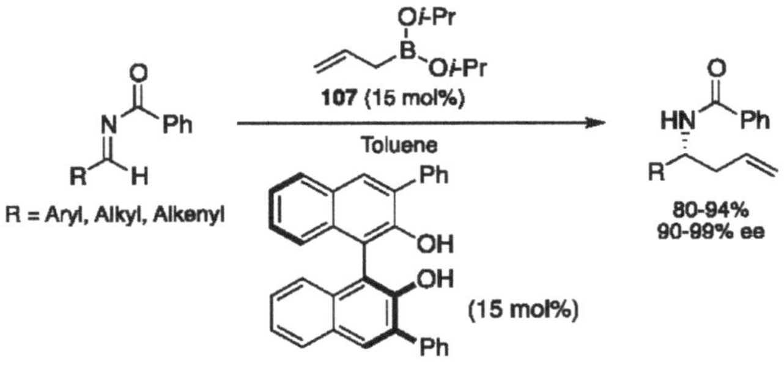

(25)

The corresponding *cis*- and *trans*-crotylboronate reagents afford a high diastereo- and enantioselectivity in the process. Quite interestingly, the *anti* product is formed with both reagents. This is in sharp contrast with what is usually observed in uncatalyzed processes. A boat-type transition state is proposed to be favored when the *cis*-crotylboronate is employed, whereas the expected chair-like transition structure is postulated with the *trans*-isomer.

As previously discussed, allylindium species formed in situ from allyl bromide and In(0) are known to add efficiently to *N*-acylhydrazones. Chiral ureas, added in catalytic amounts, have been shown to activate the imine electrophile via hydrogen bonding and allowed for enantioinduction (26) [247]:

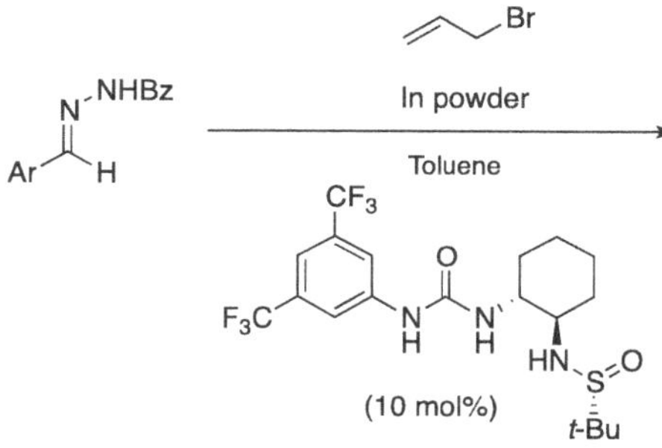

(26)

3 Conclusion

The section described above illustrated that the addition of nonstabilized nucleophiles to imine derivatives is one of the most reliable ways to form α-chiral amines. Although chiral auxiliary-based approaches are less atom-economic, they still constitute some of the most widely used methods for the synthesis of a variety of enantioenriched α-chiral amines. In the last decade, tremendous progress has been achieved in the field of asymmetric catalysis in general, and the knowledge acquired during that time is partly due to the continuous efforts devoted to the development of more efficient asymmetric addition reactions to imine derivatives. The catalytic enantioselective addition of sp^3, sp^2, or sp carbon-centered nonstabilized nucleophiles is now available for most nucleophiles (except acetylene) although addition reactions to ketoimine derivatives to generate tertiary carbinylamine still represent a considerable challenge in certain cases.

References

1. Blaser H-U, Spindler F (2009) Org React 74:1
2. Shibasaki M, Kanai M, Mita T (2008) Org React 70:1
3. Mukherjee S, Yang JW, Hoffmann S, List B (2007) Chem Rev 107:5471
4. Verkade JMM, van Hemert LJC, Quaedflieg PJLM, Rutjes FPJT (2008) Chem Soc Rev 37:29
5. Lebel H, Leogane O (2005) Org Lett 7:4107
6. Yamada K, Tomioka K (2008) Chem Rev 108:2874
7. Charette AB, Boezio AA, Cote A, Moreau E, Pytkowicz J, Desrosiers JN, Legault C (2005) Pure Appl Chem 1259
8. Appel R, Mayr H (2011) J Am Chem Soc 133:8240
9. Yamamoto Y, Komatsu T, Maruyama K (1984) J Am Chem Soc 106:5031
10. Cainelli G, Giacomini D, Panunzio M, Zarantonello P (1992) Tetrahedron Lett 33:7783
11. Franz T, Hein M, Veith U, Jager V, Peters EM, Peters K, Vonschnering HG (1994) Angew Chem Int Ed 33:1298
12. Reetz MT, Jaeger R, Drewlies R, Hubel M (1991) Angew Chem Int Ed 30:103
13. Solladié-Cavallo A, Bonne F (1996) Tetrahedron Asymmetry 7:171
14. Claremon DA, Lumma PK, Phillips BT (1986) J Am Chem Soc 108:8265
15. Baker WR, Condon SL (1993) J Org Chem 58:3277
16. Remuzon P, Dussy C, Jacquet JP, Roty P, Bouzard D (1996) Tetrahedron Asymmetry 7:1181
17. Thiam M, Chastrette F (1990) Tetrahedron Lett 31:1429
18. Cativiela C, Díaz-de-Villegas MD, Gálvez JA (1997) Tetrahedron Asymmetry 8:1605
19. Alexakis A, Lensen N, Mangeney P (1991) Tetrahedron Lett 32:1171
20. Alexakis A, Lensen N, Tranchier JP, Mangeney P, Feneaudupont J, Declercq JP (1995) Synthesis 1038
21. Alexakis A, Lensen N, Tranchier JP, Mangeney P (1992) J Org Chem 57:4563
22. Yamamoto Y, Ito W (1988) Tetrahedron 44:5415
23. Hanessian S, Yang RY (1996) Tetrahedron Lett 37:5273
24. Miyabe H, Ushiro C, Naito T (1997) Chem Commun 1789
25. Dondoni A, Junquera F, Merchan FL, Merino P, Tejero T (1992) Tetrahedron Lett 33:4221
26. Dondoni A, Franco S, Merchan FL, Merino P, Tejero T (1993) Tetrahedron Lett 34:5475
27. Dondoni A, Franco S, Merchan F, Merino P, Tejero T (1993) Synlett 78
28. Dondoni A, Franco S, Junquera F, Merchan FL, Merino P, Tejero T, Bertolasi V (1995) Chem Eur J 1:505
29. Dondoni A, Franco S, Merchan FL, Merino P, Tejero T (1993) Tetrahedron Lett 34:5479
30. Dondoni A, Junquera F, Merchan FL, Merino P, Tejero T (1994) Synthesis 1450
31. Dondoni A, Junquera F, Merchan FL, Merino P, Tejero T (1995) Chem Commun 2127
32. Merino P, Anoro S, Castillo E, Merchan F, Tejero T (1996) Tetrahedron Asymmetry 7:1887
33. Merino P, Franco S, Merchan FL, Tejero T (1997) Tetrahedron Asymmetry 8:3489
34. Dondoni A, Perrone D (1997) Tetrahedron Lett 38:499
35. Merino P, Castillo E, Merchan FL, Tejero T (1997) Tetrahedron Asymmetry 8:1725
36. Merino P, Lanaspa A, Merchan FL, Tejero T (1997) Tetrahedron Asymmetry 8:2381
37. Merino P, Lanaspa A, Merchan FL, Tejero T (1997) Tetrahedron Lett 38:1813
38. Badorrey R, Cativiela C, DiazdeVillegas MD, Galvez JA (1997) Tetrahedron 53:1411
39. Charette AB, Mellon C (1998) Tetrahedron 54:10525
40. Thiam M, Chastrette F (1992) Bull Chem Soc Fr 129:161
41. Alexakis A, Tranchier JP, Lensen N, Mangeney P (1995) J Am Chem Soc 117:10767
42. Matsubara S, Ukita H, Kodama T, Utimoto K (1994) Chem Lett 831
43. Yamamoto Y, Nishii S, Maruyama K, Komatsu T, Ito W (1986) J Am Chem Soc 108:7778
44. Beuchet P, Lemarrec N, Mosset P (1992) Tetrahedron Lett 33:5959
45. Wang DK, Dai LX, Hou XL (1995) Tetrahedron Lett 36:8649
46. Wang DK, Dai LX, Hou XL, Zhang Y (1996) Tetrahedron Lett 37:4187
47. Gao YA, Sato F (1995) J Org Chem 60:8136

48. Alvaro G, Savoia D, Valentinetti MR (1996) Tetrahedron 52:12571
49. Wakchaure VN, Mohanty RR, Shaikh AJ, Nugent TC (2007) Eur J Org Chem 959
50. Takahashi H, Suzuki Y, Inagaki H (1982) Chem Pharm Bull 30:3160
51. Suzuki Y, Takahashi H (1983) Chem Pharm Bull 31:31
52. Suzuki Y, Takahashi H (1983) Chem Pharm Bull 31:2895
53. Takahashi H, Chida Y, Suzuki T, Yanaura S, Suzuki Y, Masuda C (1983) Chem Pharm Bull 31:1659
54. Takahashi H, Suzuki Y, Hori T (1983) Chem Pharm Bull 31:2183
55. Takahashi H, Chida Y, Suzuki T, Onishi H, Yanaura S (1984) Chem Pharm Bull 32:2714
56. Takahashi H, Chida Y, Higashiyama K, Onishi H (1985) Chem Pharm Bull 33:4662
57. Ukaji Y, Watai T, Sumi T, Fujisawa T (1991) Chem Lett 1555
58. Higashiyama K, Fujikura H, Takahashi H (1995) Chem Pharm Bull 43:722
59. Hashimoto Y, Takaoki K, Sudo A, Ogasawara T, Saigo K (1995) Chem Lett 235
60. Bhuyan PJ, Prajapati D, Sandhu JS (1993) Tetrahedron Lett 34:7975
61. Giammaruco M, Taddei M, Ulivi P (1993) Tetrahedron Lett 34:3635
62. Bellucci C, Cozzi PG, Umanironchi A (1995) Tetrahedron Lett 36:7289
63. Tanaka H, Inoue K, Pokorski U, Taniguchi M, Torii S (1990) Tetrahedron Lett 31:3023
64. Bocoum A, Savoia D, Umanironchi A (1993) Chem Commun 1542
65. Basile T, Bocoum A, Savoia D, Umanironchi A (1994) J Org Chem 59:7766
66. Loh TP, Ho DSC, Xu KC, Sim KY (1997) Tetrahedron Lett 38:865
67. Dembélé YA, Belaud C, Hitchcock P, Villiéras J (1992) Tetrahedron Asymmetry 3:351
68. Nyzam V, Belaud C, Zammattio F, Villieras J (1996) Tetrahedron Asymmetry 7:1835
69. Hashimoto Y, Kobayashi N, Kai A, Saigo K (1995) Synlett 961
70. Saudan LA, Bernardinelli G, Kündig EP (2000) Synlett 483
71. Kohara T, Hashimoto Y, Saigo K (1999) Tetrahedron 55:6453
72. Bertrand MP, Coantic S, Feray L, Nouguier R, Perfetti P (2000) Tetrahedron 56:3951
73. Sun XX, Worthy AD, Tan KL (2011) Angew Chem Int Ed 50:8167
74. Simon J, Chelain E, Brigaud T (2012) Org Lett 14:604
75. Murali RVNS, Chandrasekhar S (2012) Tetrahedron Lett 53:3467
76. Job A, Janeck CF, Bettray W, Peters R, Enders D (2002) Tetrahedron 58:2253
77. Enders D, Schubert H, Nübling C (1986) Angew Chem Int Ed 25:1109
78. Denmark SE, Weber T, Piotrowski DW (1987) J Am Chem Soc 109:2224
79. Denmark SE, Edwards JP, Nicaise O (1993) J Org Chem 58:569
80. Kim YH, Choi JY (1996) Tetrahedron Lett 37:5543
81. Takahashi H, Tomita K, Otomasu H (1979) Chem Commun 668
82. Takahashi H, Tomita K, Noguchi H (1981) Chem Pharm Bull 29:3387
83. Takahashi H, Inagaki H (1982) Chem Pharm Bull 30:922
84. Takahashi H, Suzuki Y (1983) Chem Pharm Bull 31:4295
85. Denmark SE, Edwards JP, Weber T, Piotrowski DW (2010) Tetrahedron Asymmetry 21:1278
86. Friestad GK, Korapala CS, Ding H (2006) J Org Chem 71:281
87. Friestad GK, Ding H (2001) Angew Chem Int Ed 40:4491
88. Sugiura M, Kobayashi S (2005) Angew Chem Int Ed 44:5176
89. Cook GR, Maity BC, Kargbo R (2004) Org Lett 6:1741
90. Friestad GK, Qin J (2000) J Am Chem Soc 122:8329
91. Friestad GK, Qin J (2001) J Am Chem Soc 123:9922
92. Fernández M, Alonso R (2003) Org Lett 5:2461
93. Friestad GK, Marié JC, Deveau AM (2004) Org Lett 6:3249
94. Friestad GK (2005) Eur J Org Chem 3157
95. Friestad GK, Draghici C, Soukri M, Qin J (2005) J Org Chem 70:6330
96. Friestad GK, Marié JC, Suh YS, Qin J (2006) J Org Chem 71:7016
97. Korapala CS, Qin J, Friestad GK (2007) Org Lett 9:4243
98. Friestad GK, Ji A (2008) Org Lett 10:2311

99. Friestad GK, Banerjee K (2009) Org Lett 11:1095
100. Friestad GK, Ji A, Korapala CS, Qin J (2011) Org Biomol Chem 9:4039
101. Friestad GK, Ji A, Baltrusaitis J, Korapala CS, Qin J (2012) J Org Chem 77:3159
102. Dieter RK, Datar R (1993) Can J Chem 71:814
103. Gallagher PT, Lightfoot AP, Moody CJ, Slawin AMZ (1995) Synlett 445
104. Brown DS, Gallagher PT, Lightfoot AP, Moody CJ, Slawin AMZ, Swann E (1995) Tetrahedron 51:11473
105. Gallagher PT, Hunt JCA, Lightfoot AP, Moody CJ (1997) J Chem Soc-Perkin Transactions 1:2633
106. Moody CJ, Lightfoot AP, Gallagher PT (1997) J Org Chem 62:746
107. Moody CJ, Lightfoot AP, Gallagher PT (1997) Synlett 659
108. Cao ZP, Liu ZQ, Liu YL, Du HF (2011) J Org Chem 76:6401
109. Hua DH, Miao SW, Chen JS, Iguchi S (1991) J Org Chem 56:4
110. Moreau P, Essiz M, Mérour JY, Bouzard D (1997) Tetrahedron Asymmetry 8:591
111. Koriyama Y, Nozawa A, Hayakawa R, Shimizu M (2002) Tetrahedron 58:9621
112. Wipf P, Nunes RL, Ribe S (2002) Helvetica Chimica Acta 85:3478
113. Li GG, Wei HX, Whittlesey BR, Batrice NN (1999) J Org Chem 64:1061
114. Li SW, Batey RA (2004) Chem Commun 1382
115. Liu GC, Cogan DA, Ellman JA (1997) J Am Chem Soc 119:9913
116. Cogan DA, Liu GC, Ellman J (1999) Tetrahedron 55:8883
117. Cogan DA, Ellman JA (1999) J Am Chem Soc 121:268
118. Shaw AW, deSolms SJ (2001) Tetrahedron Lett 42:7173
119. Ellman JA, Owens TD, Tang TP (2002) Accounts Chem Res 35:984
120. Mukade T, Dragoli DR, Ellman JA (2003) J Comb Chem 5:590
121. Weix DJ, Ellman JA (2003) Org Lett 5:1317
122. Reynolds TE, Binkley MS, Scheidt KA (2008) Org Lett 10:5227
123. Ding CH, Chen DD, Luo ZB, Dai LX, Hou XL (2006) Synlett 1272
124. Lettan RB, Scheidt KA (2005) Org Lett 7:3227
125. Weix DJ, Shi YL, Ellman JA (2005) J Am Chem Soc 127:1092
126. Beenen MA, Weix DJ, Ellman JA (2006) J Am Chem Soc 128:6304
127. Bolshan Y, Batey RA (2005) Org Lett 7:1481
128. Kong JR, Cho CW, Krische MJ (2005) J Am Chem Soc 127:11269
129. Patterson AW, Ellman JA (2006) J Org Chem 71:7110
130. Yang TK, Chen RY, Lee DS, Peng WS, Jiang YZ, Mi AQ, Jong TT (1994) J Org Chem 59:914
131. Yan L, Guishu Y, Yaozhong J, Tengkuia Y (1995) Synth Commun 25:1551
132. Robak MT, Herbage MA, Ellman JA (2010) Chem Rev 110:3600
133. Watanabe K, Ito K, Itsuno S (1995) Tetrahedron Asymmetry 6:1531
134. Itsuno S, Watanabe K, Ito K, ElShehawy AA, Sarhan AA (1997) Angew Chem Int Ed 36:109
135. Chataigner I, Zammattio F, Lebreton J, Villieras J (1998) Synlett 275
136. Itsuno S, Yokoi A, Kuroda S (1999) Synlett 1987
137. Chen G-M, Ramachandran PV, Brown HC (1999) Angew Chem Int Ed 38:825
138. Sugiura M, Hirano K, Kobayashi S (2004) J Am Chem Soc 126:7182
139. Ramachandran PV, Burghardt TE (2005) Chem Eur J 11:4387
140. Wu TR, Chong JM (2006) J Am Chem Soc 128:9646
141. Canales E, Hernandez E, Soderquist JA (2006) J Am Chem Soc 128:8712
142. Chataigner I, Zammattio F, Lebreton J, Villieras J (2008) Tetrahedron 64:2441
143. Panek JS, Jain NF (1994) J Org Chem 59:2674
144. Schaus JV, Jain N, Panek JS (2000) Tetrahedron 56:10263
145. Berger R, Rabbat PMA, Leighton JL (2003) J Am Chem Soc 125:9596
146. Berger R, Duff K, Leighton JL (2004) J Am Chem Soc 126:5686
147. Hanessian S, Yang RY (1996) Tetrahedron Lett 37:8997
148. Nakamura M, Hirai A, Nakamura E (1996) J Am Chem Soc 118:8489

149. Cook GR, Kargbo R, Maity B (2005) Org Lett 7:2767
150. Kargbo R, Takahashi Y, Bhor S, Cook GR, Lloyd-Jones GC, Shepperson IR (2007) J Am Chem Soc 129:3846
151. Tomioka K, Inoue I, Shindo M, Koga K (1990) Tetrahedron Lett 31:6681
152. Tomioka K, Inoue I, Shindo M, Koga K (1991) Tetrahedron Lett 32:3095
153. Inoue I, Shindo M, Koga K, Tomioka K (1993) Tetrahedron Asymmetry 4:1603
154. Inoue I, Shindo M, Koga K, Tomioka K (1994) Tetrahedron 50:4429
155. Inoue I, Shindo M, Koga K, Kanai M, Tomioka K (1995) Tetrahedron Asymmetry 6:2527
156. Jones CA, Jones IG, North M, Pool CR (1995) Tetrahedron Lett 36:7885
157. Jones CA, Jones IG, Mulla M, North M, Sartori L (1997) J Chem Soc Perkin Trans 1:2891
158. Soai K, Hatanaka T, Miyazawa T (1992) Chem Commun 1097
159. Zhang XM, Gong LH, Mi AQ, Cui X, Jiang YH, Choi MCK, Chan ASC (2001) Tetrahedron Lett 42:6369
160. Denmark SE, Nakajima N, Nicaise OJC (1994) J Am Chem Soc 116:8797
161. Itsuno S, Sasaki M, Kuroda S, Ito K (1995) Tetrahedron Asymmetry 6:1507
162. Denmark SE, Nicaise OJC (1996) Chem Commun 999
163. Denmark SE, Nakajima N, Stiff CM, Nicaise OJC, Kranz M (2008) Adv Synth Catal 350:1023
164. Denmark SE, Nakajima N, Nicaise OJC, Faucher AM, Edwards JP (1995) J Org Chem 60:4884
165. Denmark SE, Stiff CM (2000) J Org Chem 65:5875
166. Beresford KJM (2002) Tetrahedron Lett 43:7175
167. Dahmen S, Bräse S (2002) J Am Chem Soc 124:5940
168. Hermanns N, Dahmen S, Bolm C, Bräse S (2002) Angew Chem Int Ed 41:3692
169. Zhang HL, Liu H, Cui X, Mi AQ, Jiang YZ, Gong LZ (2005) Synlett 615
170. Liu H, Zhang HL, Wang SJ, Mi AQ, Jiang YZ, Gong LZ (2005) Tetrahedron Asymmetry 16:2901
171. Fujihara H, Nagai K, Tomioka K (2000) J Am Chem Soc 122:12055
172. Nagai K, Fujihara H, Kuriyama M, Yamada K, Tomioka K (2002) Chem Lett 8
173. Soeta T, Nagai K, Fujihara H, Kuriyama M, Tomioka K (2003) J Org Chem 68:9723
174. Wang MC, Xu CL, Zou YX, Liu HM, Wang DK (2005) Tetrahedron Lett 46:5413
175. Boezio AA, Charette AB (2003) J Am Chem Soc 125:1692
176. Boezio AA, Pytkowicz J, Côté A, Charette AB (2003) J Am Chem Soc 125:14260
177. Côté A, Boezio AA, Charette AB (2004) Proc Natl Acad Sci U S A 101:5405
178. Côté A, Boezio AA, Charette AB (2004) Angew Chem Int Ed 43:6525
179. Côté A, Charette AB (2005) J Org Chem 70:10864
180. Desrosiers JN, Côté A, Charette AB (2005) Tetrahedron 61:6186
181. Charette AB, Côté A, Desrosiers JN, Bonnaventure I, Lindsay VNG, Lauzon C, Tannous J, Boezio AA (2008) Pure Appl Chem 881
182. Bonnaventure I, Charette AB (2008) J Org Chem 73:6330
183. Côté A, Charette AB (2008) J Am Chem Soc 130:2771
184. Lauzon C, Charette AB (2006) Org Lett 8:2743
185. Shi M, Wang CJ (2003) Adv Synth Catal 345:971
186. Shi M, Lei ZY, Xu Q (2006) Adv Synth Catal 348:2237
187. Wang CJ, Shi M (2003) J Org Chem 68:6229
188. Shi M, Zhang W (2003) Tetrahedron Asymmetry 14:3407
189. Pizzuti MG, Minnaard AJ, Feringa BL (2008) J Org Chem 73:940
190. Porter JR, Traverse JF, Hoveyda AH, Snapper ML (2001) J Am Chem Soc 123:984
191. Porter JR, Traverse JF, Hoveyda AH, Snapper ML (2001) J Am Chem Soc 123:10409
192. Akullian LC, Snapper ML, Hoveyda AH (2003) Angew Chem Int Ed 42:4244
193. Akullian LC, Porter JR, Traverse JE, Snapper ML, Hoveyda AH (2005) Adv Synth Catal 347:417
194. Fu P, Snapper ML, Hoveyda AH (2008) J Am Chem Soc 130:5530

195. Nishimura T, Yasuhara Y, Hayashi T (2006) Org Lett 8:979
196. Patel SJ, Jamison TF (2003) Angew Chem Int Ed 42:1364
197. Patel SJ, Jamison TF (2004) Angew Chem Int Ed 43:3941
198. Ngai MY, Barchuk A, Krische MJ (2007) J Am Chem Soc 129:12644
199. Skucas E, Kong JR, Krische MJ (2007) J Am Chem Soc 129:7242
200. Lou S, Schaus SE (2008) J Am Chem Soc 130:6922
201. Nishimura T, Yasuhara Y, Hayashi T (2006) Org Lett 8:979
202. Hayashi T, Ishigedani M (2001) Tetrahedron 57:2589
203. Hayashi T, Kawai M, Tokunaga N (2004) Angew Chem Int Ed 43:6125
204. Kuriyama M, Soeta T, Hao XY, Chen O, Tomioka K (2004) J Am Chem Soc 126:8128
205. Duan HF, Jia YX, Wang LX, Zhou QL (2006) Org Lett 8:2567
206. Jagt RBC, Toullec PY, Geerdink D, de Vries JG, Feringa BL, Minnaard ADJ (2006) Angew Chem Int Ed 45:2789
207. Nakagawa H, Rech JC, Sindelar RW, Ellman JA (2007) Org Lett 9:5155
208. Trincado M, Ellman JA (2008) Angew Chem Int Ed 47:5623
209. Kurihara K, Yamamoto Y, Miyaura N (2009) Adv Synth Catal 351:260
210. Tokunaga N, Otomaru Y, Okamoto K, Ueyama K, Shintani R, Hayashi T (2004) J Am Chem Soc 126:13584
211. Otomaru Y, Tokunaga N, Shintani R, Hayashi T (2005) Org Lett 7:307
212. Huffman MA, Yasuda N, Decamp AE, Grabowski EJJ (1995) J Org Chem 60:1590
213. Frantz DE, Fassler R, Carreira EM (1999) J Am Chem Soc 121:11245
214. Kauffman GS, Harris GD, Dorow RL, Stone BRP, Parsons RL, Pesti JA, Magnus NA, Fortunak JM, Confalone PN, Nugent WA (2000) Org Lett 2:3119
215. Wu TR, Chong JM (2006) Org Lett 8:15
216. Wei CM, Li CJ (2002) J Am Chem Soc 124:5638
217. Wei CM, Mague JT, Li CJ (2004) Proc Natl Acad Sci U S A 101:5749
218. Ji JX, Wu J, Chan ASC (2005) Proc Natl Acad Sci U S A 102:11196
219. Benaglia M, Negri D, Dell'Anna G (2004) Tetrahedron Lett 45:8705
220. Orlandi S, Colombo F, Benaglia M (2005) Synthesis 1689
221. Colombo F, Benaglia M, Orlandi S, Usuelli F, Celentano G (2006) J Org Chem 71:2064
222. Bisai A, Singh VK (2006) Org Lett 8:2405
223. Liu B, Huang L, Liu JT, Zhong Y, Li XS, Chan ASC (2007) Tetrahedron Asymmetry 18:2901
224. Hatano M, Asai T, Ishihara K (2008) Tetrahedron Lett 49:379
225. Irmak M, Boysen MMK (2008) Adv Synth Catal 350:403
226. Koradin C, Polborn K, Knochel P (2002) Angew Chem Int Ed 41:2535
227. Koradin C, Gommermann N, Polborn K, Knochel P (2003) Chem Eur J 9:2797
228. Gommermann N, Koradin X, Polborn K, Knochel P (2003) Angew Chem Int Ed 42:5763
229. Gommermann N, Knochel P (2004) Chem Commun 2324
230. Gommermann N, Knochel P (2005) Synlett 2799
231. Gommermann N, Knochel P (2005) Chem Commun 4175
232. Gommermann N, Knochel P (2006) Chem Eur J 12:4380
233. Knopfel TF, Aschwanden P, Ichikawa T, Watanabe T, Carreira EM (2004) Angew Chem Int Ed 43:5971
234. Aschwanden P, Stephenson CRJ, Carreira EM (2006) Org Lett 8:2437
235. Traverse JF, Hoveyda AH, Snapper ML (2003) Org Lett 5:3273
236. Nakamura H, Nakamura K, Yamamoto Y (1998) J Am Chem Soc 120:4242
237. Nakamura K, Nakamura H, Yamamoto Y (1999) J Org Chem 64:2614
238. Fernandes RA, Stimac A, Yamamoto Y (2003) J Am Chem Soc 125:14133
239. Fernandes RA, Yamamoto Y (2004) J Org Chem 69:735
240. Fernandes RA, Yamamoto Y (2004) J Org Chem 69:3562
241. Gastner T, Ishitani H, Akiyama R, Kobayashi S (2001) Angew Chem Int Ed 40:1896
242. Fang XM, Johannsen M, Yao SL, Gathergood N, Hazell RG, Jørgensen KA (1999) J Org Chem 64:4844

243. Ferraris D, Dudding T, Young B, Drury WJ, Lectka T (1999) J Org Chem 64:2168
244. Hamada T, Manabe K, Kobayashi S (2003) Angew Chem Int Ed 42:3927
245. Wada R, Shibuguchi T, Makino S, Oisaki K, Kanai M, Shibasaki M (2006) J Am Chem Soc 128:7687
246. Lou S, Moquist PN, Schaus SE (2007) J Am Chem Soc 129:15398
247. Tan KL, Jacobsen EN (2007) Angew Chem Int Ed 46:1315

Top Curr Chem (2014) 343: 75–102
DOI: 10.1007/128_2013_482

Published online: 25 October 2013

Transition Metal-Catalyzed Enantioselective Hydrogenation of Enamides and Enamines

Qi-Lin Zhou and Jian-Hua Xie

Abstract Transition metal-catalyzed enantioselective hydrogenation of enamides and enamines is one of the most important methods for the preparation of optically active amines. This review describes the recent developments of highly efficient catalytic asymmetric hydrogenation of enamides, and enamines. It specifically focuses on the substrates because hydrogenation of enamides and enamines is highly dependent on the substrates although the chiral metal catalysts play a significant role.

Keywords Catalysis · Chiral amines · Enamides · Enamines · Enantioselective hydrogenation

Contents

Q.-L. Zhou (✉) and J.-H. Xie
State Key Laboratory and Institute of Elemento-Organic Chemistry, Nankai University, Tianjin 300071, China
e-mail: qlzhou@nankai.edu.cn; jhxie@nankai.edu.cn

1 Introduction

Chiral amines are an important class of organic compounds which can be used as resolving reagents, chiral auxiliaries, and intermediates for the synthesis of a variety of biologically active molecules including natural and unnatural products. Catalytic enantioselective hydrogenation of enamides, which have an acyl group at the nitrogen atom, and enamines, which have no acyl group at the nitrogen atom, with chiral transition metal complexes bearing chiral ligands was found to be one of the most efficient and convenient methods for the preparation of chiral amines and their derivatives.

In 1972, Kagan and Dang [1] reported the first example of enantioselective hydrogenation of *N*-acetyl enamides using a rhodium catalyst containing ligand DIOP for the synthesis of protected chiral amines (Scheme 1). Since then, the catalytic enantioselective hydrogenation of enamides for the preparation of optically pure amines and their derivatives has received considerable attention. As a result, a number of efficient transition metal catalysts with chiral ligands have been developed for the hydrogenation of a wide range of enamides, offering protected chiral primary amines, with good to excellent enantioselectivities [2–4]

In contrast, only limited progress had been made with the hydrogenation of simple enamines at the start of this century although it has advantages such as no need for the introduction and subsequent removal of the *N*-acyl protecting group. In 1994, Lee and Buchwald reported a highly enantioselective asymmetric hydrogenation of enamines [5]. By using a chiral *ansa*-titanocene catalyst **2**, they achieved as high as 98% *ee* for the hydrogenation of *N*,*N*-dialkyl α-arylethenamines. Börner and co-workers used chiral rhodium–diphosphine complexes for this hydrogenation and obtained the chiral tertiary amines in moderate enantioselectivity (up to 72% *ee*) [6]. The *N*-acetyl group in the enamides is considered indispensable for the substrate to form a chelate complex with a transition metal catalyst (Fig. 1), giving good reactivity and enantioselectivity [7]. However, due to the lack of *N*-acyl group, simple enamines such as *N*,*N*-dialkyl/aryl substituted enamines coordinate to the metal center of the catalyst only in a configurationally unstable η^2-fashion [8]. This makes the enantiocontrol of the catalyst in the asymmetric hydrogenation of *N*,*N*-dialkyl/aryl enamines very difficult.

Recently, with the revival of monodentate chiral phosphorus ligands [9], the asymmetric hydrogenation of simple enamines has received increasing attention. Several successful chiral catalysts such as rhodium and iridium complexes containing chiral monophosphonites and monophosphoramidites have been developed for the asymmetric hydrogenation of a wide range of acyclic and cyclic *N*,*N*-dialkyl/aryl enamines [10]. These represent new progress in the catalytic asymmetric hydrogenation of enamines.

It is worth noting that the catalytic asymmetric hydrogenation of enamides and enamines has been successfully applied to the industrial production of chiral drugs and agrochemicals [11]. This further underlines the importance of asymmetric hydrogenation of enamides and enamines.

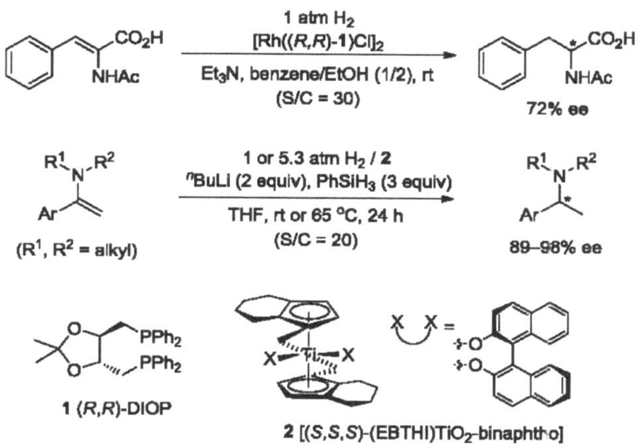

Scheme 1 Early examples of asymmetric hydrogenation of enamides and enamines

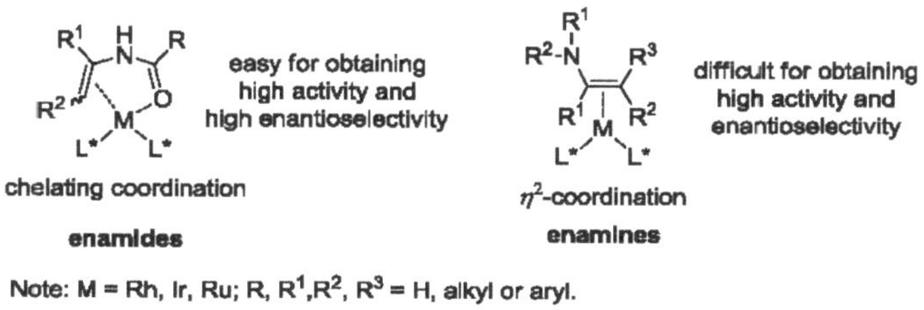

Fig. 1 Coordination models of enamide and enamine to the metal center of catalyst

In this chapter we intend to give an overview of the transition metal-catalyzed enantioselective hydrogenation of enamides (not including *N*-acetyl enamide esters and related compounds) and enamines in the synthesis of chiral amines with special emphasis on recent developments.

2 Enantioselective Hydrogenation of Enamides

Since Kagan and Dang reported that rhodium catalyst containing chiral diphosphine ligand DIOP was efficient for enantioselective hydrogenation of *N*-acetyl enamides [1], a number of chiral rhodium catalysts bearing chiral ligands have been developed. With these chiral rhodium catalysts, as well as several chiral ruthenium and iridium catalysts, a wide range of acyclic β-unsubstituted and β-substituted enamides and cyclic enamides were hydrogenated to protected chiral amines with good to excellent enantioselectivities.

2.1 Acyclic β-Unsubstituted Enamides

2.1.1 Acyclic α-Arylethenamides

Acyclic *β*-unsubstituted enamides such as *α*-arylethenamides **3** (Table 1) are the most investigated substrates in the asymmetric hydrogenation of enamides. Since the first highly enantioselective hydrogenation of this type of enamides reported by Burk et al. [12] in 1996 using rhodium catalysts of diphosphine ligands Me-DuPhos (**5a**) and Me-BPE (**6**) (Fig. 2), a number of rhodium catalysts containing different chiral phosphorus ligands have been developed (Figs. 2, 3, and 4). Most of these catalysts are highly efficient for the asymmetric hydrogenation of *α*-arylethenamides **3** (Table 1).

The 1,2-diphosphine ligand **10** derived from D-mannitol has high enantioselectivity (95–99% *ee*, Table 1) [13] for the asymmetric hydrogenation of *α*-arylethenamides **3**, and was better than the 1,2-diphosphine ligand Binaphane (**9**, 89.5–90.0% *ee*) [14]; however, the 1,2-diphosphine ligand **11** containing a six-membered ring showed higher activity, catalyzing the reaction under 1 atm of H_2 pressure [15]. For DIOP-type ligands **12**, the introduction of two alkyl substitutes at the *α*-position of the diphenylphosphino groups is necessary for obtaining high enantioselectivity. With the catalyst of Rh-**12a** and Rh-**12b**, *α*-arylethenamides **3** were hydrogenated to chiral amides **4** in 97.6 to >99% *ee* [16] and 99% *ee* [17], respectively. The 1,4-diphosphine BDPMI (**13**) bearing an imidazolidin-2-one backbone is another prominent DIOP-type ligand, showing 97.8 to >99% *ee* in the hydrogenation of *α*-arylethenamides **3** [18]. The reactions could also be performed under 1 atm of H_2 pressure at room temperature.

Chan et al. [19] introduced aminophosphine ligands BDPAP (**7**) and H_8-BDPAP (**8**) to rhodium-catalyzed hydrogenation of *α*-arylethenamides **3** and obtained excellent enantioselectivities (up to 99% *ee*). The *ortho* substituted BINAPO ligand Ph-*o*-BINAPO (**14**, 94.1–96.3% *ee*) [20] and the aminophosphine ligand **15** (97.6–99.9% *ee*) based on a 4,4′,6,6′-tetrakistrifluoromethylbiphenyl scaffold [21] are also highly efficient.

The chiral rhodium catalysts containing chiral ferrocene-based diphosphorus ligands such as ligands **16** and **17** were efficient for hydrogenation of *α*-arylethenamides **3**. The phosphinephosphoramidite ligand PPFAPhos (**16**) having a ferrocenyl backbone and an axially chiral binaphthyl moiety gave 98.7–99.6% *ee* [22]. When the catalyst loading was decreased to as low as 0.02 mol% (S/C = 5,000, S/C: the ratio of substrate to catalyst), high enantioselectivity was still maintained. The replacement of the binaphthyl moiety of the ligand **16** with a more rigid H_8-binaphthyl moiety leads to slightly lower enantioselectivity (95.5–99.0% *ee*) [23]. The rhodium catalyst of fluorinated phosphinoferrocenylaminophosphine ligand **17** was very stable to air and moisture both as a solid and in solution. The reactions can be performed in air or water-containing solvent with excellent enantioselectivities (98.3–99.7% *ee*) [24]. The phosphine-phosphoramidite ligand **18** with an (*S*)-*α*-phenylethylamine and an (*S*)-binaphthyl

Table 1 Asymmetric hydrogenation of α-arylethenamides **3** with rhodium catalysts of diphosphine ligands

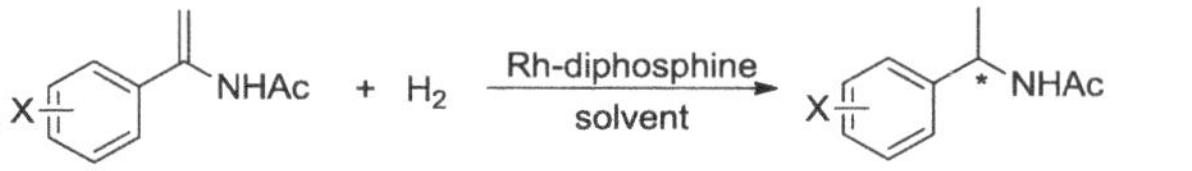

3 **4**

X at *para*- or *meta*-positions

Ligand	Reaction conditions	*ee* (%)	References
5a	0.2 mol% $[Rh(COD)_2]OTf$, 4 atm, MeOH, 20°C, 15 h	94.9–97.5	[12]
6	0.2 mol% $[Rh(COD)_2]OTf$, 4 atm, MeOH, 20°C, 15 h	74.8–97.8	[12]
7	0.5 mol% $[Rh(COD)_2]BF_4$, 1 atm, THF, 5°C, 0.5 h	90.1–96.1	[19]
8	0.5 mol% $[Rh(COD)_2]BF_4$, 1 atm, THF, 5°C, 0.5 h	96–99	[19]
9	1 mol% $[Rh(COD)_2]PF_6$, 1.36 atm, CH_2Cl_2, rt, 24 h	89.5–90.0	[14]
10	1 mol% $[Rh(COD)_2]PF_6$, 10 atm, MeOH, rt, 24 h	95–99	[13]
11	1 mol% $[Rh(NBD)_2]SbF_6$, 1 atm, CH_2Cl_2, rt, 24 h	95 to >99	[15]
12a	2 mol% $[Rh(COD)_2]SbF_6$, 10 atm H_2, MeOH, rt, 48–60 h	97.6 to >99	[16]
12b	1 mol% $[Rh(COD)_2]BF_4$, 1.3 atm H_2, CH_2Cl_2, rt, 10 h	99	[17]
13	1 mol% $[Rh(COD)_2]BF_4$, 1 atm H_2, CH_2Cl_2, rt, 12 h	97.8 to >99	[18]
14	1 mol% $[Rh(COD)_2]PF_6$, 3 atm H_2, MeOH, rt, 12 h	94.1–96.3	[20]
15	0.5 mol% $[Rh(COD)_2]BF_4$, 10 atm H_2, toluene, 5°C, 0.5–1 h	97.6–99.9	[21]
16	0.1 mol% $[Rh(COD)_2]BF_4$, 10 atm H_2, CH_2Cl_2, rt, 1 h	98.7–99.6	[22]
17	0.2 mol% $[Rh(COD)_2]BF_4$, 20 atm H_2, THF, 5°C, 30 h	98.3–99.7	[24]
18	1 mol% $[Rh(COD)_2]BF_4$, 10 atm H_2, CH_2Cl_2, rt, 12 h	98.5–99.9	[25]
19	1 mol% $[Rh(COD)_2]BF_4$, 10 atm H_2, CH_2Cl_2, rt, 24 h	99.4–99.9	[26]
20	1 mol% $[Rh(NBD)_2]SbF_6$, 13 atm H_2, MeOH, rt, 12 h	≥99	[27]
21	1 mol% $[Rh(NBD)_2]PF_6$, 1 atm H_2, MeOH, rt, 1–11 h	≥99	[28]
22	1 mol% $[Rh(NBD)_2]BF_4$, 6.8 atm H_2, CH_2Cl_2, 0°C, 12 h	98 to >99	[29]
23	0.1 mol%$[Rh(COD)_2]SbF_6$, 3 atm H_2, MeOH, rt, 0.2–18 h	85–99.9	[30]
24	0.1 mol%$[Rh(COD)_2]SbF_6$, 3 atm H_2, MeOH, rt, 0.5–5 h	92.9–99.9	[30]
25	0.1 mol%$[Rh(COD)_2]SbF_6$, 3 atm H_2, MeOH, rt, 0.7–4 h	90–99.6	[30]
26	0.1 mol%$[Rh(COD)_2]SbF_6$, 3–4 atm H_2, MeOH, rt, 18 h	43–92.8	[31]
27	0.5–0.1 mol%$[Rh(COD)_2]SbF_6$, 3–4 atm H_2, MeOH, rt, 18 h	96.1–98.6	[31]
28	1 mol% $[Rh(NBD)_2]BF_4$, 1 atm H_2, MeOH, rt, 2 min	99.3	[33]

moiety also gave 98.5–99.9% *ee* for the hydrogenation of α-arylethenamides **3** [25]. This result indicates that the ferrocenyl moiety in the ligand **16** can be replaced by a simple phenyl moiety. This conclusion was approved by THNAPhos (**19**), a phosphine-phosphoramidite ligand containing a more rigid tetrahedronaphthyl moiety [26]. With a matched combination of configurations, the ligand **19** gave 99.3–99.7% *ee* for the hydrogenation of α-arylethenamides **3**.

Although the *P*-chiral diphosphine ligands are unstable and difficult to synthesize, many *P*-chiral diphosphines were recently developed and showed high enantioselectivity for the hydrogenation of α-arylethenamides **3**. For examples, the *P*-chiral diphosphine ligand **20** (TangPhos) with two five-membered rings in the backbone gave excellent enantioselectivity (99.3% *ee*) at very low catalyst loading (*ca* 0.01 mol%, S/C = 10,000) [27]. The similar *P*-chiral diphosphine ligand DisquareP* (**21**) [28] with two four-membered rings in the backbone and

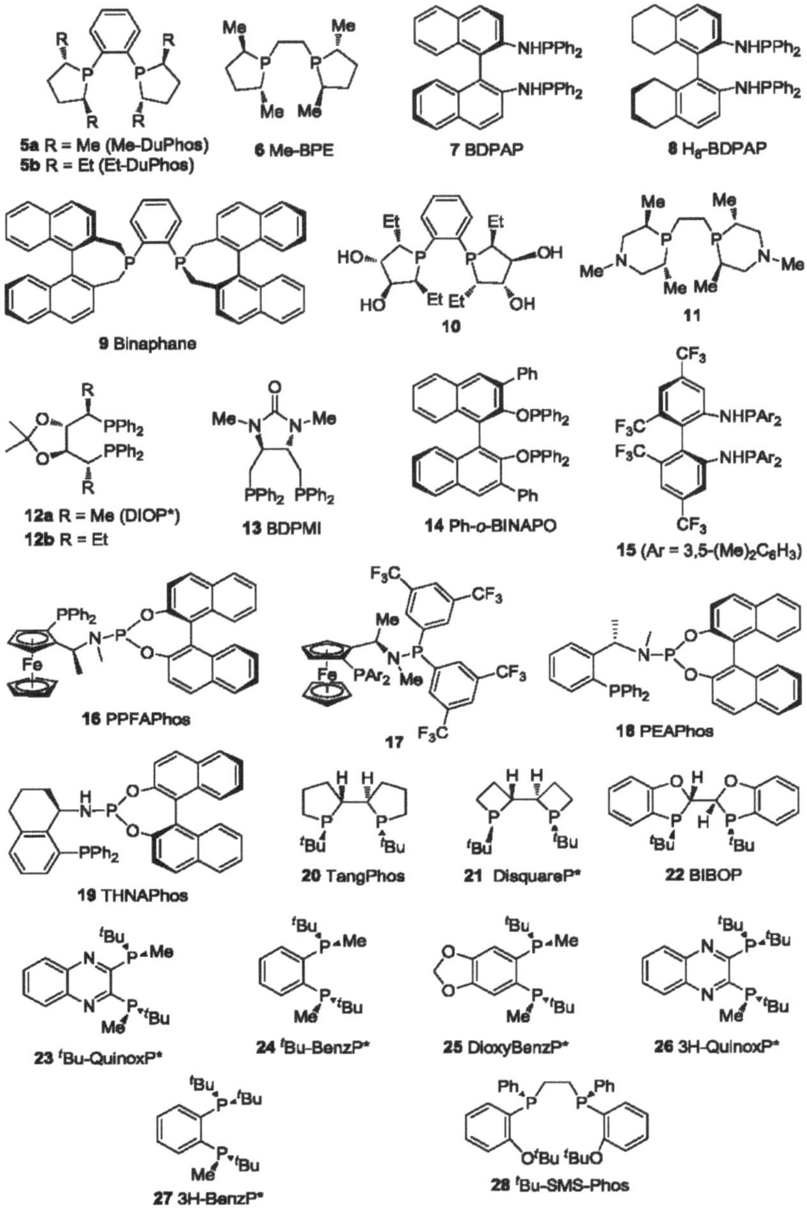

Fig. 2 Selected efficient chiral diphosphorus ligands

BIBOP (**22**) [29] with a rigid bisdihydrobenzo[*d*]-[1,3]-oxaphosphole structure are also highly enantioselective for the hydrogenation of α-arylethenamides **3**. Recently, the *P*-chiral diphosphine ligands QuinoxP* (**23**), BenzP* (**24**),

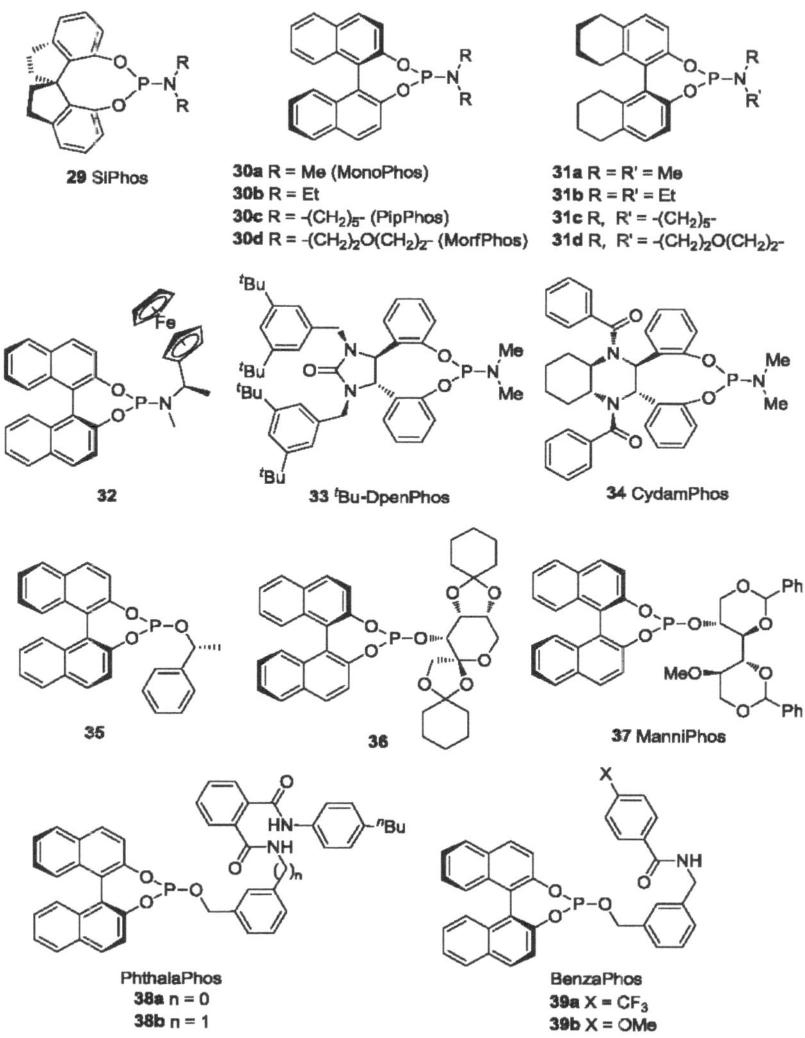

Fig. 3 Selected efficient chiral monophosphorus ligands

DioxyBenzP* (**25**) [30], 3H-QuinoxP* (**26**), 3H-BenzP* (**27**) [31], and ^{t}Bu-SMS-Phos (**28**) [32–35] were reported to be stable and efficient for the asymmetric hydrogenation of α-arylethenamides **3**. The rhodium complexes of these ligands can catalyze the hydrogenation of α-arylethenamides **3** under mild reaction conditions (S/C up to 10,000, 3 atm H_2, room temperature) in good to excellent enantioselectivities (up to 99.9% *ee*).

Chiral monodentate phosphorus ligands were used in the enantioselective hydrogenation of enamides in the beginning of this century. In 2002, Zhou

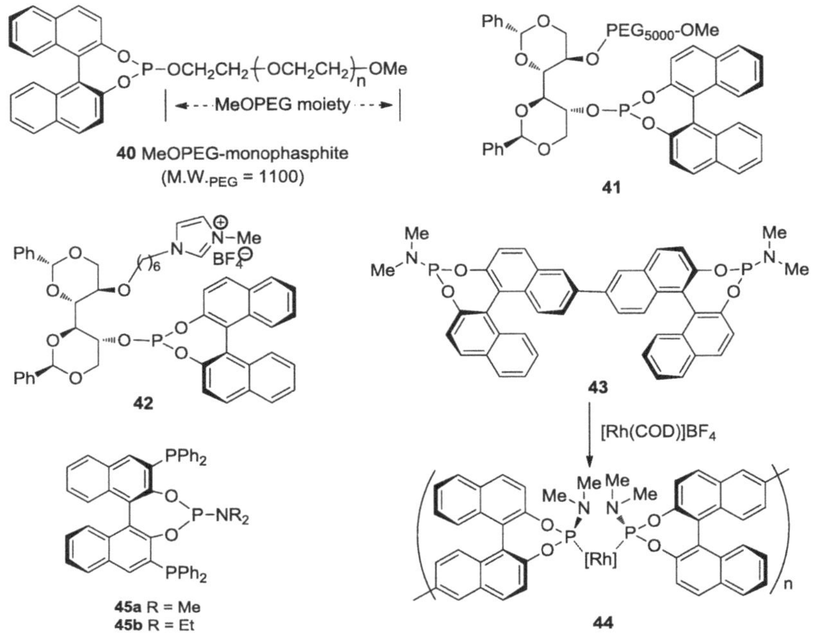

Fig. 4 Other chiral ligands and catalysts

et al. [36] reported that the chiral spiro monophosphoramidite ligand SiPhos (**29**, Fig. 3) was highly efficient for the rhodium-catalyzed asymmetric hydrogenation of α-arylethenamides **3** (91–99.7% *ee*, Table 2). Subsequently, many chiral monodentate phosphorus ligands were developed and used for the enantioselective hydrogenation of α-arylethenamides **3**.

Chiral monophosphoramidite ligands **30** (Fig. 3) based on a binaphthyl backbone exhibited good to high enantioselectivities for the hydrogenation of α-arylethenamides **3**. The ligand **30a** with a dimethylamino group gave good enantioselectivity (90–96% *ee*) only at low temperature (−20°C) [37]. Higher enantioselectivities (95–99.6% *ee*) could be obtained at room temperature by using ligands Monophos-Et (**30b**) [38], PipPhos (**30c**), and MorfPhos (**30d**) [39]. The monophosphoramidite ligand **31a** with a more rigid octahydro binaphthyl backbone gave higher enantioselectivities (92–99% *ee*) than the ligand **30a**, indicating that a rigid backbone in the monophosphoramidite ligand is beneficial to the enantioselectivity of reaction [40]. The monophosphoramidite ligand **32** containing a chiral ferrocenyl moiety gave even higher enantioselectivities (98–99.5% *ee*) [41].

The monophosphoramidite ligands based on other chiral backbones are also highly efficient for this transformation. For example, the monophosphoramidite ligands ^{t}Bu-DpenPhos (**33**) [42] derived from optically pure 1,2-di (2-dimethoxyphenyl)-1,2-ethylenediamine and CydamPhos (**34**) [43] derived

Table 2 Asymmetric hydrogenation of α-arylethenamides **3** with rhodium catalysts of monophosphorus ligands

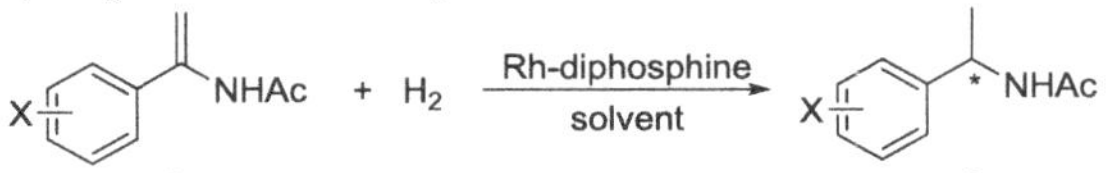

3
4
X at *para*- or *meta*-positions

Ligand	Reaction conditions	*ee* (%)	References
29	1 mol% $[Rh(COD)_2]BF_4$, 50 atm, toluene, 5°C, 12 h	91–99.7	[36]
30a	1 mol% $[Rh(COD)_2]BF_4$, 20 atm, CH_2Cl_2, −20°C, 8 h	90–96	[37]
30b	1 mol% $[Rh(COD)_2]BF_4$, 20 atm H_2, THF, 5°C to rt, 4–6 h	95–99.6	[38]
30c	2 mol% $[Rh(COD)_2]BF_4$, 25 atm H_2, CH_2Cl_2, rt, 8 h	≥99	[39]
31a	1 mol% $[Rh(COD)_2]BF_4$, 20 atm H_2, THF, −10°C, 8 h	92–99	[40]
32	1 mol% $[Rh(COD)_2]BF_4$, 10 atm H_2, CH_2Cl_2, rt, 20 h	98–99.5	[41]
33	1 mol% $[Rh(COD)_2]BF_4$, 40 atm H_2, CH_2Cl_2, rt, 2 h	96.1–99.8	[42]
34	1 mol% $[Rh(COD)_2]BF_4$, 10 atm H_2, CH_2Cl_2, rt, 2 h	91.8–98.4	[43]
35	0.2 mol% $[Rh(COD)_2]BF_4$, 60 atm H_2, CH_2Cl_2, 30°C, 20 h	94.8–95.8	[44]
36	1 mol% $[Rh(COD)_2]BF_4$, 10 atm H_2, CH_2Cl_2, rt, 12 h	95.0–98.5	[45]
37	1 mol% $[Rh(COD)_2]BF_4$, 10 atm H_2, CH_2Cl_2, 20°C, 12 h	99.5–99.9	[46]
38b	1 mol% $[Rh(COD)_2]BF_4$, 5 atm, CH_2Cl_2, rt, 20 h	98	[47]
39b	1 mol% $[Rh(COD)_2]BF_4$, 5 atm, CH_2Cl_2, rt, 20 h	>99	[48]
40	1 mol% $[Rh(COD)_2]BF_4$, 10 atm, CH_2Cl_2, rt, 20 h	96–99	[49]
41	0.5 mol% $[Rh(COD)_2]OTf$, 13.6 atm, CH_2Cl_2, rt,14 h	96.4	[50]
42	1 mol% $[Rh(COD)_2]BF_4$, 10 atm H_2, CH_2Cl_2, rt, 1 h	>99	[51]
43	1 mol% $[Rh(COD)_2]BF_4$, 40 atm H_2, toluene, 25°C, 10 h	97.3	[52]
45a	1 mol% $[Rh(COD)_2]BF_4$, 1–5 atm H_2, CH_2Cl_2, rt, 24 h	91–99.6	[55]
45b	1 mol% $[Rh(COD)_2]BF_4$, 5 atm H_2, CH_2Cl_2, rt, 24 h	94.2–99.0	[55]

from *trans*-1,2-diaminocyclohexane provided 98–99.5% *ee* and 91.8–98.4% *ee*, respectively, for the hydrogenation of α-arylethenamides **3**.

Chiral monophosphites are another type of ligand which have been widely applied in the asymmetric hydrogenation of α-arylethenamides **3**. The binaphthyl-based monophosphite ligand **35** containing a (*R*)-1-phenylethoxy group afforded 94.8–95.8% *ee* under 60 atm of H_2 pressure [44]. The monophosphite ligand **36** containing a "D-fructose" moiety gave better enantioselectivities (95.0–98.5% *ee*) than ligand **35** [45] The ligand ManniPhos (**37**) with a "D-mannitol" moiety was found to be the most efficient monophosphite ligands, providing 99.5–99.9% *ee* for a series of α-arylenamides substrates **3** (except the *ortho*-chloro substituted substrate, 79.1% *ee*) [46]. Moreover, by using ManniPhos ligand, the rhodium catalyst loading can be lowered to 0.02 mol% (S/C = 5,000). The chiral supramolecular ligands derived from PhthalaPhos (**38**) [47] and BenzaPhos (**39**) [48] showed higher enantioselectivities (>99% *ee*) than the structurally related monodentate phosphite ligands such as ligand **35**.

Recycling of chiral catalysts is another topic in the research of enantioselective hydrogenation of α-arylethenamides **3** using monodentate phosphorus ligands. In 2006, Zheng et al. reported that the PEGpoly (ethyleneglycol) monomethyl ether-derived polymer monophosphite **40** (MeOPEG-monophosphites, Fig. 4) gave high enantioselectivity (96–99% *ee*) for the hydrogenation of several α-arylethenamides **3**, and the rhodium catalyst containing MeOPEG-monophosphite ligand **40** can be recycled four times without seriously diminishing enantioselectivity [49]. Polymer-supported monophosphite ligand **41** also showed as high as 96.4% *ee* for the hydrogenation of the α-arylenamide **3** (X = H) [50]. In 2011, Huang and Xia developed chiral ionic phosphite ligand **42** and found that the rhodium complex of ligand **42** gave >99% *ee* for the hydrogenation of α-arylenamides **3**. This catalyst was reused ten times in mixed solvents of [bmin]BF_4 and toluene (v/v = 1:1) at 1 mol% catalyst loading [51]. The bis-MonoPhos ligand **43** was reported by Ding et al. [52], which was derived by linking two MonoPhos ligands **30a** with a covalent bond. The Bis-MonoPhos ligand **43** was capable of self-assembly by reacting with the catalyst precursor $[Rh(COD)_2]BF_4$ to form a polymer catalyst **44**, which showed as high as 97.3% *ee* for the hydrogenation of α-phenylethenamide (**3**, X = H) under 40 atm H_2 at 25°C in toluene. The chiral rhodium catalysts containing dendrimer-supported monophosphoramidites [53] and the non-covalent (hydrogen bonding and metal coordination) self-assembled chiral rhodium catalysts of monodentate phosphoramidites [54] were also efficient for the asymmetric hydrogenation of α-arylenthenamides **3** and could be reused.

However, most of the aforementioned chiral ligands and their chiral rhodium catalysts are efficient only for the *N*-acetyl α-arylethenamides **3**, which have *para* or *meta* substituents. Interestingly, the triphosphorus ligands **45** have high enantioselectivities for the α-arylenthenamides **3**, which have substituents in any position of the aromatic ring [55].

The asymmetric rhodium-catalyzed hydrogenation of α-arylenthenamides **3** have been successfully applied to the preparations of chiral drugs or potential pharmaceuticals (Fig. 5) The rhodium catalysts of *P*-chiral diphosphine ligands QuinoxP* (**23**), BenzP* (**24**), and DioxyBenzP* (**25**) have been used for the synthesis of SDZ-ENA-713 (a precursor of acetylcholinesterase inhibitor), CGP-55845 (a GAVA-B antagonist), Cospitant (a potent neurokinin 1 receptor antagonist), and Cinacalcet hydrochloride (a calcimimetic agent for treatment of hyperparathyroidism and hypercalcemia) [30]. The enantioselectivities of the hydrogenations were 85–99.9% *ee* and the catalyst loadings were 0.1–0.01 mol%. The rhodium catalyst of ligand Et-DuPhos (**5b**) was also used for the preparation of AZ-23 (95.3% *ee*), a novel agent with potential therapeutic reagent for neuroblastoma and multiple other cancer indications, and AZD1480 (>99% *ee*), a potent Jak2 inhibitor [56]. Compound **46**, a potent orally bioavailable bradykinin B_1 antagonist, was prepared with Rh-TangPhos (Rh-**20**) catalyzed asymmetric hydrogenation (99.5% *ee*) [57]. The compound **47**, a potent melanocortin (MC4) receptor agonist was also prepared by hydrogenation of corresponding enamides (97.1% *ee*) [58].

2.1.2 Acyclic α-Alkylethenamides

In addition to α-arylethenamides, α-alkylethenamides such as α-*tert*-butylethenamide (**48a**) and α-1-adamantylethenamide (**48b**) can also be hydrogenated to the

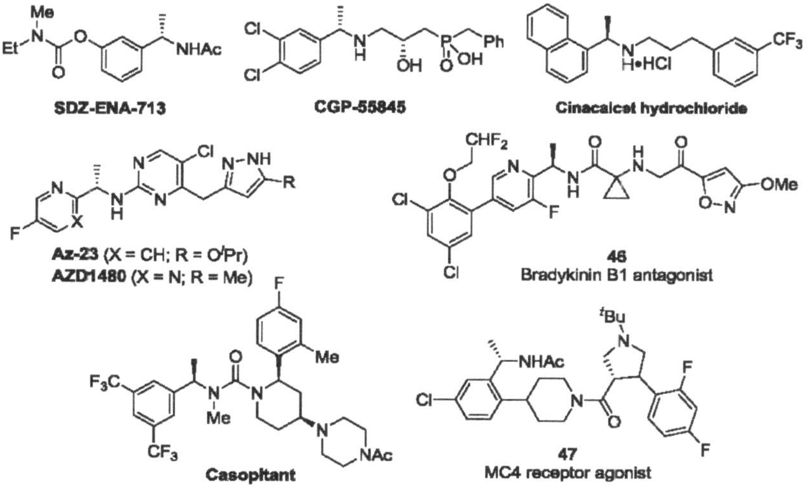

Fig. 5 Chiral drugs or potential pharmaceuticals made by asymmetric hydrogenation of enamides

corresponding chiral amides **49a** and **49b** in good to excellent enantioselectivities by using chiral rhodium catalysts containing chiral diphosphine ligands, in particular the *P*-chiral diphosphines (Scheme 2).

The rhodium catalyst of *P*-chiral ligand ^{t}Bu-BisP* (**50**) gave 99% *ee* for the hydrogenations of both α-alkylethenamides **48a** and **48b** with 0.1 mol% of catalyst loading (S/C = 1,000) [59]. The ligand ^{t}Bu-MiniPhos (**51**) showed the same *ee* value for the hydrogenation of α-1-adamantylethenamide (**48b**). Other *P*-chiral ligand such as DisquareP* (**21**, 93% *ee* for **48a**) [28], QuinoxP* (**23**, 99% *ee* for **48a**, 96.3% *ee* for **48b**), BenzP* (**24**, 96.6% *ee* for **48a**), DioxyBenP* (**25**, 95.1% *ee* for **48a**) [30], and trichickenfootPhos (**52**, 99% *ee* for **48a**) [60] also gave good to high enantioselectivities for the hydrogenation of α-alkylethenamides **48**. Furthermore, the 1,2-diphosphine ligand MalPhos (**53**) having a maleic anhydride backbone offered 96% *ee* for the hydrogenation of α-*tert*-butyl enamide (**48a**) [61]. However, chiral monodentate phosphorus ligands were less efficient for the hydrogenation of α-alkylethenamides **48**. The catalyst Rh-**30c** gave only a moderate *ee* value (82% *ee*) for the hydrogenation of α-*tert*-butylethenamide (**48a**) [39].

2.1.3 Other Acyclic β-Unsubstituted Enamides

Changing the *N*-acetyl group of α-arylethenamides to other protecting groups such as *N*-trifluoroacetyl and *N*-phthaloyl sometime improved the enantioselectivity in rhodium-catalyzed asymmetric hydrogenation of enamides. For example, the

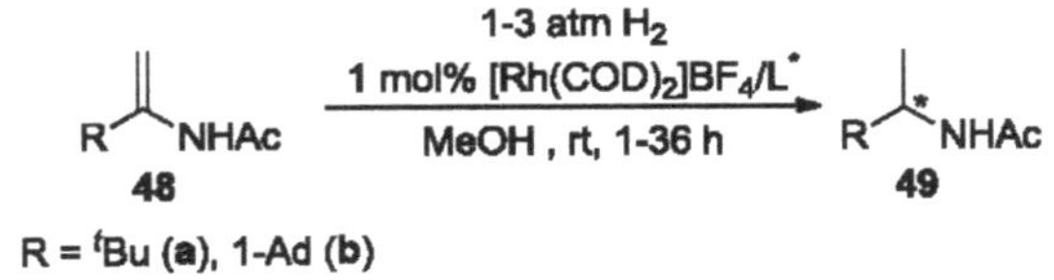

L*		21[a]	23[b]	24[b]	25[b]	50	51	52	53[b]
ee (%)	48a	93	99	96.6	95.1	99	99	99	96
	48b		96.3			99			

Note: [a] $[Rh(NBD)_2]PF_6$ as a precursor.
[b] 0.1 mol% $[Rh(COD)_2]SbF_6$ as a precursor.
[c] CH_2Cl_2 as a solvent.

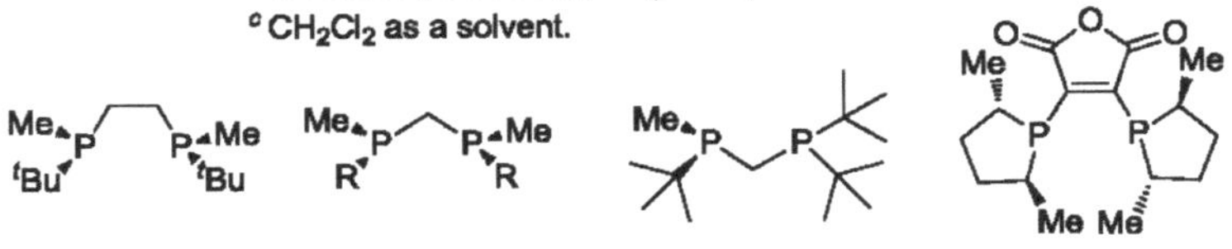

50 ^{t}Bu-BisP* **51** ^{t}Bu-MiniPhos **52** trichickenfootPhos **53** MalPhos

Scheme 2 Hydrogenation of α-alkylethenamides **48**

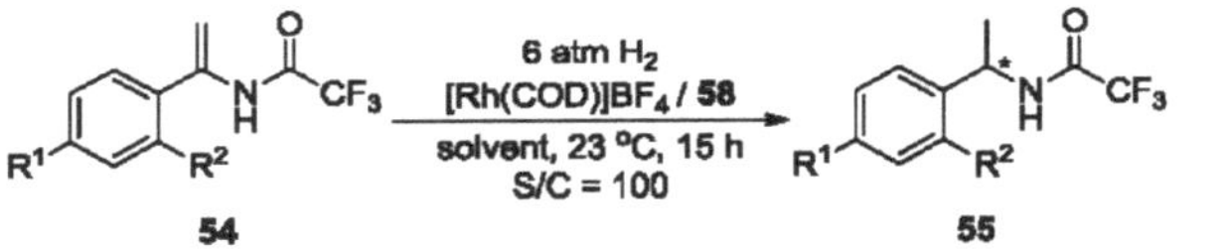

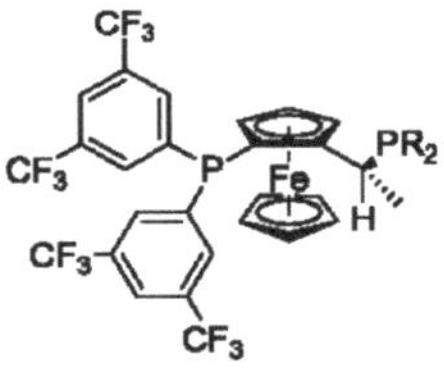

58a R = Cy
58b R = 3,5-$(Me)_2$-C_6H_3

54	R^1	R^2	L*	ee (%)
54a	CF_3	H	58a	97.1[a]
54b	Br	F	58b	91.8[b]

[a] ^{i}PrOH as a solvent
[b] 1,2-dichlorobenzene as a solvent

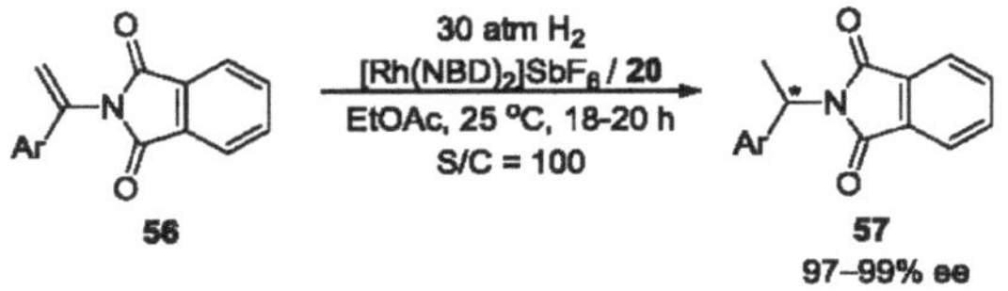

Scheme 3 Enantioselective hydrogenation of enamides **54** and **56**

hydrogenation of the *N*-trifluoroacetyl enamides **54** catalyzed by the rhodium catalyst of Josiphos-type ligand **58** offered chiral amides **55** with high enantioselectivities (Scheme 3) [62]. The *N*-phthaloyl enamines **56** with *meta* and *para*-substituents can be hydrogenated to chiral amides **57** in 97–99% *ee* with the catalyst Rh-TangPhos (Rh-**20**) [63].

2.2 Acyclic β-Substituted Enamides

The acyclic β-substituted enamides were generally prepared as *E*/*Z* mixtures and the separation of these mixtures is demonstrated to be problematic. Fortunately, the rhodium catalysts of diphosphine ligands such as DuPhos (**5**), BPE (**6**), and Binaphane (**9**) reported by Burk et al. [12] and Zhang et al. [14] can catalyze the hydrogenations of *E*/*Z* mixtures in high enantioselectivities. Recently, several new rhodium catalysts containing chiral diphosphine and monophosphorus ligands were reported to be efficient for the hydrogenation of the *E*/*Z* mixtures of acyclic β-substituted enamides. The substrates of this reaction have been expanded from β-alkyl substituted α-arylenamides to MOM-protected β-hydroxy substituted α-arylenamides and β-aryl substituted α-alkylenamides.

2.2.1 Acyclic β-Alkyl Substituted α-Arylenamides

Chiral rhodium catalysts containing DIOP-type 1,4-diphosphine and BPE-type 1,2-diphosphine ligands have recently been found to be popular in enantioselective hydrogenation of the *Z*/*E* mixtures of acyclic β-alkyl substituted α-arylenamides (Table 3).

The 1,4-diphosphine ligand DIOP* (**12a**) gave excellent enantioselectivities (97.3–99% *ee*, Table 3) for the hydrogenation of the *Z*/*E* mixture of β-alkyl substituted α-arylenamides **59** (X = H, R = Me) with 2 mol% catalyst (S/C = 50) under 10 atm of H_2 [16]. The ligand **12b** showed 98% *ee* for the same reaction under milder conditions (1.3 atm H_2) with 1 mol% of catalyst (S/C = 100) [17]. The 1,4-diphosphine ligand **61** and **62** (SK-Phos, Fig. 6) also offered 95–98% *ee* and 94–98% *ee*, respectively, in this reaction [64]. The most efficient 1,4-diphosphine ligand for this transformation is Me-BDPMI (**13**), providing >99% *ee* for several *Z*/*E* mixtures of β-alkyl substituted α-arylenamides **59** under mild conditions (1 atm H_2, room temperature) [18]. The 1,2-diphosphine ligand **63**, an analogue of Binaphane (**9**), was also efficient for the hydrogenation of the *Z*/*E* mixtures of enamides **59**, providing chiral amides **60** with 93.1–99.5% *ee* [65]. The more rigid 1,2-diphosphine ligand **11** gave better enantioselectivities (97 to >99% *ee*) under mild conditions (1 atm H_2) [15]. Furthermore, the *P*-chiral 1,2-diphosphane ligand TangPhos (**20**, 98–99% *ee*) [27] and the phosphine-phosphoramidite ligand THNAPhos (**19**, 99.1% *ee*) [26] also gave good to excellent enantioselectivity for the hydrogenation of the *Z*/*E* mixtures of β-alkyl substituted α-arylenamides **59**.

Several monodentate phosphorus ligands were demonstrated to be highly efficient for the rhodium-catalyzed enantioselective hydrogenation of the *Z*/*E* mixtures of *N*-acetyl α-arylenamides **59**. The ligand **64** derived from an achiral catechol and a chiral amine gave 93% *ee* for the hydrogenation of the *Z*/*E* mixtures of β-substituted enamide **59** (X = H, R = Et, *Z*/*E* = 2:3), which was the calculated average of the enantioselectivities for the *Z* and *E* isomers of substrate [66]. Higher

Table 3 Hydrogenation of enamides **59** with rhodium catalysts of phosphorus ligands

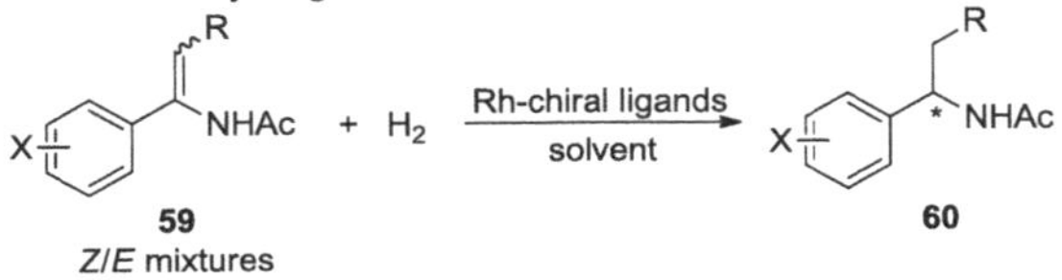

Ligand	Reaction conditions	*ee* (%)	References
11	1 mol% $[Rh(NBD)_2]SbF_6$, 1 atm, CH_2Cl_2, rt, 24 h	97 to >99	[15]
12a	2 mol% $[Rh(COD)_2]SbF_6$, 10 atm H_2, MeOH, rt, 48–60 h	97.3 to >99	[16]
12b	1 mol% $[Rh(COD)_2]BF_4$, 1.3 atm H_2, CH_2Cl_2, rt, 10 h	98	[17]
13	1 mol% $[Rh(COD)_2]BF_4$, 1 atm H_2, CH_2Cl_2, rt, 12 h	$\geq$99	[18]
19	1 mol% $[Rh(COD)_2]BF_4$, 10 atm H_2, CH_2Cl_2, rt, 24 h	99.1	[26]
20	1 mol% $[Rh(NBD)_2]SbF_6$, 1.3 atm H_2, MeOH, rt, 12 h	98–99	[27]
36	1 mol% $[Rh(COD)_2]BF_4$, 10 atm H_2, CH_2Cl_2, rt, 12 h	96.7	[45]
37	1 mol% $[Rh(COD)_2]BF_4$, 10 atm H_2, CH_2Cl_2, 20°C, 12 h	99.2	[46]
45a	1 mol% $[Rh(COD)_2]BF_4$, 5 atm H_2, CH_2Cl_2, rt, 24 h	96–99	[55]
61	1 mol% $[Rh(NBD)_2]SbF_6$, 3 atm H_2, MeOH, rt, 24 h	94–98	[64]
62	1 mol% $[Rh(NBD)_2]SbF_6$, 3 atm H_2, MeOH, rt, 24 h	95–98	[64]
63	1 mol% $[Rh(NBD)_2]SbF_6$, 2.7 atm H_2, MeOH, rt, 24 h	93.1–99.5	[65]
64	1 mol% $[Rh(COD)_2]BF_4$, 25 atm H_2, CH_2Cl_2, rt, 16 h	93	[66]
65	1 mol% $[Rh(COD)_2]BF_4$, 10 atm H_2, CH_2Cl_2, rt, 20 h	94–97	[41]
66	1 mol% $[Rh(COD)_2]BF_4$, 10 atm H_2, CH_2Cl_2, 20°C, 12 h	97.3–99.0	[67]

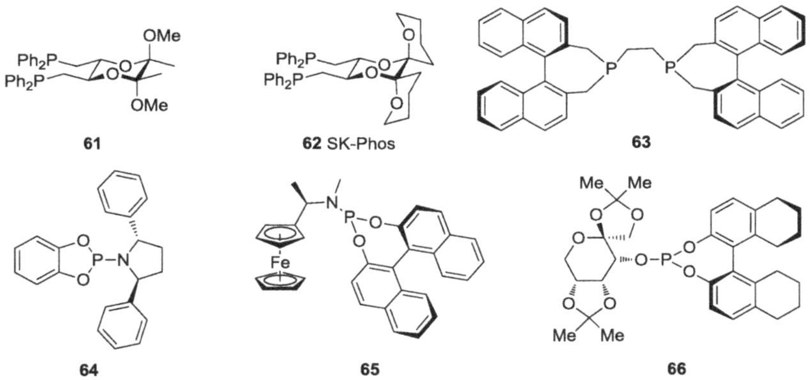

Fig. 6 Chiral phosphorus ligands for hydrogenation of α-arylenamides **59**

enantioselectivities were obtained with monophosphoramidite ligand **65** containing both ferrocenyl and binaphthyl backbones under similar conditions [41]. The chiral monophosphite ligands were better than chiral monophosphoramidite ligands in terms of enantioselectivity for the hydrogenation of the *Z*/*E* mixtures of enamides **59**. The monophosphite ligand **36** with a "D-fructose" moiety gave 96.7% *ee* for the *Z*/*E* mixtures of enamide **59** (X = H, R = Me) [45]. However, as high as 99.2% *ee* was obtained with the "D-mannitol" based monophosphite ligand ManniPhos (**37**)

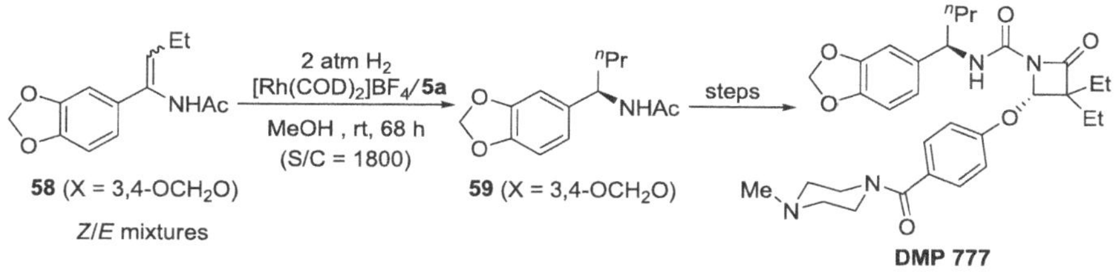

Scheme 4 Enantioselective synthesis of DMP 777

[46]. The H_8-binaphthyl-based monophosphite ligand **66**, an analogue of the ligand **36**, also gave 97.3 and 99.0% *ee* for the hydrogenation of enamides **59** with the R group Me and Et, respectively [67]. The triphosphorus ligand **45a** based on a binaphthyl backbone showed up to 99% *ee* for the asymmetric hydrogenation of β-alkyl substituted α-arylenamides **59** [55].

By using the Rh-Me-DuPhos (Rh-**5a**) catalyzed asymmetric hydrogenation of the *Z*/*E* mixtures of β-ethyl substituted α-arylenamides **59** (X = 3,4-OCH_2O) as a key step, Storace et al. developed an efficient large-scale process for the preparation of chiral prodrug DMP 777, a human leukocyte elastase inhibitor (Scheme 4) [68].

2.2.2 Acyclic β-Methoxymethoxy Substituted α-Arylenamides

The catalytic enantioselective hydrogenation of protected β-hydroxy enamides is a practical way to generate optically pure β-amino alcohols. The rhodium catalyst of DIOP-type ligand **61** was found to be efficient for the hydrogenation of the *Z*/*E* mixtures of acyclic β-methoxymethoxy α-arylenamides **67**, yielding chiral amides **68** in 96 to >99% *ee* (Scheme 5) [64]. These results are comparable to or better than those obtained from BICP and Me-DuPhos ligands [69].

2.2.3 Acyclic β-Substituted α-Alkylenamides

The catalytic asymmetric hydrogenation of acyclic β-substituted α-alkylenamides is still a challenging task. To date, only a few chiral catalysts are capable of efficient hydrogenation of such enamides (Scheme 6). In 2009, Zhang et al. reported that the rhodium catalyst of TangPhos (**20**) is highly enantioselective for the hydrogenation of the (*Z*)-isomers of β-aryl substituted α-alkylenamides **69** [70]. Under the conditions of 30 atm H_2 in EtOAc at room temperature for 20 h the catalyst Rh-**20** gave 96.6 to >99.9% *ee* for (*Z*)-α-alkylenamides **69**. However, very low enantioselectivity (<50% *ee*) was obtained for the (*E*)-isomer of α-alkylenamide **69**. Recently, Zhou and co-workers found that the rhodium complexes of chiral spiro monophosphite ShiP ligands and spiro monophosphine SCTIP ligands were efficient for the hydrogenation of (*Z*) and (*E*)-isomers of enamides **69**, respectively [71]. With the catalyst Rh-**71** the (*Z*)-isomer of enamides **69** can be hydrogenated to

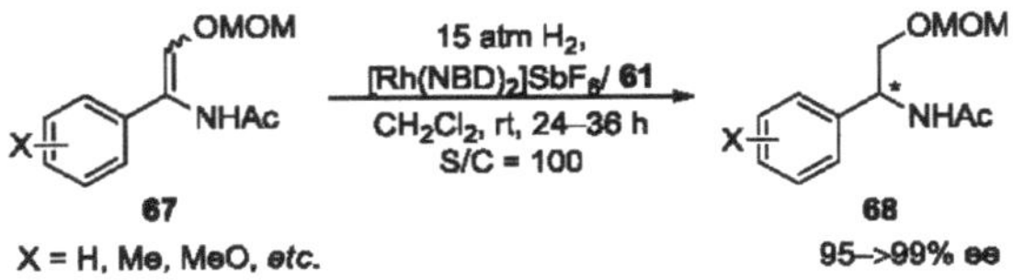

Scheme 5 Enantioselective hydrogenation of enamides **67**

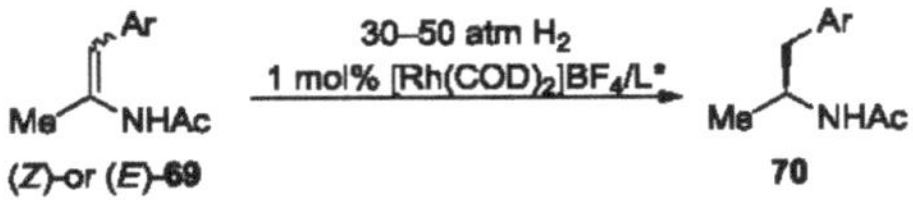

L*	69	conditions	%ee
20	(Z)-69	30 atm, EtOAc, rt, 20 h	96.6–>99.9% ee (9 examples)
71	(Z)-69	50 atm, CH_2Cl_2/toluene, rt, 12 h	90–95% ee (8 examples)
72	(E)-69	50 atm, CH_2Cl_2, rt, 12 h	81–97% ee (13 examples)

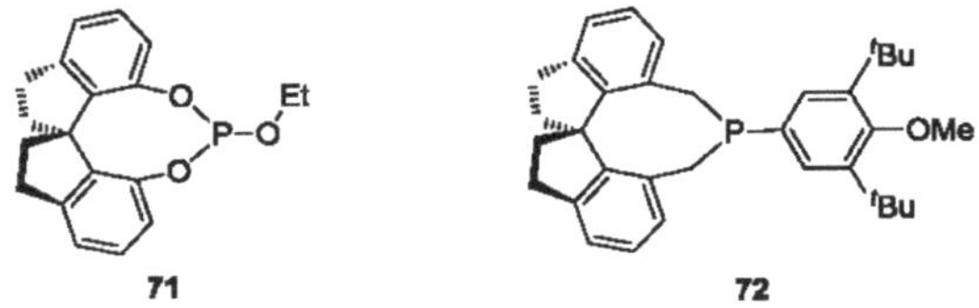

Scheme 6 Enantioselective hydrogenation of acyclic β-substituted α-alkylenamides **69**

chiral amides **70** in full conversion with 90–95% *ee* under 50 atm H_2. Better results were achieved by the catalyst Rh-**72** for the hydrogenation of the (*E*)-isomer of enamides **69**, providing the chiral amides **70** with up to 97% *ee* under similar reaction conditions. The products of this hydrogenation **70** are important intermediate for the synthesis of pharmaceuticals such as (*S*)-amphetamine, (*R*,*R*)-formoterol, (*R*)-selegiline, and (*R*)-tamsulosin.

Wallace et al. reported that the rhodium complex of Josiphos type ligand **75** was efficient for the asymmetric hydrogenation of β-disubstituted α-alkylenamide **73** (Scheme 7) [72]. Under 10 atm H_2 at 40°C in TFE (2,2,2-trifluoroethanol) for 17 h (S/C = 2,000) the catalyst Rh-**75** gave chiral amide **74** with two adjacent chiral centers in 90% yield with 96% *ee*. With chiral amide **74**, a cannabinoid-1 receptor (CB1R) inverse agonist, taranabant, was conveniently synthesized.

2.3 Cyclic Enamides

The enantioselective hydrogenation of cyclic enamides is an efficient way to access optically pure chiral cyclic amines or nitrogen-containing heterocyclic compounds, which have wide applications in the synthesis of pharmaceuticals. The asymmetric hydrogenation of cyclic enamides has been extensively studied. The rhodium complex of chiral diphosphine ligand *o*-Ph-HexMeO-BIPHEP (**82**) was reported

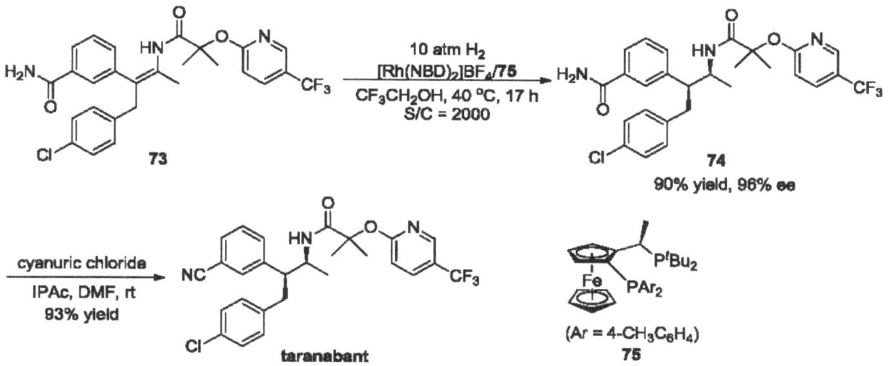

Scheme 7 Enantioselective hydrogenation of β-disubstituted α-alkylenamide **73**

to be an efficient catalyst for the hydrogenation of cyclic enamides **76** and **78a**, derived from 1-indanone and α-tetralone (Scheme 8) [73]. This catalyst offered cyclic amides **77** and **79a** with 96 and 98% *ee*, respectively, under low hydrogen pressure (1.3 atm). However, only moderate *ee* value (66% *ee*) was obtained for the enamide **78b** derived from 4-chromanone. The catalyst Rh-**82** has also been successfully applied in the enantioselective hydrogenation of a racemic cyclic carbamate **80** for the synthesis of a chiral precursor of sertraline ((1*S*,4*S*)-**81**).

Rhodium catalysts of chiral monophosphorus ligands such as SiPhos (**29**), PipPhos (**30c**), and ManniPhos (**37**) showed excellent enantioselectivity for the hydrogenation of cyclic enamides. The catalyst Rh-**29** gave 88–95% *ee* for the hydrogenation of enamide **76** and its analogues (with 5-Br or 6-MeO group on the phenyl ring) [74]. However, the catalyst Rh-**30c** was found efficient for these three types of cyclic enamides and gave up to 98% *ee* [39]. The catalyst Rh-**37** provided 96% *ee* for the hydrogenation of cyclic enamide **76** [46]. Recently, the supramolecular rhodium catalysts derived from monophosphite ligands **39a** and **39b** were reported to be highly enantioselective for the hydrogenation of cyclic enamide **78a**, providing chiral cyclic amide **79a** in full conversion with 99% *ee* [48].

The cyclic enamide **83** derived from β-tetralone can be hydrogenated to chiral amide **84** by chiral ruthenium and rhodium catalysts (Scheme 9). In the presence of 0.5 mol% of catalyst $[Ru(O_2CCF_3)_2((R)\text{-}\mathbf{87})]$ the enamide **83** was hydrogenated to amide **84** in 94% *ee* under 10 atm of H_2 pressure in MeOH [75]. A higher enantioselectivity (96% *ee*) was obtained by changing the protecting group *N*-acetyl to *N*-benzoyl of the substrate. The catalyst Rh-SupraPhos (Rh-**90**), generated from self-assembly of chiral monophosphite ligand and an achiral phosphine ligand with a rhodium precursor $[Rh(COD)_2]BF_4$, gave as high as 94% *ee* for the hydrogenation of cyclic enamide **83** under 12 atm H_2 pressure [76].

The ruthenium catalyst of ligand SynPhos (**88**) was highly enantioselective for the hydrogenation of cyclic enamides derived from 3-chromanones (Scheme 9) [77]. The *N*-acyl group of the substrates has a significant effect on

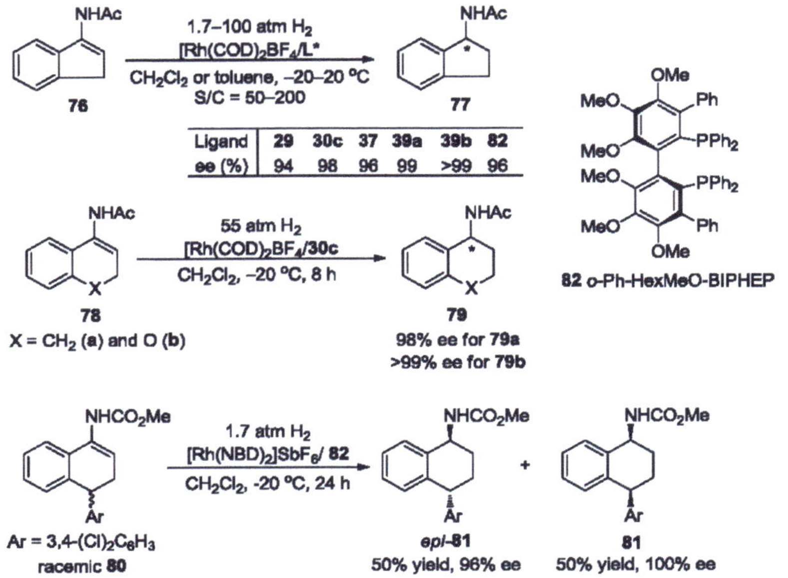

Scheme 8 Enantioselective hydrogenation of cyclic enamides **76**, **78**, and **80**

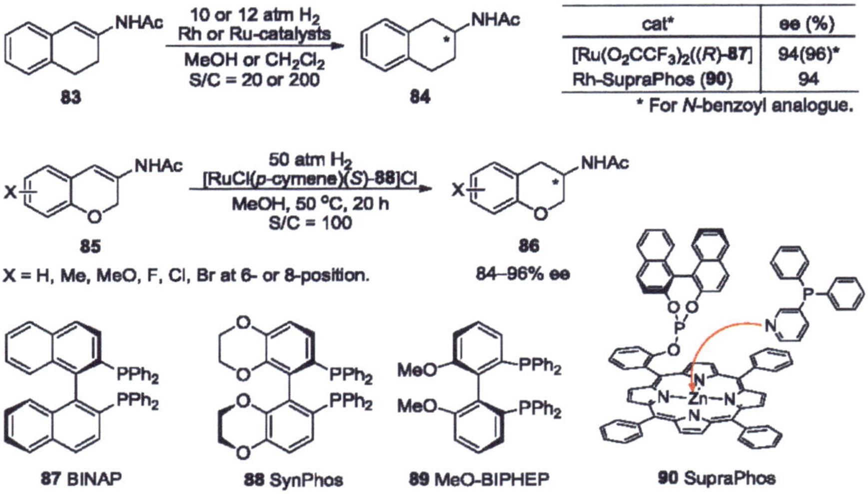

Scheme 9 Enantioselective hydrogenation of cyclic enamides **83** and **85**

enantioselectivity, and the enamide with *N*-acetyl protecting group gave higher enantioselectivity. Under the optimal conditions, a series of *N*-acetyl cyclic enamides **85** with substituents at the 6- or 8-positions were hydrogenated by catalyst Ru-**88** to chiral amides **86** with 84–96% *ee*. In contrast, the ruthenium

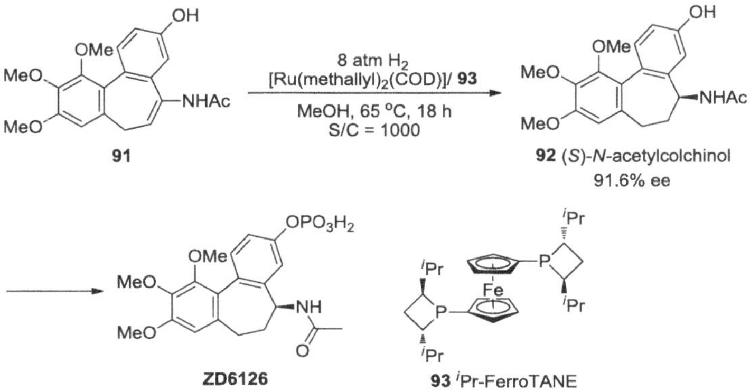

Scheme 10 Enantioselective synthesis of ZD6126

catalyst of ligand MeO-BIPHEP (**89**) gave a higher enantioselectivity for the *N*-benzoyl enamide **85** (X = H, 92% *ee*) than the *N*-acetyl enamide **85** (X = H) (72% *ee*) [76]. Recently, the catalytic asymmetric hydrogenation of methyl 6,8-difluoro-2*H*-chromen-3-ylcarbamate, a *N*-methyloxycarbonyl analogue of cyclic enamide **85** (X = 6,8-$(F)_2$), with chiral ruthenium catalysts of diphosphine ligands was successfully applied to the preparation of Etamicastat, a novel peripherally selective dopamine β-hydroxylase (DBH) inhibitor [78].

The ruthenium complex of the ferrocene-based ligand ^{i}Pr-FerroTANE (**93**) showed good enantioselectivity and reactivity in the hydrogenation of cyclic enamide **91** (Scheme 10) [79]. When the hydrogenation was performed at S/C of 1,000 under 8 atm H_2 at 65°C, the chiral amide **92** (*N*-acetylcolchinol), a key intermediate for the synthesis of the drug ZD6126, was obtained in 91.6% *ee* with 100% conversion. In contrast, the ruthenium or rhodium catalysts containing other chiral ligands such as DuPhos (**5**) and BPE (**6**) yielded the chiral amide **92** with only moderate enantioselectivities (<76% *ee*).

The tetrasubstituted cyclic enamide **94** derived from 2-methyl-1-indanone can be hydrogenated to chiral amide *cis*-**95** in 96% *ee* with rhodium catalyst of ligand *o*-Ph-HexMeO-BIPHEP (**82**, Scheme 11) [73]. However, this catalyst gave low enantioselectivity (37% *ee*) for the hydrogenation of the tetrasubstituted cyclic enamide derived from 2-methyl-1-tetralone. These results are similar to those obtained by rhodium catalyst of rigid diphosphine ligand Me-PennPhos [80]. The hydrogenation of tetrasubstituted cyclic enamides such as **96** derived from 1-benzylic substituted 2-tetralones has been studied with ruthenium catalysts of ligands Me-DuPhos (**5a**) and Me-BPE (**6**), but only moderate enantioselectivities (52–72% *ee*) were obtained [81]. Recently, the chiral dinuclear rhodium catalyst **101** bridged by two anionic sulfonamidophosphoramidite ligands **100** was reported to be highly enantioselective (>99% *ee*) for the hydrogenation of enamide **96**, although with only 56% conversion at S/C = 100 [82]. The fluorine-containing tetrasubstituted cyclic enamide **98** could be hydrogenated to chiral cyclic amide **99** with 99% yield in 99.1% *ee* by using the rhodium catalyst of Josiphos ligand **102**

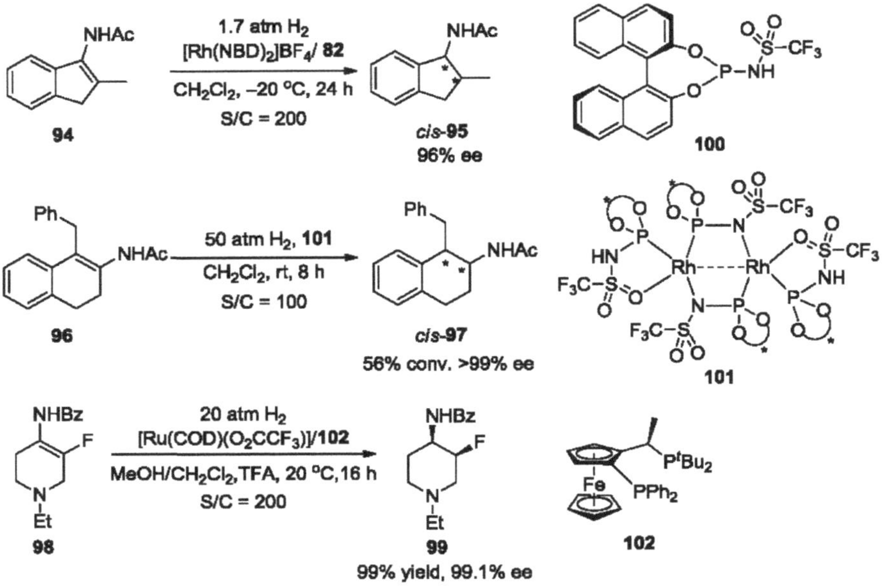

Scheme 11 Enantioselective hydrogenation of tetrasubstituted cyclic enamides

[83]. This highly efficient asymmetric hydrogenation has been successfully applied to the preparation of enantiomerically pure fluoropiperidine on the multi-kilogram scale.

In 1986, Noyori et al. [84] reported that the catalyst Ru-BINAP (**87**) can catalyze the enantioselective hydrogenation of *N*-acyl-1-alkylidenetetrahydroisoquinolines, a type of cyclic enamides with an exo carbon-carbon double bond, to chiral tetrahydroisoquinolines. However, this catalyst is only efficient for the (*Z*)-isomer of substrate, whereas the (*E*)-isomer is recovered intact. The subsequent studies were limited to 1-methylenetetrahydroisoquinolines, which have no *Z*/*E* isomers, such as **103a** and **103b** (Scheme 12). The rhodium catalysts of diphosphine ligands TangPhos (**20**) [27] and **61** [64] were efficient for the hydrogenation of enamide **103b**, providing *N*-acetyl salsolidine (**104b**) in 97 and 94% *ee*, respectively. Recently, the rhodium catalysts of 1,2-diphosphine ligand MalPhos (**53**) and its benzamide derivatives **109a** and **109b** showed high enantioselectivity for the hydrogenation of enamide **103a** to produce tetrahydroisoquinoline **104a** in 92, 94, and 95% *ee*, respectively [61]. With these chiral rhodium catalysts, the exocyclic enamide **105** derived from tetrahydro-*β*-carboline was hydrogenated to chiral dihydro-*β*-carboline **106** in up to 97% *ee*.

The exocyclic enamides **107** can be synthesized as a single (*Z*)-isomer from dihydrobenzoxazine, and was studied in the rhodium-catalyzed asymmetric hydrogenation [85]. The catalyst Rh-Me-DuPhos (Rh-**5a**) was found to be a choice among the catalysts screened. Under optimal conditions a series of exocyclic enamides **107** were hydrogenated to the benzomorpholine products **108** in high

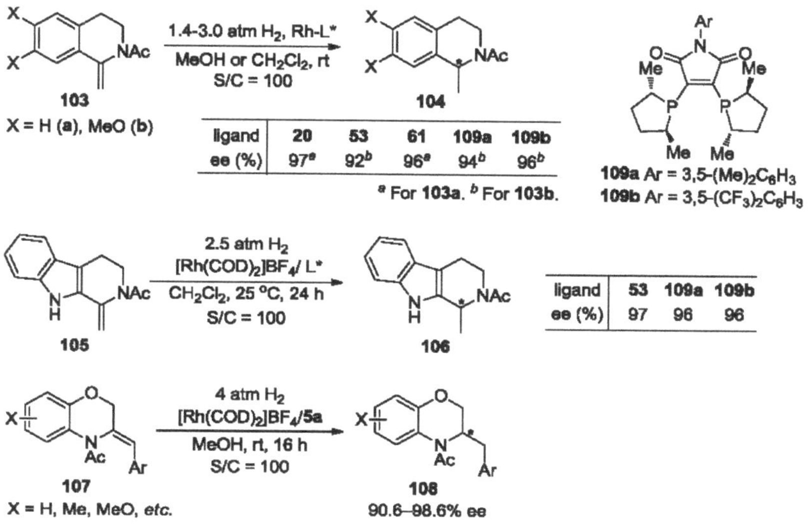

Scheme 12 Enantioselective hydrogenation of exocyclic enamides

enantioselectivities (90.6–98.6% *ee*) regardless of the substituent on the aromatic ring.

3 Enantioselective Hydrogenation of Enamines

Compared with the enamides (*N*-acyl enamines), unprotected enamines are challenging substrates in catalytic asymmetric hydrogenation. The lack of a chelating acyl group of substrate for the metal of the catalyst makes the enantiocontrol in the hydrogenation of carbon-carbon double bond of enamines very difficult. To date, very few efficient chiral catalysts have been developed for the enantioselective hydrogenation of acyclic and cyclic unprotected enamines.

3.1 Acyclic Enamines

3.1.1 Acyclic α-Arylethenamines

Chiral iridium complexes of phosphine-oxazoline ligands and phosphoramidite ligands were recently reported to be efficient catalysts for the hydrogenation of acyclic α-arylethenamines **110** (Scheme 13). The iridium catalyst of phosphine-oxazoline ligand **112** (Ir-**112**) showed moderate to good enantioselectivities (64–87% *ee*) for the hydrogenation of *N*,*N*-diethyl α-arylethenamines (**110**, R^1,

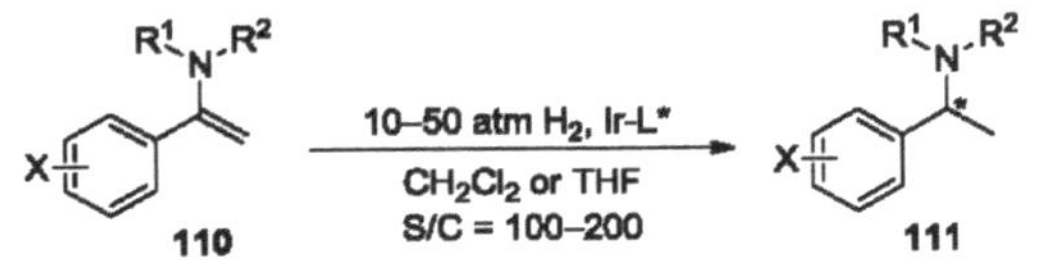

ligand	R^1, R^2	X	ee (%)
112	Et, Et	H, Me, MeO, *etc* (*p*)	64–87
113	Me, Bn	H, MeO, Cl (*p*)	50–92.5
114	Me, Ar*	H	90–91
115	$-(CH_2)_4-$	H, Me, MeO *etc.* (*p,m,o*)	52–87

* Ar = Ph, 4-MeO-Ph, 4-Cl-Ph.

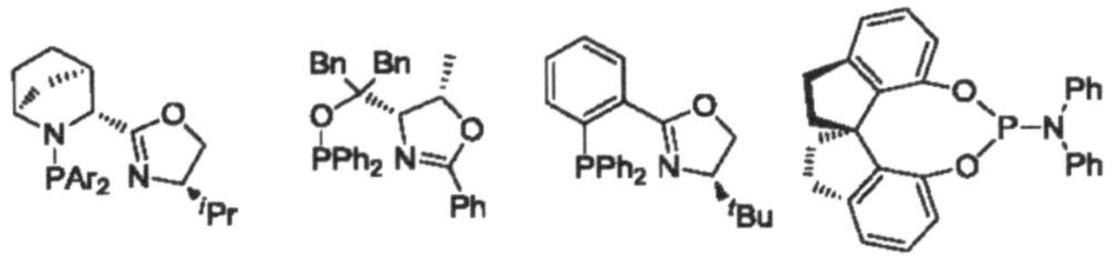

112 (Ar = 4-Me-Ph) 113 114 115

Scheme 13 Enantioselective hydrogenation of enamines **110**

R^2 = Et) under 50 atm H_2 (S/C = 200). However, low enantioselectivities (<40% *ee*) were obtained for the α-arylethenamines **110** with cyclic amino groups such as pyrrolidyl and morpholidyl [86]. The iridium catalysts with ligands Threphox (**113**) and PHOX (**114**) also offered moderate to good enantioselectivities for the hydrogenations of *N*-benzyl-*N*-methyl and *N*-aryl-*N*-methyl α-arylethenamines **110**, respectively [87] However, both catalysts showed low enantioselectivity for the hydrogenation of α-arylethenamines **110** with pyrrolidyl and *N,N*-diethyl groups.

Recently, the iridium catalyst of monodentate spiro *N,N*-diarylphosphoramidite ligand **115** (Ir-**115**) was used to catalyze the hydrogenation of *N,N*-dialkyl α-arylethenamines **110**, giving the corresponding amines in 52–87% *ee*. In the presence of 5 mol% I_2 the hydrogenation can be carried out at 10 atm H_2 [88]. All these aforementioned results showed that the enantioselective hydrogenation of simple α-arylethenamines is still a challenge, and more efficient chiral catalysts are highly desirable for this useful reaction.

3.1.2 Acyclic β-Substituted α-Arylethenamines

The β-aryl substituted α-arylethamines **116** with a pyrrolidyl group can be hydrogenated by a rhodium catalyst of chiral spiro phosphonite ligand **118** in excellent enantioselectivities (Scheme 14). In the presence of 2 mol% I_2 and 20 mol% HOAc as additives and under 10 atm H_2, a variety of β-aryl substituted α-arylethamines **116** were hydrogenated by the catalyst Rh-**118** to amines **117** in 80–99.9% *ee* [89]. Generally, the substrates with electron-donating substituents

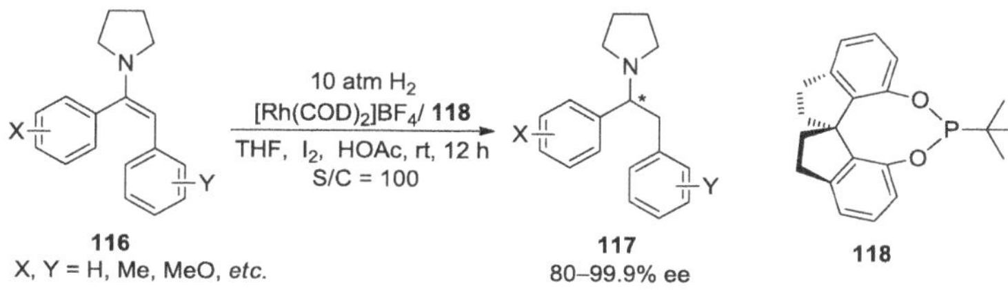

Scheme 14 Enantioselective hydrogenation of enamines **116**

(X group) on the α-phenyl ring and/or electron-withdrawing substituents (Y group) on the β-phenyl ring gave higher enantioselectivity. The highest enantioselectivity (99.9% *ee*) was achieved in the hydrogenation of the enamine having a 4-F on the β-phenyl ring of the substrate. In this reaction I_2 remarkably improved the enantioselectivity and HOAc accelerated the reaction rate.

3.2 Cyclic Enamines

Chiral cyclic tertiary amines are essential structural units in natural products and drugs, and catalytic enantioselective hydrogenation of cyclic *N,N*-dialkylenamines provides a direct approach to optically active tertiary amines. The iridium complex of spiro phosphoramidite ligand **123** (SiPhos-pe) was found to be a highly enantioselective catalyst for this transformation (Scheme 15). In combination with I_2 the catalyst Ir-**123** in situ generated from [Ir(COD)Cl]$_2$ and ligand **123** gave 82–97% *ee* for the hydrogenation of cyclic enamines **119** under mild conditions (1 atm H_2, room temperature, S/C = 100), and the isoquinoline alkaloid crispine A was prepared by the Ir-**123**-catalyzed hydrogenation of tricyclic enamines **121** (X = MeO) in 90% *ee* [90]. The addition of I_2 in this catalytic system is crucial for obtaining full conversion and high enantioselectivity.

The cyclic enamines **124** derived from benzocyclic ketones such as 1-indanone and α-tetralone were also hydrogenated in high enantioselectivity by chiral iridium catalysts containing chiral spiro monophosphorus ligands (Scheme 16). The iridium catalyst of ligand **115** gave 87% *ee* for hydrogenation of cyclic enamine **124** ($n = 2$, X = MeO) derived from 6-methoxy-3,4-dihydronaphthalenone [87]. Comparable enantioselectivities (80–89% *ee*) were obtained by iridium catalyst of ligand **118** in this reaction [88].

Tetrahydroisoquinolines, widely existing in plants and several tissues in mammalian species, is a common structural motif of numerous alkaloids. The enantioselective hydrogenation of the easily obtained *N*-alkyl-1-alkylidenetetrahydroisoquinolines **126** is a direct method for the synthesis of optically pure *N*-alkyltetrahydroisoquinolines. With catalyst Ir-**123** and under ambient hydrogen pressure (1 atm H_2) a series of enamines **126** were hydrogenated to chiral *N*-alkyl tetrahydroisoquinolines **127** in high yields with up to 98% *ee* (Scheme 16) [91].

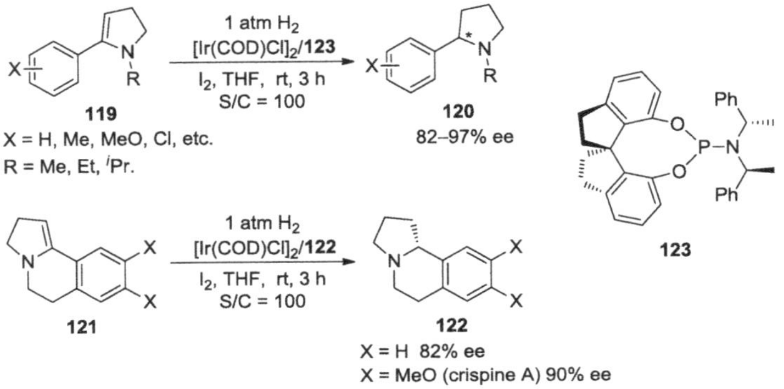

Scheme 15 Enantioselective hydrogenation of enamines **119** and **121**

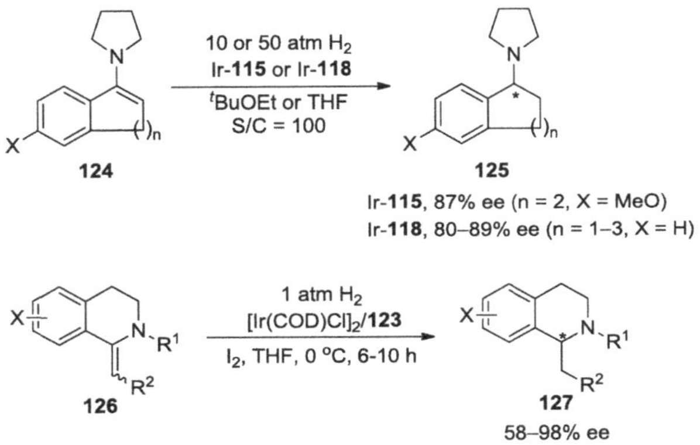

Scheme 16 Enantioselective hydrogenation of enamines **124** and **126**

4 Conclusion and Outlook

As can be seen from the contents of this review, although significant progress has been made over the past decade in the field of transition metal-catalyzed hydrogenation of enamides and enamines, the highly enantioselective hydrogenation of both enamides and enamines is still a challenge. The existing drawbacks such as the low enantioselectivity for the *E*/*Z* isomeric mixture of acyclic enamides and enamines and the poisoning effect of the amines products toward the catalysts need to be addressed. Several recent breakthroughs on the highly efficient hydrogenations of the simple enamines provided new strategies for finding efficient catalysts. The catalytic enantioselective hydrogenation of enamides and enamines will continue to be an exciting and promising area of asymmetric catalysis.

References

1. Kagan HB, Dang TP (1972) J Am Chem Soc 94:6429
2. de Vries JG, Elsevier CJ (2007) The handbook of homogeneous hydrogenation. Wiley VCH, Weinheim
3. Nugent TC, El-Shazly M (2010) Adv Synth Catal 352:753
4. Xie JH, Zhou QL (2012) Acta Chim Sinica 70:1427
5. Lee NE, Buchwald SL (1994) J Am Chem Soc 116:5985
6. Tararov VI, Kadyrov R, Riermeier TH, Holz J, Börner A (2000) Tetrahedron Lett 41:2351
7. Ohta T, Ikegami H, Miyake T, Takaya H (1995) J Organometal Chem 502:169
8. Becalski AG, Cullen WR, Fryzuk MD, James BR, Kang GJ, Retting SJ (1991) Inorg Chem 30:5002
9. Guo HC, Ding KL, Dai LX (2004) Chin Sci Bull 49:1575
10. Xie JH, Zhu SF, Zhou QL (2012) Chem Soc Rev 41:4126
11. Blaser HU, Spindler F, Thommen M (2007) In: de Vries JG, Elsevier CJ (eds) The handbook of homogeneous hydrogenation. Wiley VCH, Weinheim, p 1279
12. Burk MJ, Wang YM, Lee JR (1996) J Am Chem Soc 118:5142
13. Li W, Zhang Z, Xiao D, Zhang X (2000) J Org Chem 65:3489
14. Xiao D, Zhang Z, Zhang X (1999) Org Lett 1:1679
15. Yan Y, Zhang X (2006) Tetrahedron Lett 47:1567
16. Li W, Zhang X (2000) J Org Chem 65:5871
17. Yan YY, RajanBabu TV (2000) Org Lett 2:4137
18. Lee SG, Zhang YJ, Song CE, Lee JK, Choi JH (2002) Angew Chem Int Ed 41:847
19. Zhang FY, Pai CC, Chan ASC (1998) J Am Chem Soc 120:5808
20. Zhou YG, Zhang X (2002) Chem Commun 1124
21. Wang CJ, Gao F, Liang G (2008) Org Lett 10:4711
22. Hu XP, Zheng Z (2004) Org Lett 6:3585
23. Zeng QH, Hu XP, Duan ZC, Liang XM, Zheng Z (2005) Tetrahedron Asymmetry 16:1233
24. Li X, Jia X, Xu L, Kok SHL, Yip CW, Chan ASC (2005) Adv Synth Catal 347:1904
25. Huang JD, Hu XP, Duan ZC, Zeng QH, Yu SB, Deng J, Wang DY, Zheng Z (2006) Org Lett 8:4367
26. Qiu M, Hu XP, Wang DY, Deng J, Huang JD, Yu SB, Duan ZC, Zheng Z (2008) Adv Synth Catal 350:1413
27. Tang W, Zhang X (2001) Angew Chem Int Ed 41:1612
28. Imamoto T, Oohara N, Takahashi H (2004) Synthesis 1353
29. Tang W, Qu B, Capacci AG, Rodriguez S, Wei X, Haddad N, Narayanan B, Ma S, Grinberg N, Yee NK, Krishnamurthy D, Senanayake CH (2010) Org Lett 12:176
30. Imamoto T, Tamura K, Zhang Z, Horiuchi Y, Sugiya M, Yoshida K, Yanagisawa A, Gridnev ID (2012) J Am Chem Soc 134:1754
31. Zhang Z. Tamura K, Mayama D, Sugiya M, Imamoto T (2012) J Org Chem 77:4184
32. Stephan M, Šterk D, Mohar B (2009) Adv Synth Catal 351:2779
33. Zupančič B, Mohar B, Stephan M (2010) Org Lett 12:1296
34. Zupančič B, Mohar B, Stephan M (2010) Org Lett 12:3022
35. Stephan M, Šterk D, Zupančič B, Mohar B (2011) Org Biomol Chem 9:5266
36. Hu AG, Fu Y, Xie JH, Zhou H, Wang LX, Zhou QL (2002) Angew Chem Int Ed 41:2348
37. Jia X, Guo R, Li X, Yao X, Chan ASC (2002) Tetrahedron Lett 43:5541
38. Jia X, Li X, Xu L, Shi Q, Yao X, Chan ASC (2003) J Org Chem 68:4539
39. Bernsman H, van den Berg M, Hoen R, Minnaard AJ, Mehler G, Reetz MT, de Vries JG, Feringa BL (2005) J Org Chem 70:943
40. Li X, Jia X, Lu G, AuYeung TTL, Lam KHT, Lo WH, Chan ASC (2003) Tetrahedron Asymmetry 14:2687
41. Zeng QH, Hu XP, Duan ZC, Liang XM, Zheng Z (2006) J Org Chem 71:393
42. Liu Y, Ding K (2005) J Am Chem Soc 127:10488

43. Zhao B, Wang Z, Ding K (2006) Adv Synth Catal 348:1049
44. Reetz MT, Mehler G, Meiswinkel A, Sell T (2002) Tetrahedron Lett 43:7941
45. Huang H, Zheng Z, Luo H, Bai C, Hu X, Chen H (2003) Org Lett 5:4137
46. Huang H, Zheng Z, Luo H, Bai C, Hu X, Chen H (2004) J Org Chem 69:2355
47. Pignataro L, Boghi M, Civera M, Carboni S, Piarulli U, Gennari C (2012) Chem Eur J 18:1383
48. Pignataro L, Bovio C, Civera M, Piarulli U, Gennari C (2012) Chem Eur J 18:10368
49. Hu XP, Huang JD, Zeng QH, Zheng Z (2006) Chem Commun 293
50. Chen W, Roberts SN, Whittall J (2006) Tetrahedron Lett 47:4263
51. Zhao Y, Huang H, Shao J, Xia C (2011) Tetrahedron Asymmetry 22:769
52. Wang X, Ding K (2004) J Am Chem Soc 126:10524
53. Zhang F, Li Y, Li ZW, He YM, Zhu SF, Fan QH, Zhou QL (2008) Chem Commun 6048
54. Shi L, Wang X, Sandoval CA, Li M, Qi Q, Li Z, Ding K (2006) Angew Chem Int Ed 45:4108
55. Zhang W, Zhang X (2006) Angew Chem Int Ed 45:5515
56. Wang T, Lamb ML, Scott DA et al (2008) J Med Chem 51:4672
57. O'Shea PD, Gauvreau D, Gosselin F, Hughes G, Nadeau C, Roy A, Shultz CS (2009) J Org Chem 74:4547
58. Lefort L, Boogers JAF, Kuilman T, Vijn RJ, Janssen J, Straatman H, de Vries JG, de Vries AHM (2010) Org Process Res Dev 14:568
59. Gridnev ID, Yasutake M, Higashi N, Imamoto T (2001) J Am Chem Soc 123:5268
60. Gridnev ID, Imamoto T, Hoge G, Kouchi M, Takahashi H (2008) J Am Chem Soc 130:2560
61. Enthaler S, Erre G, Junge K, Addis D, Kadyrov R, Beller M (2008) Chem Asian J 3:1104
62. Allwein SP, McWilliams JC, Secord EA, Mowrey DR, Nelson TD, Kress MH (2006) Tetrahedron Lett 47:6409
63. Yang Q, Gao W, Deng J, Zhang X (2006) Tetrahedron Lett 47:821
64. Li W, Waldkirich JP, Zhang X (2002) J Org Chem 67:7618
65. Chi Y, Zhang X (2002) Tetrahedron Lett 43:4849
66. Hoen R, van den Berg M, Bernsmann H, Minnaard AJ, de Vries JG, Feringa BL (2004) Org Lett 6:1433
67. Huang H, Liu X, Chen H, Zheng Z (2005) Tetrahedron Asymmetry 16:693
68. Storace L, Anzalone L, Confalone PN et al (2002) Org Proc Res Dev 6:54
69. Zhu G, Casalnuovo AL, Zhang X (1998) J Org Chem 63:8100
70. Chen J, Zhang W, Geng H, Li W, Hou G, Lei A, Zhang X (2009) Angew Chem Int Ed 48:800
71. Zhu SF, Liu T, Yang S, Song S, Zhou QL (2012) Tetrahedron 68:7685
72. Wallace DJ, Campos KR, Shultz CS, Klapars A, Zewge D, Crump BR, Phenix BD, McWilliams JC, Krska S, Sun Y, Chen CY, Spindler F (2009) Org Proc Res Dev 13:84
73. Tang W, Chi Y, Zhang X (2002) Org Lett 4:1695
74. Fu Y, Guo XX, Zhu SF, Hu AG, Xie JH, Zhou QL (2004) J Org Chem 69:4648
75. Renaud JL, Dupau P, Hay AE, Guingouain M, Dixneuf PH, Bruneau C (2003) Adv Synth Catal 345:230
76. Jiang XB, Lefort L, Goudriaan PE, de Vires AHM, van Leeuwen PWNM, de Vries JG, Reek JNH (2006) Angew Chem Int Ed 45:1223
77. Wu Z, Ayad T, Ratovelomenana-Vidal V (2011) Org Lett 13:3782
78. Beliaev A, Wahnon J, Russo D (2012) Org Proc Res Dev 16:704
79. Lennon IC, Ramsden JA, Brear CJ, Broady SD, Muir JC (2007) Tetrahedron Lett 48:4623
80. Zhang Z, Zhu G, Jiang Q, Xiao D, Zhang X (1999) J Org Chem 64:1774
81. Dupau P, Bruneau C, Dixneuf PH (2001) Adv Synth Catal 343:331
82. Patureau FW, de Boer S, Kuil M, Meeuwissen J, Breuil PAR, Siegler MA, Spek AL, Sandee AJ, de Bruin B, Reek JNH (2009) J Am Chem Soc 131:6683
83. Stumpf A, Reynolds M, Sutherlin D, Babu S, Bappert E, Spindler F, Welch M, Gaudino J (2011) Adv Synth Catal 353:3367
84. Noyori R, Ohta M, Hsiao Y, Kitamura M, Ohta T, Takaya H (1986) J Am Chem Soc 108:7117
85. Zhou YG, Yang PY, Han XW (2005) J Org Chem 70:1679

86. Cheruku P, Church TL, Trifonova A, Wartmann T, Andersson PG (2008) Tetrahedron Lett 49:7290
87. Baeza A, Pfaltz A (2009) Chem Eur J 15:2266
88. Yan PC, Xie JH, Zhou QL (2010) Chin J Chem 28:1736
89. Hou GH, Xie JH, Wang LX, Zhou QL (2006) J Am Chem Soc 128:11774
90. Hou GH, Xie JH, Yan PC, Zhou QL (2009) J Am Chem Soc 131:1366
91. Yan PC, Xie JH, Hou GH, Wang LX, Zhou QL (2009) Adv Synth Catal 351:3243

Top Curr Chem (2014) 343: 103–144
DOI: 10.1007/128_2013_483

Published online: 6 November 2013

Asymmetric Hydrogenation of Imines

Wei Li and Xumu Zhang

Abstract Homogeneous catalytic asymmetric hydrogenation has evolved into a significantly useful methodology for the prepartion of enantiopure compounds. Significant advances in asymmetric hydrogenation of imines have enabled a straightforward and powerful approach to various chiral amines. An overview of the many variants of the chiral transition metal catalysts, illustrations of synthetic applications, and discussions of emerging and future strategies from a mechanistic perspective are presented.

Keywords Asymmetric · Catalysis · Hydrogenation · Imines · Transition metal

Contents

W. Li
Institute of Chemical Biology and Drug Discovery, Stony Brook University, Stony Brook, NY 11794-3400, USA

X. Zhang (✉)
Departments of Chemistry and Chemical Biology, Department of Medicinal Chemistry, Rutgers, the State University of New Jersey, 160 Frelinghuysen Rd, Piscataway, NJ 08854, USA
e-mail: xumu@rci.rutgers.edu

Abbreviations

BARF	(Tetrakis(3,5-bis(trifluoromethyl)phenyl)borate)
COD	Cyclooctadiene
Cp	Cyclopentadienyl
Cp*	1,2,3,4,5-Pentamethylcyclopentadienyl
DCE	Dichloroethane
D-DTTA	Di-*p*-toluoyl-D-tartaric acid
DPEN	1,2-Diphenylethane-1,2-diamine
MEA	2-Methyl-5-ethylaniline
Py	Pyridine
SPO	Secondary phosphine oxide
TBME	*tert*-Butyl methyl ether
TFE	Trifluoroethanol
THF	Tetrahydrofuran
TMI	2,3,3-Trimethylindolenine
TOF	Turnover frequency
TON	Turnover number
Ts	Tosyl (p-$CH_3C_6H_4SO_2$)

1 Introduction

Chiral amines are an important class of organic compounds that can serve as key intermediates for the synthesis of a variety of biologically active molecules. Reducing prochiral imines to obtain chiral imines represents an attractive strategy that enables a straightforward approach to valuable synthetic building blocks.

Extensively studied since the late 1960s, homogeneous catalytic asymmetric transformations in organic syntheses constitute mainstream synthetic chemistry. In particular, as an important division of asymmetric catalysis, advances in asymmetric hydrogenation have provided an important driving force for basic research on synthetic organic methodology and industrial processes. Its simplistic nature, environmental friendliness, and cost effectiveness have undoubtedly made asymmetric hydrogenation one of the most studied methodologies in the last 50 years [1–4]. Catalytic enantioselective hydrogenation of imines by chiral transition metal complexes was found to be one of the most efficient and convenient methods for preparing chiral amines and their derivatives [5–7].

A broad range of chiral catalysts have been developed in the hope that the chirality from the chiral catalyst could be effectively transferred to the reactant, providing the desired chiral amines or their derivatives with excellent enantioselectivities. One very notable and motivating example is the first and most successful industrial application of asymmetric imine hydrogenation producing chiral herbicide (*S*)-metolachlor – more than 10,000 tons per year – by the chiral iridium–Xyliphos [8].

Compared with the success achieved in olefin and ketone hydrogenation, the method developed for asymmetric imine reduction was less useful [9, 10]. As a result of the properties of the C = N bond: (1) relatively low catalytic activities were usually observed, probably due to the strong coordination of the amine product to the transition metal catalyst; (2) the amine adduct of the catalyst loses activity toward hydrogenation and thus results in low turnover numbers and yields; (3) the inseparable and sometimes interconvertible *E*/*Z* isomers of imines can also result in poor enantioselectivity; (4) imine-enamine tautomerization could also erode the enantioselectivities. Generally, the nature of the substrates, particularly the substituent directly attached to the N-atom, significantly influences the properties and thus the reduction capability of the C = N functional groups.

2 Transition Metal Catalysts for Asymmetric Hydrogenation of Imines

2.1 Rh-Catalyzed Asymmetric Hydrogenation of Imines

In the early stage of asymmetric hydrogenation, the combinations of rhodium and chiral phosphine ligands were demonstrated as powerful catalysts in the hydrogenation of C = C double bonds. Thus, rhodium was one of the first transition metals investigated in the form of chiral complexes with chiral phosphorus ligands in the enantioselective hydrogenation of imines, initiated by the pioneering work reported by Scorrano et al. [11]. However, only a few rhodium-based catalyst systems have been reported to provide high enantioselectivities in the hydrogenation of C = N double bonds (Scheme 1).

In 1988 Cullen and Fryzuk et al. reported another piece of pioneering work on Rh-catalyzed hydrogenation of *N*-benzylimines in which (*R*)-Cycphos was applied to achieve up to 91% *ee* in the presence of KI as an additive [12].

Subsequently, Bakos et al. reported high enantioselectivities of the same *N*-benzylimine substrates using Rh complexes of BDPP and monosulfonated-BDPP (84% *ee* and 94% *ee*, respectively) [13]. In 1996, Buriak and Osborn introduced AOT (bis(2-ethylhexyl) sulfosuccinate) and 15-crown-5 to enhance the performance of Rh–BDPP catalyst in the two-phase reaction system [14].

Börner and co-workers described their studies of a variety of chiral diphosphine ligands such as DIOP derivatives and chiral diphosphinite ligands such as BDPCH in Rh-catalyzed enantioselective *N*-benzylimine reduction [15]. The diphosphinite

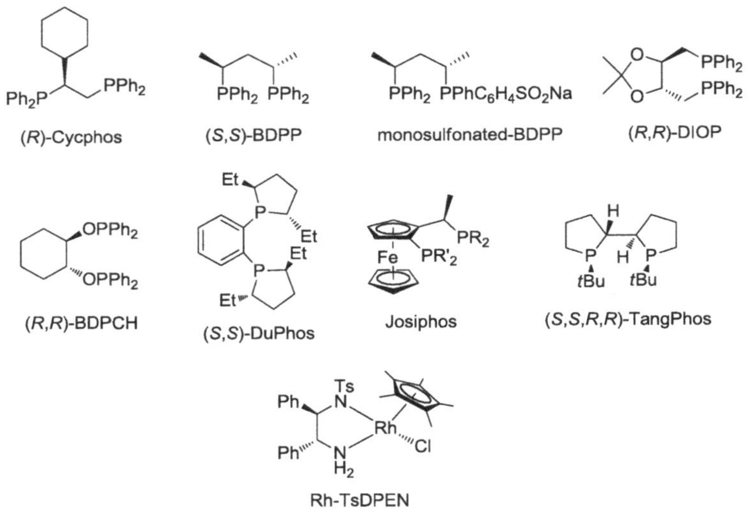

Scheme 1 Chiral phosphorus ligands in Rh catalyst and chiral Rh complex

ligand BDPCH was found to be superior (up to 71% *ee*) to those diphosphine ligands (up to 41% *ee*), when employed in the [RhL*(COD)]BF_4 complex. When screening diphosphine ligands, they found that the reaction was highly sensitive to the size of the chelate ring formed with the metal and the substituents on the phosphorus.

Using the cationic Rh(I)–DuPhos complex, Burk and co-workers successfully used asymmetric hydrogenation to reduce a series of *N*-benzoylhydrazones containing aryl and methyl substituents, and achieved corresponding products in 88–97% *ee* (Scheme 2) [16]. For a similar type of hydrazine substrates, the recent application of electron-rich but ferrocene-based ligands such as Josiphos by Yoshikawa and Tan showed promising results (64–92% *ee*, Scheme 2) [17].

Zhang and co-workers reported an example of excellent enantioselectivity of the Rh catalyst for the hydrogenation of imino esters, using electron-rich and structurally rigid bisphospholane ligand TangPhos [18]. The substrate scope included α-aryl (up to 95% *ee*) imino esters with various substituents and α-alkyl substrate (94% *ee*), and the excellent enantioselectivities and low catalyst loading (0.1–1 mol%) made this method potentially useful for the preparation of chiral glycines (Scheme 3).

Xiao and Li demonstrated an example of Rh(III) catalyst by applying asymmetric transfer hydrogenation catalyst Rh(III)-TsDPEN to asymmetric direct hydrogenation of cyclic imines [19]. The authors studied the dramatic anion effect on both enantioselectivities and kinetics and found this Cp*Rh(TsDPEN)Cl catalyst to be highly effective (up to 99% *ee*) in the presence of $AgSbF_6$ in imine hydrogenation, affording various corresponding products of tetrahydroisoquinoline and tetrahydro-β-carboline derivatives (Scheme 4).

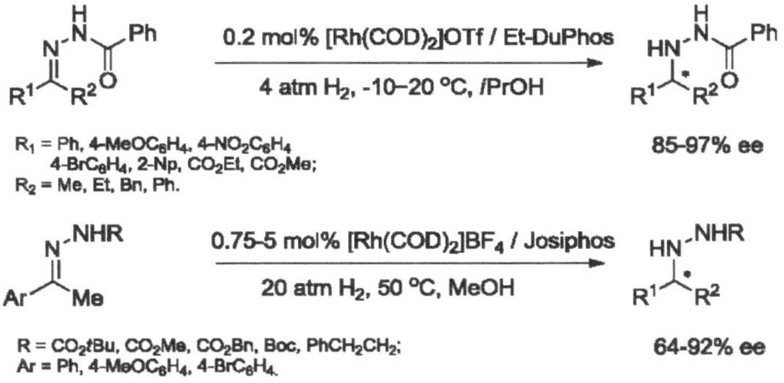

Scheme 2 Rh-catalyzed hydrogenation of *N*-benzoylhydrazones and hydrazine substrates

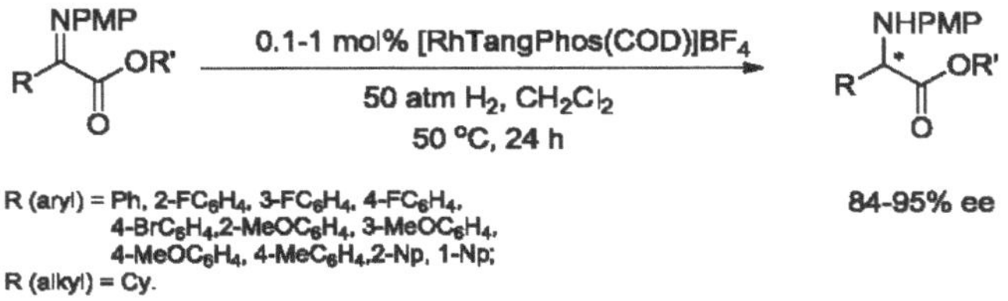

Scheme 3 Rh-catalyzed hydrogenation of imino esters

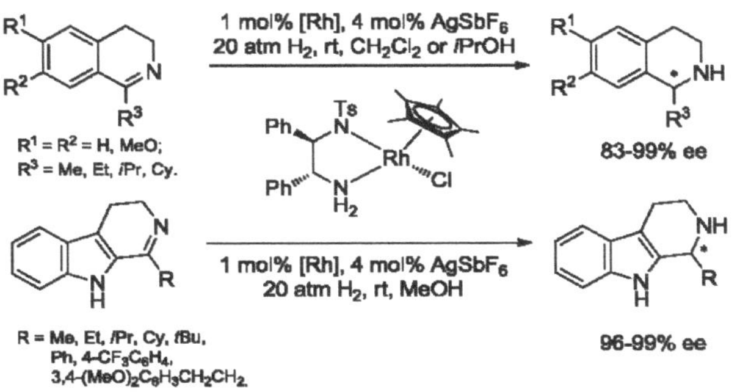

Scheme 4 Rh-catalyzed hydrogenation of cyclic imines

2.2 *Ti-Catalyzed Asymmetric Hydrogenation of Imines*

One of very few examples of a chiral titanium catalyst was involved in Buchwald's study of the chiral titanocene catalyst system, which has high levels of reactivity and enantioselectivity in the hydrogenation of a wide range of both acyclic and cyclic imines [20]. Treatment of the precatalyst chiral titanocene binolate or chloride by

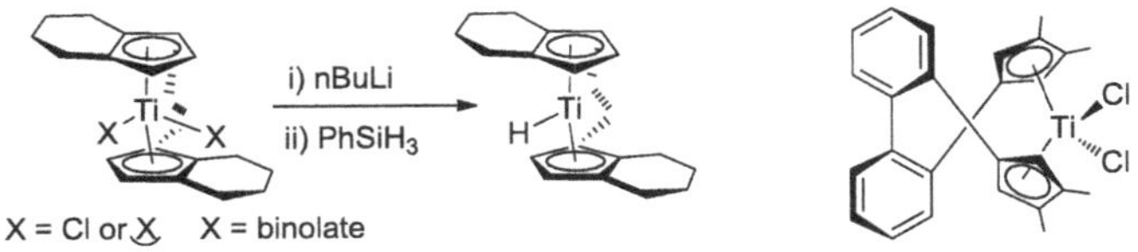

Scheme 5 Chiral titanocene catalysts in enantioselective imine hydrogenation

*n*BuLi and $PhSiH_3$ was necessary to generate the catalytic species (Scheme 5). For a series of 2-substituted cyclic imines, remarkably high enantioselectivities were achieved (95–99% *ee*) with 1 mol% of Ti catalyst under varied hydrogen pressures. However, the fact is that the acyclic imine substrates were prepared as an *E*/*Z* isomeric mixture and found to be significantly adverse to enantioselectivity, resulting in lower *ee*s (53–78% *ee*) when the same Ti catalyst was applied. Moreover, detailed kinetic and mechanistic studies were performed [21].

Brintzinger et al. developed another titanocene with a different bridge linker and investigated it in the imine hydrogenation. Similar activation by *n*BuLi was necessary to generate Ti hydride species and up to S/C = 1,000 and 98% *ee* were obtained for the reduction of cyclic imine [22].

2.3 Ru-Catalyzed Asymmetric Hydrogenation of Imines

The outstanding successes of the Ru–BINAP catalyst system in the asymmetric hydrogenation of C = C and C = O functional groups led to investigations of its performance and utility in asymmetric imine hydrogenations.

Oppolzer and co-workers reported the application of optically pure dimeric Ru complex $Ru_2Cl_4(BINAP)_2(Et_3N)$ in the hydrogenation of the cyclic imine to provide the chiral sultam [23]. Subsequently, Charette and co-workers reported the $Ru(OAc)_2$BINAP complex's application in the hydrogenation of activated *N*-tosylimines with moderate to good enantioselectivities (Scheme 6) [24].

However, Ru catalysts have proved to be more successful in imine hydrogenation after Noyori's Nobel Prize-winning Ru–diphosphine–diamine motif was discovered to be outstandingly powerful in ketone hydrogenations. After their combinatorial screening of chiral diphosphines and chiral diamines, Cobley and Henschke reported optimized catalysts for different types of substrates studied, $RuCl_2$[(*R*,*R*)-Et-Duphos][(*R*,*R*)-DACH] for *N*-arylimine (92% *ee*), $RuCl_2$[(*S*)-Tol-BINAP][(*S*,*S*)-DPEN] for *N*-benzylimine (62% *ee*), and $RuCl_2$[(*S*)-MeO-BIPHEP][(*S*,*S*)-ANDEN] for cyclic imine 2,2,3-trimethylindolenine (88% *ee*), respectively. They also evaluated the effectiveness of this Ru catalyst in the hydrogenation of heteroaromatic compounds (Scheme 7) [25].

Ohkuma et al. described a new Ru–diphosphine–diamine catalyst system of $RuBr_2$[(*S*,*S*)-Xyl-Skewphos][(*S*,*S*)-DPEN], achieving excellent reactivity (up to

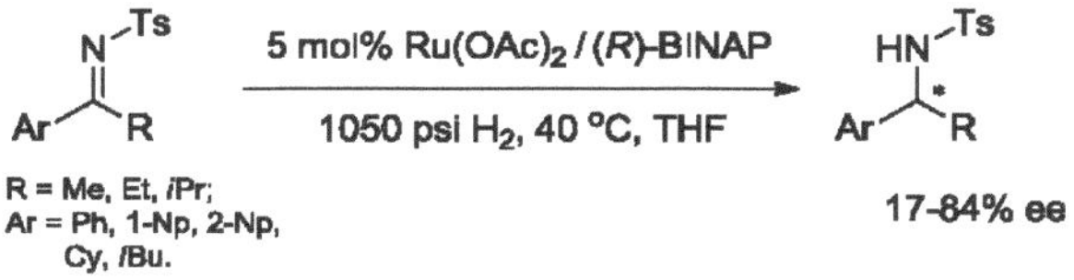

Scheme 6 Enantioselective hydrogenation of *N*-tosyl imines by Ru-BINAP catalyst

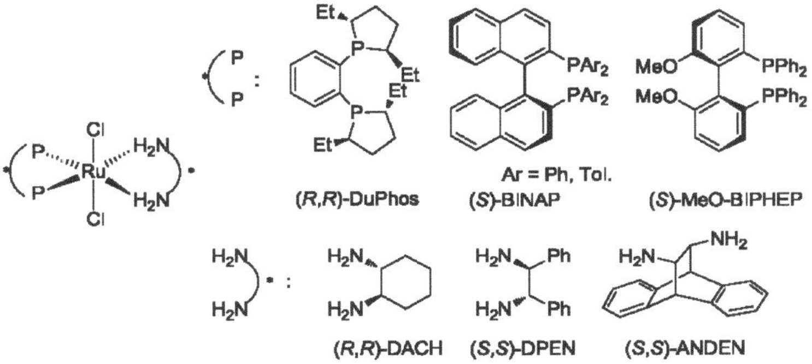

Scheme 7 Chiral *trans*-$RuCl_2$(diphosphine)(diamine) catalysts

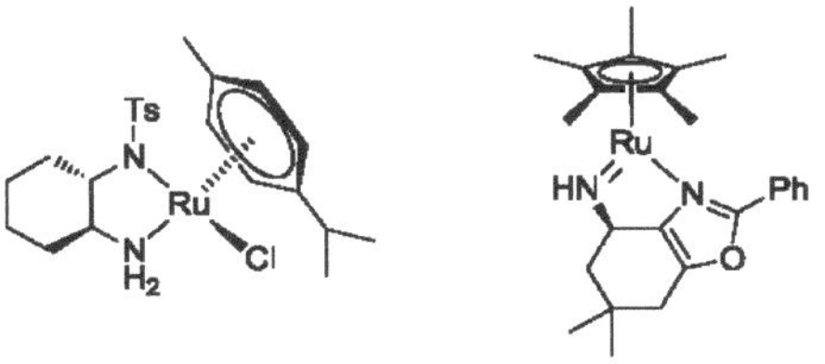

Scheme 8 Other Ru complexes in enantioselective imine hydrogenation

18,000 TON) and enantioselectivity (up to 99% *ee*) in the hydrogenation of *N*-arylimines, particularly of *N*-2′-methoxyphenyl imines [26].

More recently, several Ru-based complexes have been synthesized and found to be effective in imine hydrogenations, such as the transfer hydrogenation catalyst Ru (II)–*N*-sulfonylated diamine complex reported by Ikariya and co-workers. Up to 74% *ee* was achieved in the hydrogenation study of *N*-benzylimine [27]. The positive additive effect of silver salt $AgSbF_6$ was realized, and the excessive amount of silver salt increased enantioselectivities in the similar Ru, Rh, Ir complexes (Scheme 8).

Other recently developed Ru complexes showed only moderate enantioselectivity. For example, the chiral ruthenium amine–imine complex, reported by Andersson and co-workers, afforded up 51% *ee* in the hydrogenation of *N*-arylimine (Scheme 8) [28].

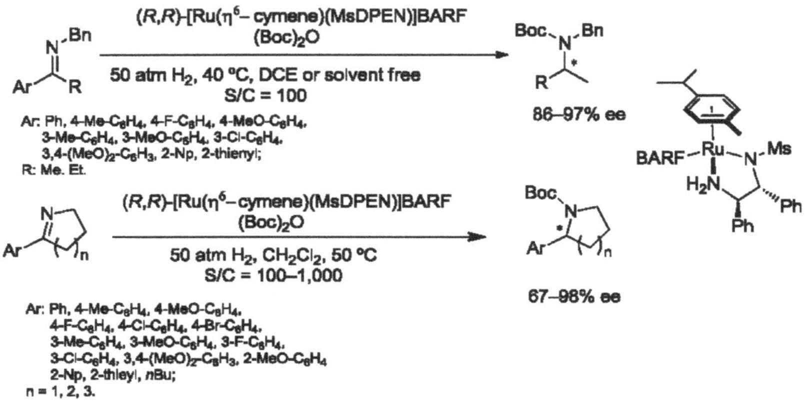

Scheme 9 Enantioselective hydrogenation of imines by [Ru(η^6-cymene)(MsDPEN)BARF catalyst

Fan et al. reported on the application of a similar complex of [Ru(η^6-cymene)(MsDPEN)BARF to asymmetric hydrogenation of acyclic *N*-benzyl ketimines under more environmentally friendly solvent-free conditions at low catalyst loadings (as low as 0.05 mol%) [29]. The same catalyst system also showed excellent efficacy in the hydrogenation of cyclic imines (Scheme 9) [30].

2.4 Ir-Catalyzed Asymmetric Hydrogenation of Imines

Although initially slow to recognize the potential of the Ir catalyst in asymmetric hydrogenation, the pioneering asymmetric Ir-catalyzed imine hydrogenation emerged more than a decade after Crabtree and Morris's first demonstration of highly efficient iridium complex in hydrogenation in a nonasymmetric version in 1976 [31]. However, iridium-based catalysts have been the most popular catalysts in the area of asymmetric hydrogenation of imines. In 1990, Spindler and co-workers described the diphosphinoiridium catalyst for the enantioselective hydrogenation of *N*-arylimine. Under somewhat milder conditions than those in Rh-catalyzed imine hydrogenations, 20 atm of H_2 and at 0°C, 84% *ee* was achieved using (*S*,*S*)-BDPP and $[Ir(COD)Cl]_2$ precursor [32]. The crucial positive effect of iodide to this Ir catalyst was revealed at the same time (Scheme 10).

Subsequently, the historic successful development of industrial synthesis of (*S*)-metolachlor [8], which we could not possibly ignore when reviewing the history of imine hydrogenation, has significantly inspired the forefront of developing highly efficient chiral catalysts, including the development of Ir-based chiral complexes and their applications to related enantioselective imine hydrogenation methodologies, and hence further pharmaceutical and fine chemical industries [33]. Blaser, Spindler, and co-workers reported this highly efficient process, achieving >1,000,000 TON (turnover number) and 79% *ee* using chiral ferrocenyl-based diphosphine Xyliphos

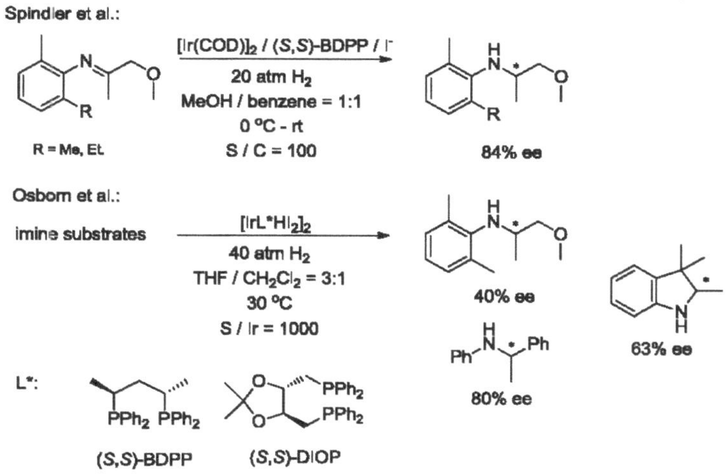

Scheme 10 Pioneering works of Ir-catalyzed enantioselective hydrogenation of imines

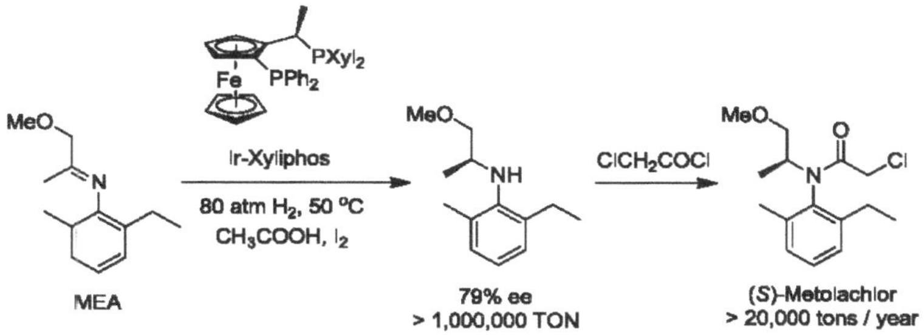

Scheme 11 Ir-catalyzed enantioselective synthesis of (*S*)-metolachlor

(Scheme 11) after the proper choice of the solvent and additive (acetic acid as the solvent and iodide as the additive were crucial to achieving high reactivity and enantiomeric excesses) [34].

2.4.1 Diphosphine Ligands in Ir-Based Catalysts

Enlightened by such seminal and significant examples, along with the fast development of asymmetric catalysis over the past two decades, the dramatically growing number of various types of chiral diphosphine ligands synthesized and their applications to the enantioselective hydrogenation of imine substrates has been leading to continuous progress and breakthroughs in this area.

Many chiral diphosphine ligands that have been studied in asymmetric reactions with other transition metals such as Ru and Rh, or have been applied to the

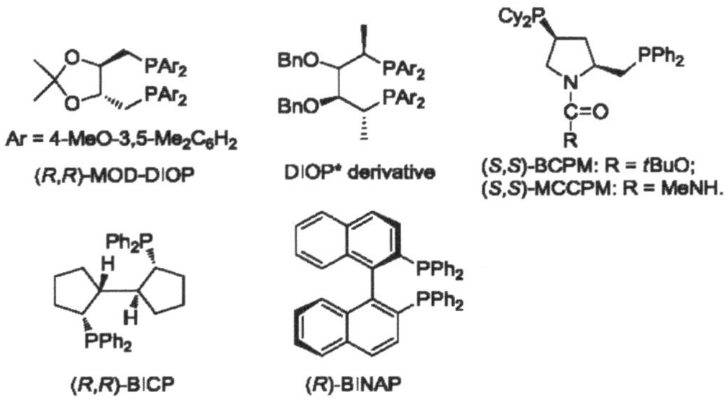

Scheme 12 Diphosphine ligands in Ir-catalyzed enantioselective hydrogenation of imines

hydrogenation of other substrate categories, such as C = C and C = O double bonds, have been investigated extensively. For example, in the form of Ir-based catalysts, the application of DIOP type ligands MOD-DIOP by Achiwa et al. achieved 81.4% *ee* for the cyclic imine 2,3,3-trimethylindolenine (TMI) [35]; the application of DIOP* ligand and its derivatives by Zhang and co-workers obtained 85.0% *ee* for TMI [36]; the application of BCPM and MCCPM ligands, also by Achiwa et al., achieved 91% *ee* and 90% *ee* for TMI [37]; the application of BICP ligand by Zhang and co-workers achieved 95% *ee* for TMI [38]; and the application of BINAP by Achiwa et al. obtained up to 86% *ee* for the isoquinoline-type imines (Scheme 12) [39].

The remarkable success of Xyliphos in industrial applications revealed the great potential of diphosphine ligands based on a ferrocene backbone, which could possibly enhance the air-stability and electron-donating ability of the ligands. Blaser, Spindler, and co-workers further extended their systematic fine-tuning investigation of Josiphos-type ligands in the hydrogenation of MEA imines and TMI to achieve optimal enantioselectivity while maintaining the extremely high TONs. Up to 83% *ee* and 94% *ee* for MEA imine and TMI substrate, respectively, were achieved in the presence of each corresponding optimal ligand containing two matched elements R and R′ groups in the planar chirality (Scheme 13) [40].

Subsequently, Reetz et al. developed another planar chiral ferrocene-based diphosphine for the asymmetric hydrogenation of TMI and achieved 79% *ee* [41]. In 2001, Zhang and co-workers developed a chiral 1,1′-bisphosphanoferrocene ligand f-Binaphane and demonstrated its excellent efficacy in hydrogenating a series of *N*-arylimines, achieving up to 98% *ee* (Scheme 14) [42]. The rigid binaphthyl moiety and the electron-rich ferrocene backbone are helpful in achieving high enantioselectivities, and the larger P–M–P bite angle could contribute to better accommodation of more sterically hindered imine substrates. These ferrocene-based ligands were later found to be outstandingly efficient for the asymmetric hydrogenation of N-H imines [43], HCl salts of unprotected β-enamine esters [44], and different cyclic imines [45–48].

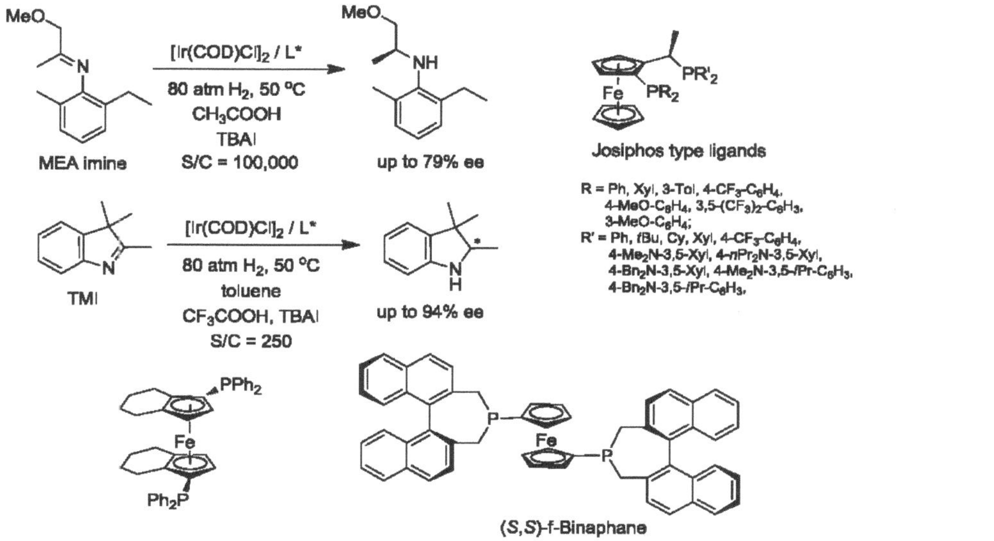

Scheme 13 Ferrocene-based diphosphine ligands in Ir-catalyzed enantioselective hydrogenation of imines

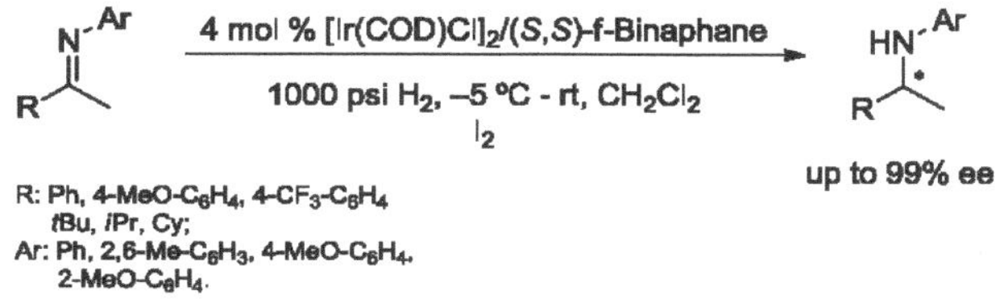

Scheme 14 Enantioselective hydrogenation of *N*-aryl imines by Ir-f-Binaphane catalyst

In another approach to probe the mechanism of the high efficacy of Ir complexes of diphosphines and the need for additives such as I_2, Dervisi et al. synthesized the diphosphine ligand ddppm from D-isomannide and its Ir complex, and displayed good to high enantioselectivities in the hydrogenation of *N*-arylimines using cationic [Ir(COD)(ddppm)]PF_6 catalyst (up to 94% *ee*, Scheme 15) [49]. It should be noted that they isolated a mixture of dimeric and trimeric Ir(III) hydride clusters, which are inactive in the hydrogenation reaction studied (Scheme 15). Such findings agreed with those iridium hydride species in the phosphine-oxazoline systems [50].

The more recently successfully developed P-chiral ligands have also brought new promising solutions for the enantioselective hydrogenation of imines. Imamoto et al. reported their discovery of the high efficacy of the P-chiral bis(trialkylphosphine) ligand *t*Bu-BisP* in the form of cationic Ir species in asymmetric hydrogenation of a wide range of *N*-arylimines, achieving 83–99% *ee* (Scheme 16). Their demonstration of the counterion effect proved that only BARF can promote

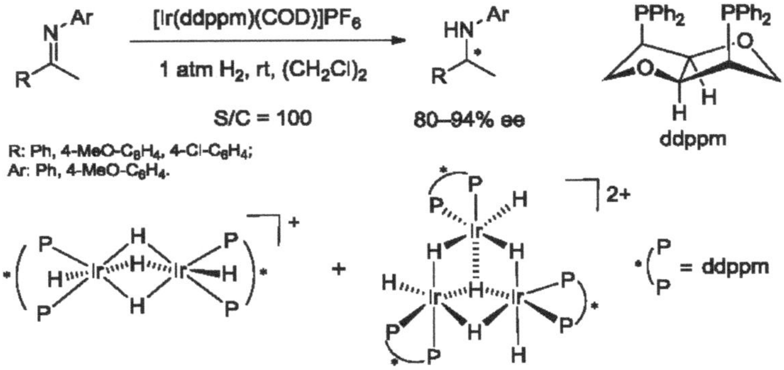

Scheme 15 Enantioselective hydrogenation of *N*-arylimines by Ir–ddppm complex and structures of formed inactive dimeric and trimeric Ir(III) hydride complexes

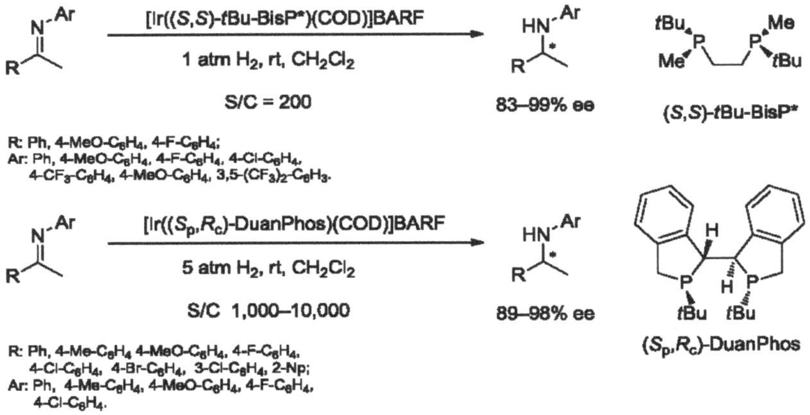

Scheme 16 Enantioselective hydrogenation of *N*-aryl imines by Ir complexes of electron-rich diphosphine ligands

the reactivity dramatically, while other counterions lead to complete loss of reactivity of the iridium catalyst [51].

Another example of a cationic Ir catalyst employing structurally rigid and electron-rich P-chiral bisphospholane DuanPhos was recently presented by Zhang and co-workers (Scheme 16) [52]. The same crucial counterion effect (BARF as the counterion) was revealed, and this more efficient Ir catalyst afforded the corresponding *N*-aryl chiral amine products at 89–98% *ee* at a very low catalyst loading under mild conditions (TON = 1,000 under 1 atm H_2, TON = 10,000 under 5 atm H_2). Such outstanding promoting and activating capability of the BARF counterion has also been more prevalently observed in the Ir-catalyzed olefin hydrogenations by catalysts containing PHOX developed by Pfaltz et al. [53] and there are many more examples [54–60].

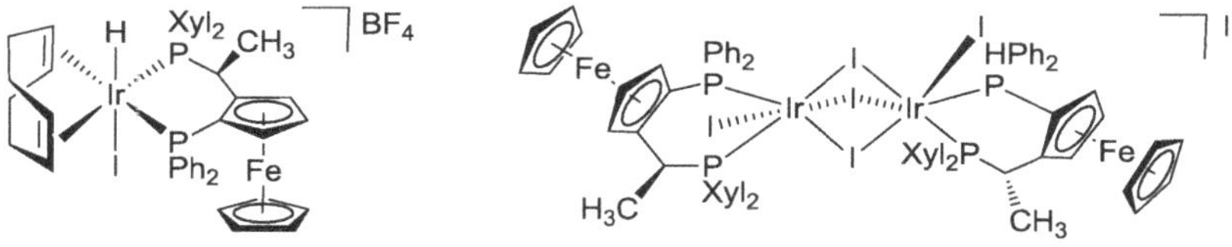

Scheme 17 Ir(III) complexes of Xyliphos

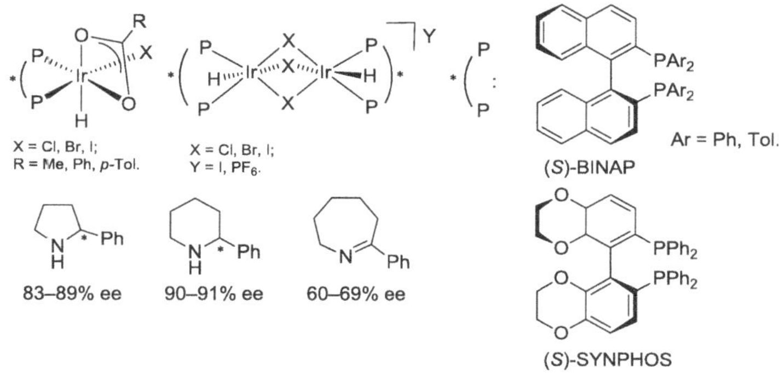

Scheme 18 Halide-carboxylate and triply halogen-bridged Ir(III) complexes and their applications in enantioselective hydrogenation of cyclic imines

Moreover, Ir(III) catalyst precursors containing diphosphine ligands have recently attracted a lot of attention. In their efforts to explain the active catalytic species in the "magic" catalyst mixture of Ir–Xyliphos from the mechanistic perspective, Dorta et al. synthesized several cationic Ir(III)–Xyliphos complexes including Ir (III)–Xyliphos iodo hydrido complex and a triply iodo-bridged dinuclear species $[Ir_2I_2(\mu\text{-}I)_3Xyliphos]I$, and thus examined the efficacy of these Ir(III) complexes in the hydrogenation of MEA imine substrate (Scheme 17) [61]. The hydrogenations applied with these Ir(III) complexes afforded 81% *ee* and 79% *ee*, respectively, although the reactivity of both was slightly lower (TOF of 1,375 h^{-1} and 2,083 h^{-1}, respectively) than in situ generated Ir–Xyliphos catalyst with I_2 as the additive (TOF of 2,778 h^{-1}).

Inspired by the investigation of the halide effect and the high reactivity of Ir(III) complexes, Genet, Mashima, and co-workers successfully synthesized mononuclear halide-carboxylate and cationic triply halogen-bridged dinuclear Ir(III) complexes of BINAP and SYNPHOS, and found that these complexes exhibited excellent enantioselectivities in cyclic imine hydrogenations (Scheme 18) [62]. As expected on the basis of the reported tendency of halide effects for iridium catalyst systems [8, 32, 35, 42], the iodide served as a better catalyst precursor in terms of catalytic activity among the dinuclear catalysts, though halogen atoms bound to the Ir(III) center did not clearly affect enantioselectivity.

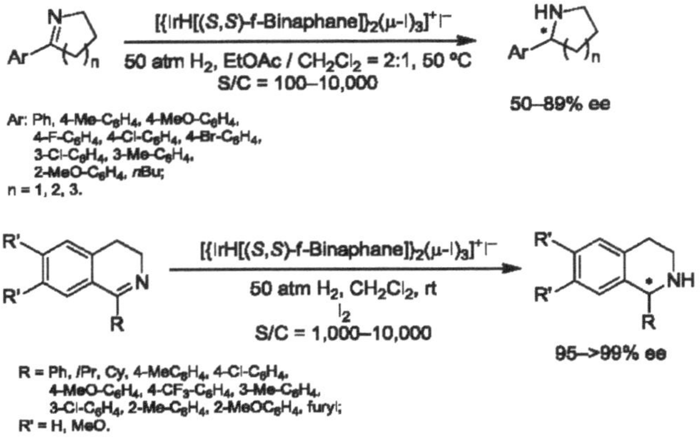

Scheme 19 Enantioselective hydrogenation of cyclic imines by $[[Ir(H)(S,S)\text{-f-Binaphane}]_2(\mu\text{-}I)_3]^+I^-$ complex

Recently, Zhang and co-workers reported the preparation of a similar triply halogen-bridged Ir(III) complex of the ferrocene-based electron-donating diphosphine ligand f-Binaphane and its application to the enantioselective hydrogenation of cyclic imines (Scheme 19) [47, 48]. By applying the Ir complex as $[[Ir(H)(S,S)\text{-f-Binaphane}]_2(\mu\text{-}I)_3]^+I^-$, various substituted arylpyrrolines, 2,3,4,5-tetrahydro-6-phenylpyridine, and 2-phenyl-4,5,6,7-tetrahydro-3*H*-azepine were efficiently (TON up to 10,000) hydrogenated with good enantioselectivities (50–89% *ee*, Scheme 19) [47]. The same Ir(III) catalyst proved to be highly effective in terms of both reactivity and enantioselectivity in the hydrogenation of a wide range of substituted 3,4-dihydroisoquinolines when affording over 95% *ee* for all substrates and up to >99% *ee* (Scheme 19) [48]. It should be noted that the usage of I_2 as the additive further enhanced the enantioselectivity of this iodo-bridged dimeric Ir(III) complex, and that substrates both bearing alkyl substituents were well-tolerated in this hydrogenation (96% *ee* and 95% *ee* for *i*Pr and Cy substituted substrates, respectively).

2.4.2 P,N Ligands in Ir-Based Catalysts

Since the Crabtree catalyst $[Ir(PCy_3)(Py)(COD)]PF_6$ showed remarkable activity in the hydrogenation of alkenes in nonasymmetric fashion [31], particularly sterically hindered tri- and even tetrasubstituted alkenes, its structure has inspired tremendous research work into chiral P,N ligands for enantioselective hydrogenation. The largest and most successful group of chiral analogues of the Crabtree catalyst is iridium complexes with phosphorus–oxazoline ligands.

The most extensively studied of these systems are the phosphino-oxazoline (PHOX) catalysts reported by Pfaltz and co-workers. In various asymmetric transformations over a broad range of substrates, good enantioselectivity has been achieved with these catalysts [10, 53, 63]. Most notably, the highest enantioselectivity

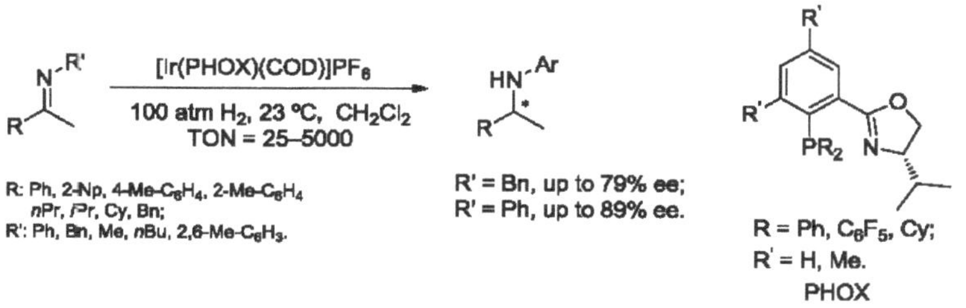

Scheme 20 Enantioselective hydrogenation of imines by Ir-PHOX catalyst

in the hydrogenation of unfunctionalized trisubstituted alkenes has been achieved with them.

In 1997, Pfaltz and co-workers pioneered research that introduced PHOX ligands into the hydrogenation of imines [63]. Under the hydrogen pressure of 100 bar in CH_2Cl_2, with 4 mol% of Ir–PHOX catalyst, the best enantioselectivity achieved for *N*-arylimines and *N*-alkylimines are 71% *ee* and 79% *ee*, respectively, whereas this system was not effective for cyclic imine substrates. Decreasing the reaction concentration further and decreasing the catalyst loading to 0.1 mol% could enhance the enantioselectivity for *N*-phenyl imine of acetophenone (Scheme 20). The counterion study suggested that replacing PF_6^- with other non-coordinating counterions such as SbF_6^-, BF_4^-, or BPh_4^- had no apparent effect. Highest *ee* values were observed in nonpolar weakly coordinating aprotic solvents such as CH_2Cl_2 and 1,2-dichloroethane. In the subsequent investigation by Leitner and Pfaltz to evaluate the efficacy of Ir–PHOX catalyst in supercritical carbon dioxide, the choice of BARF as the "CO_2-philic" counterion and the introduction of the perfluoroalkyl-substituted diphenylphosphine moiety were of eminent importance in transferring this hydrogenation reaction from conventional solvents (CH_2Cl_2) to the supercritical medium, while high levels of asymmetric induction by the Ir–PHOX catalyst were maintained [64]. Thus, >90% conversion of the imine, corresponding to >6,800 TON was achieved in spite of some enantioselectivity loss.

Based on the fundamental motif of combination of diphosphine moiety and chiral oxazoline moiety, the high modularity feature of PHOX-type ligands and the capability of steric and electronic tuning of these ligands from the substitution on the oxazoline ring greatly encouraged the syntheses of a large number of structurally similar oxazoline-derived ligands. Many of these PHOX-type ligands have been evaluated in the asymmetric hydrogenation of imines. Pfaltz and co-worker synthesized a series of derivatives such as threonine-derived phosphinite-oxazoline ligands ThrePHOX [65], SimphePHOX [66], NeoPHOX [67], and many more (Scheme 21) [68]. All of these phosphorus-oxazoline ligands have been applied in the cationic Ir complex [Ir(chiral P,N ligand)(COD)]BARF and evaluated in the asymmetric hydrogenation of imines in addition to functionalized and unfunctionalized olefins [68]. In the hydrogenation of *N*-phenylimine, most of these Ir catalysts afforded full conversion and the enantiomeric excesses ranged from 63% *ee* to 88% *ee*. At a lowered temperature of −40°C, under 5 atm of hydrogen pressure,

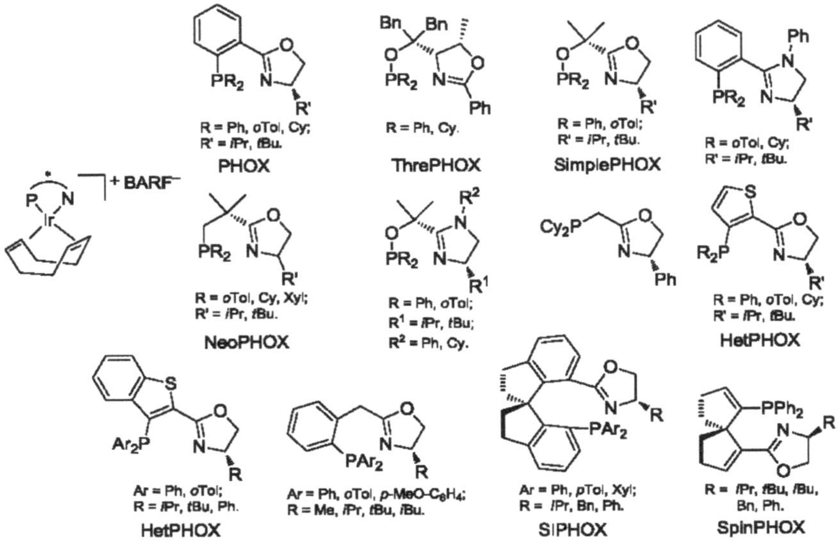

Scheme 21 Ir complexes of PHOX ligand and its analogues

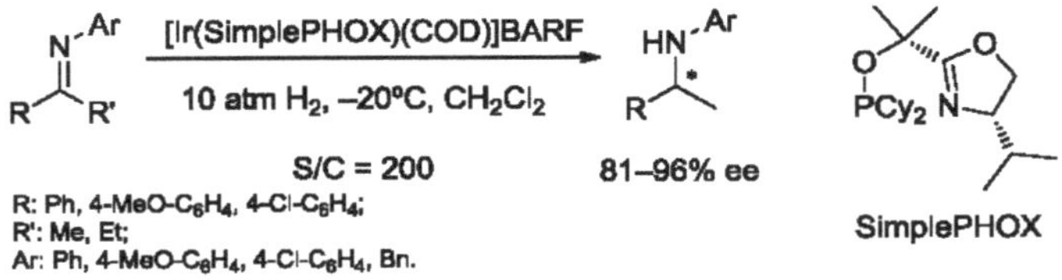

Scheme 22 Enantioselective hydrogenation of imines by Ir-SimplePHOX catalyst

up to 97% *ee* was achieved by SimplePHOX, which displayed the best efficacy for a series of *N*-arylimine substrates (91.5–96% *ee*) and one example of *N*-benzylimine (81% *ee*, Scheme 22) [68].

The development of chiral P,N ligands for the asymmetric hydrogenation of imines remained a rapidly developing field after Pfaltz's contributions. The application range of these ligands is largely complementary to chiral diphosphine ligands. For example, Cozzi et al. synthesized HetPHOX from thiophene and benzo[*b*]thiophene skeletons (Scheme 21) and achieved as high as 86% *ee* in the hydrogenation of *N*-phenylimine [69]; Hou and co-workers synthesized benzylic substituted PHOX-type ligands (Scheme 21) which gave up to 88% *ee* in the hydrogenation of *N*-arylimines [70].

In another outstanding example of phosphine–oxazoline ligand, Zhou and co-workers synthesized SIPHOX (Scheme 21) based on a spirobiindane backbone. Considering the superior rigidity and large bite angle it can provide to construct bidentate ligands, the Ir catalyst of SIPHOX can reduce a wide range of

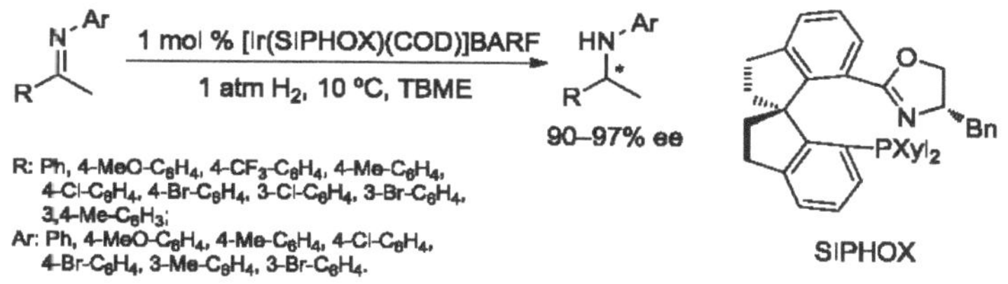

Scheme 23 Enantioselective hydrogenation of imines by Ir-SIPHOX catalyst

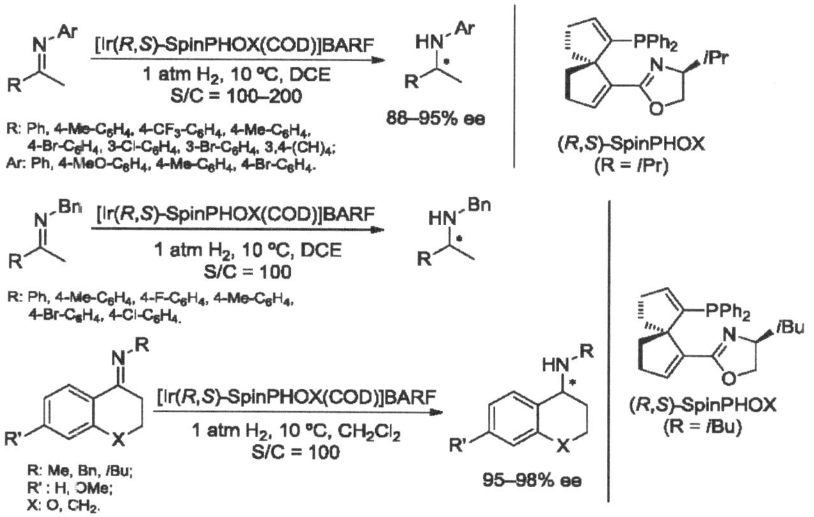

Scheme 24 Enantioselective hydrogenation of imine substrates by Ir-SpinPHOX catalyst

N-arylimines to their corresponding chiral amines with 90–97% *ee* (Scheme 23) [71] could demonstrate the excellent capability of SIPHOX for asymmetric induction in imine hydrogenation.

Taking advantage of the rigidity of the similar spiro architecture, using spiro [4,4]-1,6-nonadiene, Ding, Zhang, and co-workers synthesized SpinPHOX (Scheme 21) and achieved excellent enantioselectivities in the hydrogenation of *N*-arylimines (up to 95% *ee*), and, particularly, of very challenging *N*-alkylimines (up to 98% *ee*, Scheme 24) [72].

Furthermore, aside from oxazoline moieties, some other nitrogen donors have been also employed in P,N ligands for Ir-catalyzed imines hydrogenations. For example, Bolm et al. introduced a novel class of C_1-symmetric sulfoximines in the asymmetric hydrogenation of acyclic *N*-arylimines and achieved good to excellent enantioselectivities (69–98% *ee*, Scheme 25) [73]. Knochel et al. reported the application of synthesized ferrocenyl diphosphine with a pyridine moiety tethered

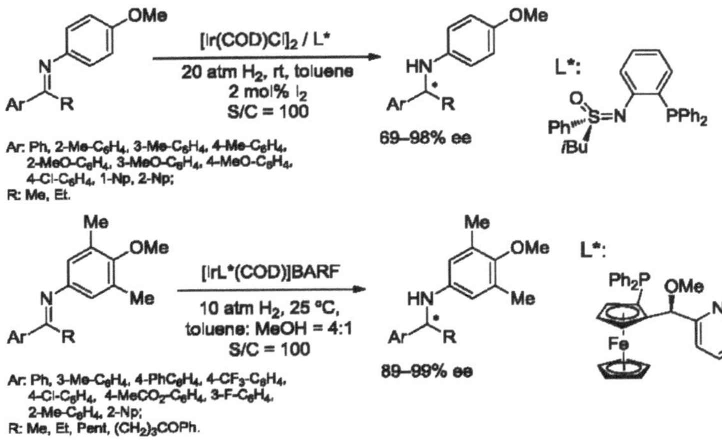

Scheme 25 Enantioselective hydrogenation of imines by Ir complex of other P,N ligands

to the backbone as the nitrogen donor in asymmetric hydrogenation of *N*-arylimines (89–99% *ee*, Scheme 25) [74].

2.4.3 Phosphite, Phosphinite, and Phosphoramidite Ligands in Ir-Based Catalysts

In sharp contrast to the fruitful discoveries of successful diphosphine and P,N ligands, efficient phosphite and phosphinite ligands have been less synthesized and reported. Castillon, Claver, and co-workers demonstrated the application of diphosphite and diphosphite ligands based on the xylofuranoside and D-glucosamine scaffolds in the hydrogenation of *N*-phenyl and *N*-benzyl imines with moderate enantiomeric excess (up to 70% *ee* for *N*-phenyl and 76% *ee* for *N*-benzyl imines) [75, 76]. Pizzano et al. reported the synthesis of chiral biphenyl-based phosphine–phosphite ligands and evaluated their efficacy in hydrogenating *N*-arylimine substrates, giving up to 85% *ee* [77].

Although the monodentate phosphoramidite ligands such as MonoPhos (Scheme 26) have been widely used in the hydrogenation of substrates of enamides [78] and olefins [79], the search for their usefulness in Ir-catalyzed imine hydrogenation have been limited. With an additional achiral nitrogen donor such as 2,6-lutidine and acridine, Faller et al. were able to prepare a cationic Ir complex of MonoPhos with BARF as the counterion. However, when the cyclic imine TMI was reduced to under 40 atm of H_2, only up to 58% *ee* was obtained [80].

More recently, Minnaard, Feringa, de Vries, and co-workers reported the application of novel BINOL-derived phosphoramidite ligands PipPhos (Scheme 26) in the hydrogenation of *N*-arylimines. Excellent capability of stereocontrol of PipPhos was observed when enantiomeric excesses ranging from 97% *ee* to >99% *ee* were

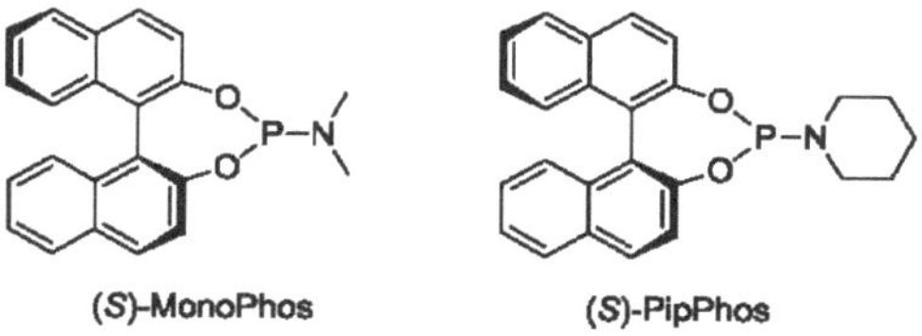

Scheme 26 Monodentate phosphoramidite ligands for enantioselective imine hydrogenation

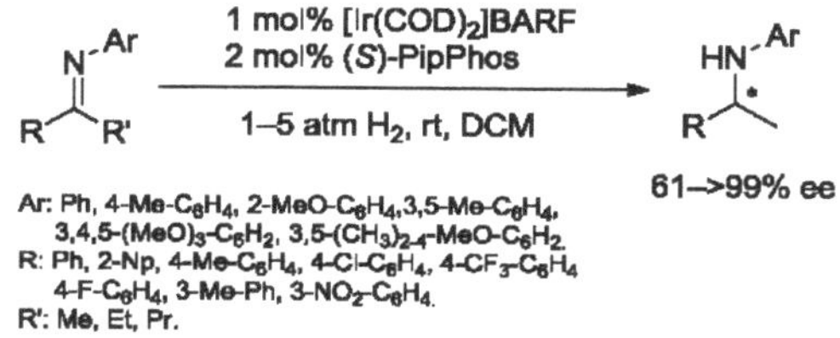

Scheme 27 Enantioselective hydrogenation of imines by Ir-PipPhos catalyst

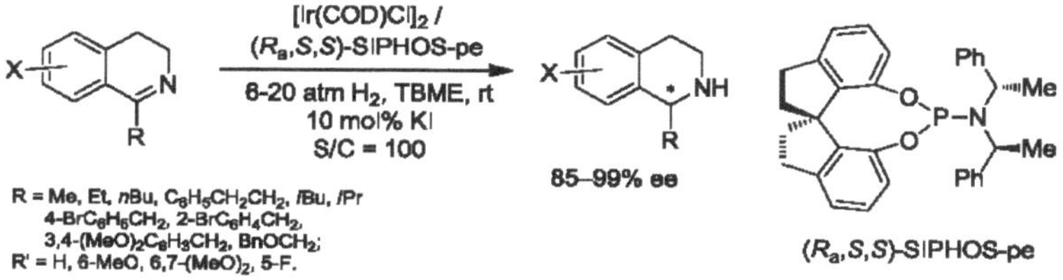

Scheme 28 Enantioselective hydrogenation of imines by Ir-SIPHOS-pe catalyst

observed for the majority of the *N*-arylimines studied, especially for those substrates bearing electron-donating groups on the *N*-phenyl ring (Scheme 27) [81]. Interestingly, the ratio of PipPhos to [Ir(COD)$_2$]BARF precursor was 2:1, while no additional nitrogen donor was necessary. The same ligand has also shown outstanding efficacies in the hydrogenation of heteroaromatic compounds such quinoline and *N*-protected indole derivatives [82, 83].

The spiro-based phosphoramidite ligand SIPHOS-pe, which has been tested and been shown to be highly efficient for the asymmetric hydrogenation of unfunctionalized cyclic enamines, has also been successfully applied in the form of an Ir complex to the asymmetric hydrogenation of 1-alkyl 3,4-dihydroisoquinolines by Zhou and co-workers [84]. In the presence of 10 mol% KI as the additive, a variety of chiral 1-alkyl tetrahydroisoquinolines were afforded by this hydrogenation with good to excellent enantioselectivities (85–99% *ee*). The synthesis of tetracyclic alkaloid (*S*)-xylopinine from 1-benzyl substitute hydrogenation product further proved the usefulness of this catalyst system (Scheme 28).

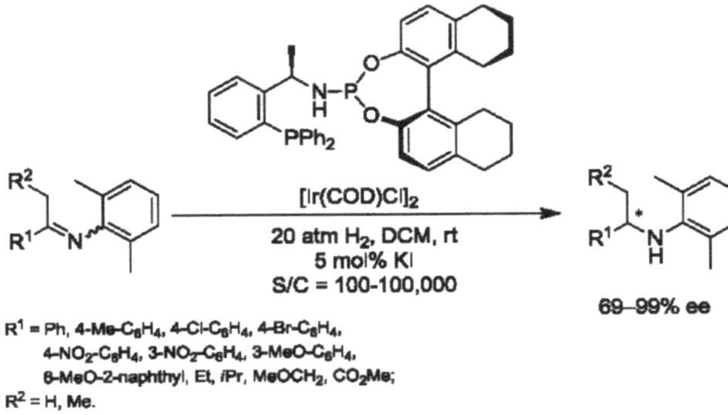

Scheme 29 Enantioselective hydrogenation of imines by Ir complex of phosphine-phosphoramidite ligand

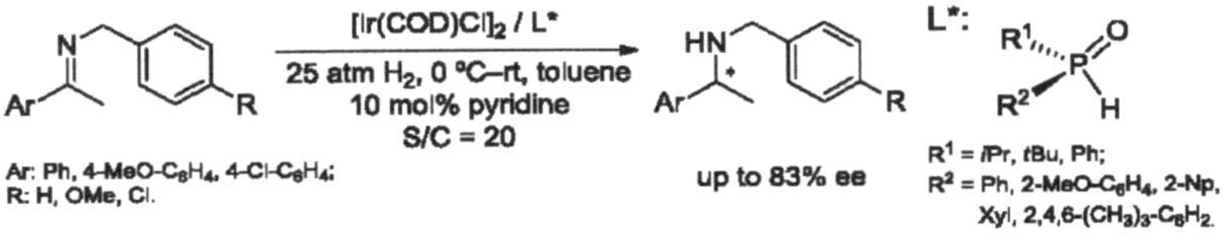

Scheme 30 Enantioselective hydrogenation of imines by Ir complex of SPO ligand

Hu and co-workers successfully applied the Ir catalyst of phosphine–phosphoramidite ligand in the asymmetric hydrogenation of a variety of sterically hindered *N*-arylimines (*N*-2,6-$(CH_3)_2$-phenyl) and demonstrated its excellent efficacy by achieving high turnover numbers (up to 100,000) and good to perfect enantioselectivities (up to 99% *ee*) in the presence of 5 mol% KI as the additive (Scheme 29) [85].

2.4.4 Other Ir Catalysts

Other than these above-mentioned types of chiral ligands, there have been investigations of other classes. Minnaard, Feringa, de Vries, and co-workers reported an interesting series of phosphinite-equivalent secondary phosphine oxide (SPO) ligands. The acidic protons in these easily synthesized ligands in their tautomeric equilibrium in solutions were expected to have an accelerating effect on Ir-catalyzed imine hydrogenations. Thus, up to 83% *ee* was achieved when the hinder *t*Bu-substituted SPO was applied in the hydrogenation of *N*-benzylimines in the presence of additional pyridine as the additive (Scheme 30). The application of structurally similar and strategically related chiral phosphoric acid together with achiral phosphines or phosphites in the asymmetric hydrogenation of *N*-benzylimines were reported by Reetz et al., achieving excellent enantioselectivities (up to 92% *ee*) [86].

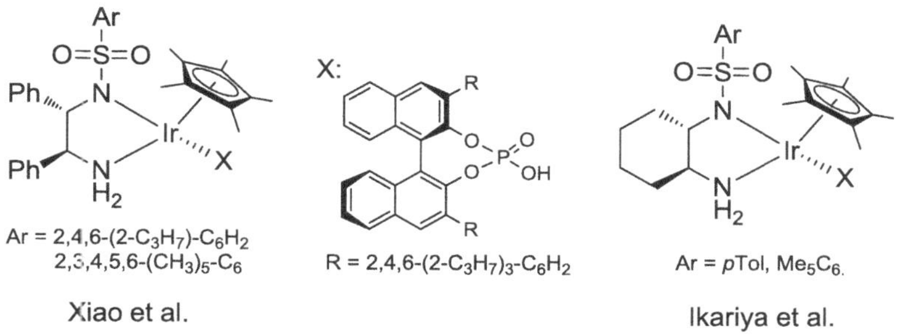

Scheme 31 Ir(Cp*)(tosyldiamine) complexes in enantioselective imine hydrogenation

Another class of Ir(III) complexes that have attracted attention are chiral Ir(Cp*) (tosyldiamine) complexes (Scheme 31) that have been developed in asymmetric transfer hydrogenations. The Xiao group and the Ikariya group have independently reported such Ir complexes and their applications in the enantioselective hydrogenation of *N*-arylimines. In acidic conditions, in the presence of a catalytic amount of chiral BINOL-based phosphoric acids, Xiao et al. obtained excellent enantioselectivities (up to 99% *ee*) for a wide range of *N*-arylimine substrates [87]. In the presence of silver salt such as $AgSbF_6$ as the additive, Ikariya et al. obtained moderate to good enantioselectivities (up to 78% *ee*) for the same category of substrates [27].

2.4.5 Additive Effects and Mechanistic Perspectives

In enantioselective hydrogenation of imines, particularly in the Ir-catalyzed hydrogenation of imines, significant additive effects have been reported and investigated to promote enantioselectivity and/or reactivity. The most notable example is undoubtedly the "magic" accelerating additive of iodide and acid in the hydrogenation of MEA imines reported by Blaser and his co-workers in their search for the ultimate solution to the industrial production process of herbicide ingredient (*S*)-metolachlor with >1,000,000 TON and 1,800,000 h^{-1} reached [8].

The pronounced halide effect was documented as early as in one of the initial homogeneous Ir-catalyzed imine hydrogenations that achieved useful *ee* by Ir–DIOP catalyst. When Bu_4NI was added as the source of iodide together with the precursor $[Ir(COD)Cl]_2$ and ligand (*S*,*S*)-DIOP (Ir/ligand/I^- = 1:1.1:2), the enantioselectivity increased to 70% *ee*, compared to 58% *ee*, which was obtained under the same conditions with the exception of the absence of the iodide additive [32]. An even stronger halogen effect was observed when 1 equiv. of iodide addition to the halogen-free catalytic system $[Ir(COD)_2]BF_4$/(*S*,*S*)-DIOP led to a dramatic increase of enantioselectivity from 4% *ee* to 68% *ee*. However, the positive effect of Cl^- and Br^- are less obvious [32]. Achiwa et al. introduced BiI_3 as another iodide source in addition to Bu_4NI in the study of hydrogenation of

Table 1 Additive effects of iodide and iodine in enantioselective hydrogenation of imines

Ir Catalyst	Additive	Substrate	S/C	Conv./ Yield (%)	% *ee*	References
$[Ir(COD)Cl]_2$/(*S*,*S*)-DIOP	Bu_4NI	MEA imine analogue	100	95	70	[32]
$[Ir(COD)Cl]_2$/(*S*,*S*)-BDPP	Bu_4NI	MEA imine analogue	100	98	84	[32]
$[Ir(COD)Cl]_2$/(*S*,*S*)-MOD-DIOP	Bu_4NI	TMI	100	100	81.4	[35]
$[Ir(COD)Cl]_2$/Josiphos	Bu_4NI	MEA imine	100,000	100	80	[40]
$[Ir(COD)Cl]_2$/Josiphos	Bu_4NI	TMI	250	100	94	[40]
$[Ir(COD)Cl]_2$/(*S*,*S*)-MCCPM	BiI_3	TMI	100	92	91	[88]
$[Ir(COD)Cl]_2$/(*S*,*S*)-MOD-DIOP	BiI_3	Cyclic imine	100	85	90	[101]
$[Ir(COD)Cl]_2$/(*S*,*S*)-f-Binaphane	I_2	*N*-Arylimine	25	100	94	[42]
$[Ir(COD)Cl]_2$/(*S*,*S*)-DIOP*	I_2	TMI	100	96.6	85	[36]
$[[Ir(H)(S,S)\text{-f-Binaphane}]_2(\mu\text{-I})_3]^+I^-$	I_2	Cyclic imine	2,000	99	>99	[48]

TMI and other imine substrates. In the presence of BiI_3, the application of a type of modified BPPM type ligand (*S*,*S*)-MCCPM showed excellent enantioselectivities in the hydrogenation of TMI imine [88]. Later, in the study of the hydrogenation of *N*-aryl imines by Zhang and co-workers, iodine was chosen as the additive to promote this catalytic process. Thus, following this pioneering work, iodine has also been studied extensively and found to have significant enhancing effects to both enantioselectivity and reactivity in the Ir-catalyzed hydrogenation of imines [36, 48] and of other closely-related substrate types, including enamines [89, 90], quinoline derivatives [91–95], quinoxalines [96], and other heteroaromatics [97–100]. Representative examples of iodide compounds and iodine as effective additives in enantioselective imine hydrogenations are summarized in Table 1.

The additive effect of iodide or iodine has been studied for different substrates, and Zhang and co-workers (Scheme 32) proposed the possible mechanistic rationale involving employing iodine as the additive [42]. This proposed mechanism was consistent with the reported synthesis of $[Ir(P\text{–}P)I_4]^-$ [102] and $[Ir(P\text{–}P)HI_2]_2$ [103], where P–P as diphosphines such as DIOP, BDDP, etc., and the demonstration that the formation of these complexes can avoid the deactivation of forming inactive hydride-bridged Ir species. These types of Ir complexes were found to be catalytically active for the hydrogenation of DMA imine, suggesting the formation of a similar active mononuclear Ir species as for the catalyst formed in situ. The formation of inactive hydride-bridged Ir species was similar to that proposed by Crabtree (Scheme 32) [104]. The isolation of dimeric and trimeric hydride-bridged Ir (III)–ddppm complexes proved to be inactive in the hydrogenation reaction of

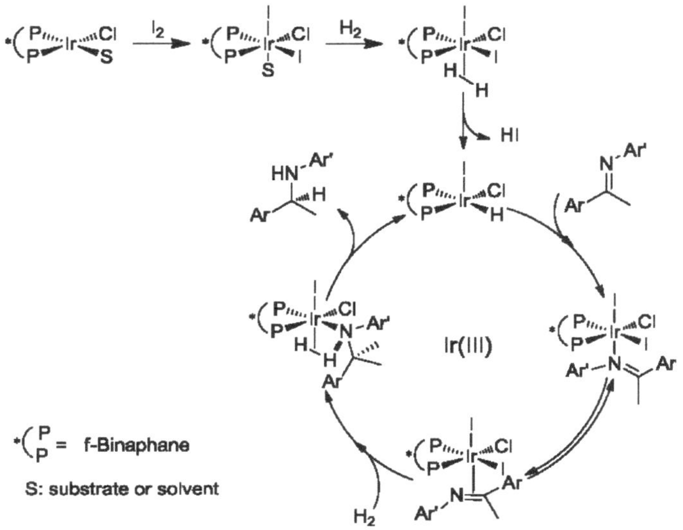

Scheme 32 Proposed mechanism for the Ir–f-Binaphane-catalyzed hydrogenation of *N*-arylimine with I_2 as the additive

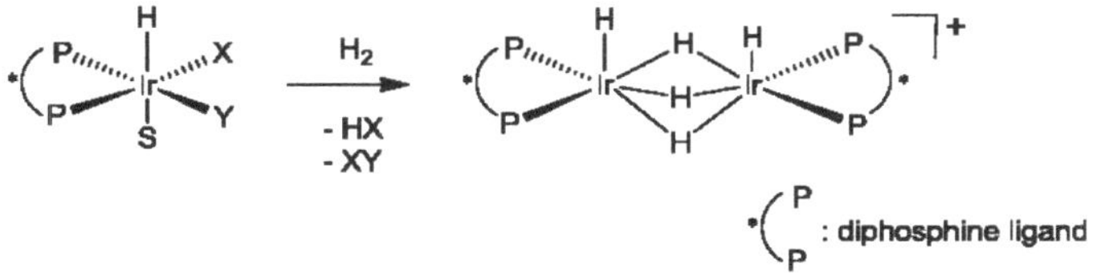

Scheme 33 Irreversible formation of an inactive dimeric Ir species

N-arylimines (Scheme 15) [49]. Furthermore, the synthesis cationic triply halogen-bridged dinuclear Ir(III) complexes of BINAP and SYNPHOS as well as the excellent enantioselectivity of these dinuclear Ir(III) complexes in the hydrogenation of cyclic imines were reported by Genet, Mashima, and co-workers (Scheme 18) [62]. The further demonstrated superior efficacy of similar preformed dinuclear Ir(III)–f-Binaphane as the catalyst in the hydrogenation of different types of cyclic imines (up to 99% *ee* and up to 10,000 TON) [47, 48] also corroborates the proposed role of iodide (or iodine) of precluding the catalytic Ir species from forming inactive hydride-bridged Ir complexes (Scheme 33).

Aside from iodide and iodine, other additives have also been surveyed extensively to find effective promoters in Ir-catalyzed imine hydrogenations. By taking advantage of different additives, enhanced enantioselectivity and reactivity have been observed toward different imine substrates (Table 2). The control experiment of replacing phthalimide *N*-methyl phthalimide excluded the possibility of an Ir–N bond, which could account for the additive effect. One possible explanation for the observed

Table 2 Other effective additives in enantioselective hydrogenation of imines

Ir Catalyst	Additive	Substrate	S/C	Conv./Yield (%)	% *ee*	References
$[Ir(COD)Cl]_2$/(*S*)-BINAP	$BnNH_2$	Cyclic imine (6-membered)	100	100	90	[106]
$[Ir(COD)Cl]_2$/(*S*,*S*)-BCPM	Phthalimide	Cyclic imine	100	95	93	[88]
$[Ir(COD)Cl]_2$/(*S*)-BINAP	F_4-Phthalimide	Cyclic imine	100	89	86	[107]
$[Ir(COD)Cl]_2$/(*S*)-BINAP	Parabanic acid	Cyclic imine	100	99	89	[39]
$[Ir(COD)Cl]_2$/(*R*,*R*)-BICP	Phthalimide	TMI	100	100	95.1	[38]
$[Ir(COD)Cl]_2$/SPO	Pyridine	$BnNH_2$	20	80	83	[105]
$[Ir(COD)Cl]_2$/(*S*)-Monophos	Acridine	TMI	100	56	58	[80]

results is that the canonical form of phthalimide involving a C–O single bond may coordinate to the Ir complex. Such a hypothesis could be supported by the result that the introduction of electron-withdrawing groups adjacent to the nitrogen atom such as the Br group in *N*-bromophthalimide provided much lower enantioselectivity. The accelerating effect of pyridine [105] and acridine [80] in Ir–monodentate systems could be explained by regarding them as the nitrogen-donor to form an untethered P,N ligation to Ir metals, but in a much less efficient way of chiral induction to form chiral amines.

2.5 Pd-Catalyzed Asymmetric Hydrogenation of Imines

2.5.1 Pd-Based Catalysts for Asymmetric Hydrogenation of Imines

In the past two decades, rapidly developing Pd-catalyzed transformations have been one of the most important methods for C–C, C–N, and C–O bond formation. The breakthrough of the asymmetric version of the powerful Pd-catalyst transformations occurred in the 1990s in enantioselective Heck reactions by Overman and Shibasaki, cross-coupling reactions by Hayashi, copolymerization of alkenes and CO by Nozaki, and nucleophilic substitutions by Trost and van Vranken [108]. In addition, the heterogeneous palladium catalyst (such as Pd/C, Lindlar catalyst) is also well known as one of the most effective catalysts in the hydrogenation of unsaturated double bonds. Nevertheless, homogeneous palladium-catalyzed asymmetric hydrogenation has been unsuccessful or ignored until 2001 [109]. Amii and Ueyama's application of Pd–BINAP catalyst in the hydrogenation of α-fluorinated iminoesters exemplified the pioneering work. Although substrate scope and reactivity

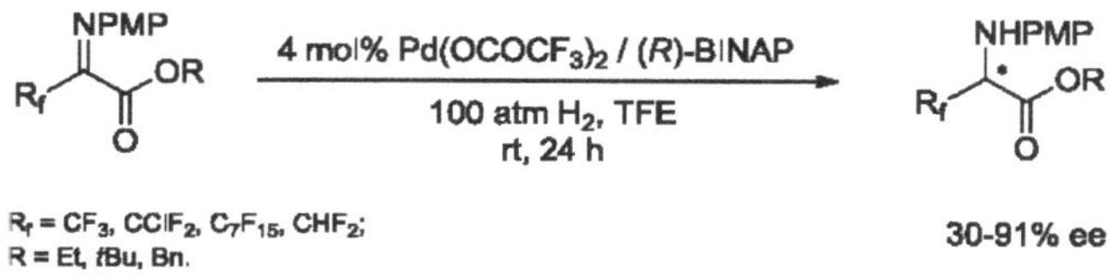

Scheme 34 Enantioselective hydrogenation of imino esters by Pd-BINAP catalyst

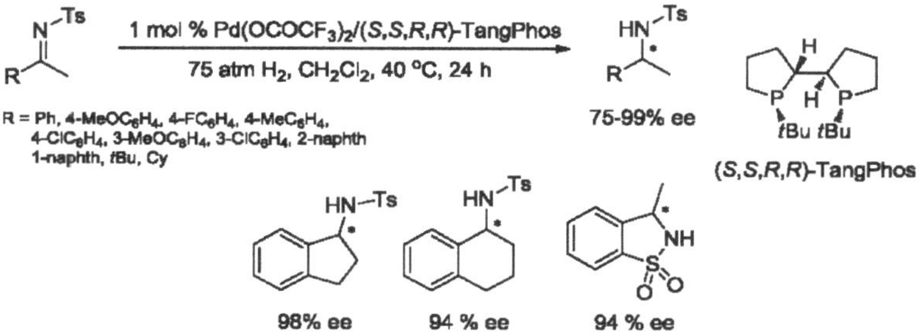

Scheme 35 Enantioselective hydrogenation of *N*-tosyl imines by Pd-TangPhos catalyst

were limited, up to 91% *ee* opened up new vistas of Pd-based catalysts in asymmetric imine hydrogenations (Scheme 34) [110]. Employing fluorinated alcohols such as CF_3CH_2OH (TFE) dramatically improved both reactivity and enantioselectivity, and gave maximum enantiomeric excess of 88% *ee*. It was suggested that the weak nucleophilicity and the low coordinative nature of TFE were responsible for the superior results obtained [111]. Moreover, TFE might activate α-fluorinated iminoesters by protonation or hydrogen bonding of the imino group [109].

In 2006, Zhang and co-workers reported the highly enantioselective catalyst in situ generated from $Pd(OCOCF_3)_2$ and TangPhos in *N*-tosylimine hydrogenation. Up to >99% *ee* was achieved for a variety of *N*-tosylimine substrates; notably, for alkyl substituent substrates bearing *t*Bu and Cy group, 98% *ee* and 75% *ee* were achieved, respectively. The weak coordinating ability of $CF_3CO_2^-$ as well as the highly electron-rich and rigid features of TangPhos played a crucial role in the high enantioselectivities. This catalyst system also proved to be effective in reducing corresponding imine to afford methyl-substituted sultam in 94% *ee* (Scheme 35) [112].

Zhou and co-workers have also tried to find homogenous asymmetric Pd-based catalysts that can be applied to hydrogenate various types of activated imines. In 2006, Zhou reported the hydrogenation of *N*-diphenylphosphinyl imines by the same $Pd(OCOCF_3)_2$ precursor and SEGPHOS with the assistance of a molecular sieve to reach high conversions (Scheme 36) [113]. In the solvent of TFE, up to 98.6% *ee* was achieved using this Pd catalyst. The reaction is strongly solvent dependent, as TFE was discovered to be the only efficient solvent. Subsequently, by screening other structurally similar atropisomeric diphosphines under the same catalytic conditions, the substrate scope was expanded to survey other imines

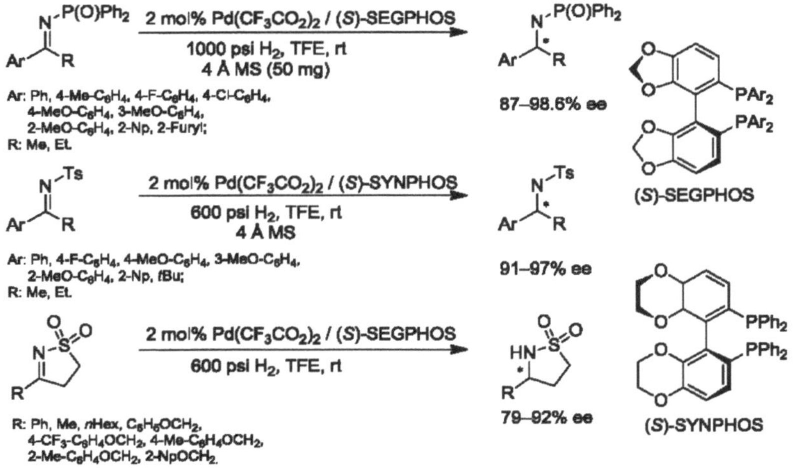

Scheme 36 Pd-catalyzed enantioselective hydrogenations of *N*-diphenylphosphinyl imines, *N*-tosyl imines and cyclic *N*-sulfonylimines

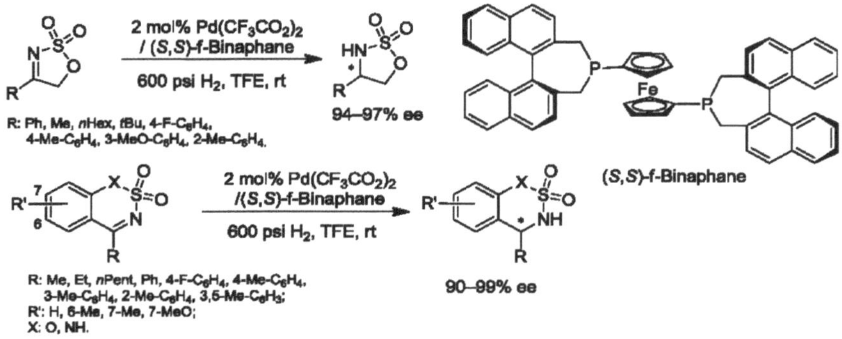

Scheme 37 Enantioselective hydrogenation of cyclic imines by Ir-f-Binaphane catalyst

such as *N*-tosylimines and cyclic *N*-sulfonylimines (Scheme 36) [114]. More recently, the groundbreaking application of similar combinations of Pd $(OCOCF_3)_2$ and atropisomeric biaryl diphosphines in the hydrogenation of unprotected indoles activated by Brønsted acids has been reported [115].

In their further extended search, Zhou et al. discovered that f-Binaphane also displays excellent efficacy in this established Pd-catalyst system in the hydrogenation of cyclic imines to chiral sulfamidates under the same conditions (TFE as the solvent) [45]. By applying 2 mol% of $Pd(CF_3COO)_2$ and the ferrocene-based ligand, a series of simple or benzo-fused imines were successfully hydrogenated with almost

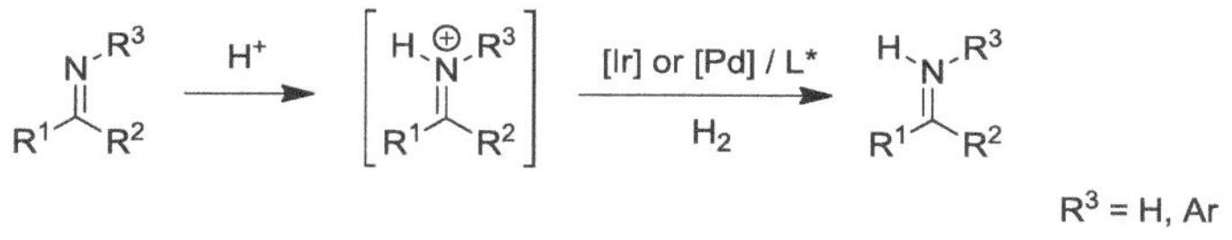

Scheme 38 "Brønsted acid strategy" to activate imines in enantioselective hydrogenation

quantitative yields and excellent enantioselectivities (90–98% *ee*, Scheme 37). To test the practicality of this catalytic system, TON of 2,000 was achieved at room temperature on the gram scale for one example of substrate with 97% isolated yield and 98% *ee*.

2.5.2 Inhibitory Effect of Amine Product and Activation Strategy

The strong donor character of the NH group of an amine, with its ability to compete for coordination at the catalytic site (catalyst poisoning), may be one factor contributing to the more difficult hydrogenation of imines. Thus, inhibition of the amine product could "sequester" the metal center [116]

To tackle the challenge of hydrogenation of non-activated imines, in which no strong electron-withdrawing groups are introduced to reduce the inhibition, the "Brønsted acid strategy" has been developed and applied to those non-activated imine substrates by introducing Brønsted acid in the reaction in Ir- and Pd-catalyzed asymmetric hydrogenations of imines (Scheme 38) [117]. In 2001, Norton demonstrated the first catalytic enantioselective hydrogenation of *N*-benzylimine through an ionic mechanism with the assistance of HBF_4 and achieved 66% *ee* [118]. Later, it was reported that the introduction of an acid activator could be achieved by preforming iminium type substrates such as N–H imines [43, 119], or by the addition of the acid activator in situ [120, 121]. Following that strategic breakthrough, groundbreaking works of Pd-catalyzed asymmetric hydrogenation of heteroaromatics, such as unprotected indoles by Zhou et al., and pyrroles by Zhou, Fan et al. were recently published [115, 122, 123].

Zhou and co-workers developed an efficient Pd-catalyzed asymmetric hydrogenation of *N*-aryl simple ketimines and tetralone-derived ketimines using a catalytic amount of Brønsted acid as an additive. Using $Pd(CF_3COO)_2$ as the Pd precursor, in TFE, C_4-TunePhos was found to be the most effective biaryl diphosphine ligand in affording the corresponding chiral amines at up to 95% *ee* (Scheme 39) [120]. A series of Brønsted acid studied in this reaction and D-DTTA was found to be the best additive in terms of enantioselectivity.

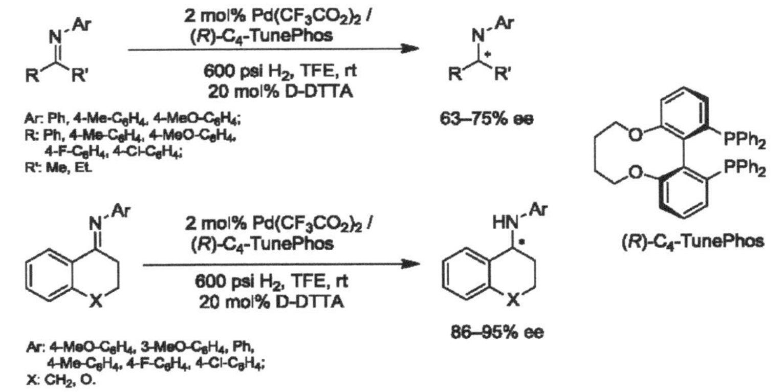

Scheme 39 Pd-catalyzed enantioselective hydrogenation of *N*-arylimines by Pd–C_4-TunePhos complex with Brønsted acid activator

3 Asymmetric Hydrogenation of Acyclic Imines

3.1 Asymmetric Hydrogenation of Activated Acyclic Imines

To address the problem of the inhibitory effect of the amine product in asymmetric hydrogenations [116], several electron-withdrawing *N*-substituents were introduced into acyclic imines, such as *N*-acylhydrazone, *N*-tosyl and *N*-diphenylphosphinyl groups (Scheme 40), which could lead to good to high enantioselectivities and reactivities in Rh-, Ir-, Ru-, and Pd-catalyzed asymmetric hydrogenation (Table 3). Aside from the reduction of inhibition, there are other advantages from the introduction of these groups that facilitate the synthetic utility: (1) these *N*-substituting groups can also act as secondary coordinating groups to metals in the catalyst to accelerate the rates of hydrogenation of these substrates, especially in the Rh-catalyzed hydrogenations [16, 124]; (2) in contrast to some other imine types, one stereoisomer (*E*/*Z*) of these activated acyclic imines is expected to dominate; (3) these *N*-substituent (protecting groups) could be readily removed to afford corresponding unprotected amines. Rh-based chiral catalysts are the pioneers for these substrates [16, 124], but examples of Pd-based chiral catalysts have recently become more dominant. The palladium catalysts of biaryl diphosphine ligands, such as BINAP analogues [114], as well as the highly electron-donating ligands, such as TangPhos [112] and ferrocene-based ligands [17], have shown extraordinary efficacies.

Moreover, the removal of the different *N*-X protecting and activating group to obtain the primary amine is easy and can easily be scaled up (Scheme 41). The N–N bond in *N*-aroylhydrazones can be readily cleaved by samarium diiodide (SmI_2) almost instantaneously at 20°C, and the deprotection proceeds with no loss of optical purity [16]. The selective deprotection of the *N*-Boc hydrazine product

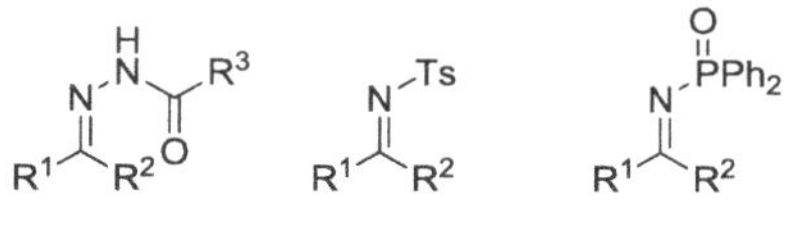

Scheme 40 *N*-Acylhydrazone, *N*-tosyl and *N*-diphenylphosphinyl imines

Table 3 Representative results of enantioselective hydrogenation of acyclic activated imine types

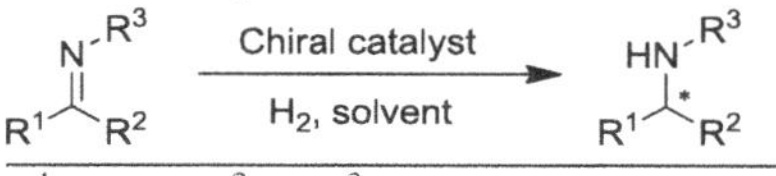

R^1	R^2	R^3	Catalyst	S/C	*ee* (%)	References
Ph	Me	NHCOPh	Rh[(*R*,*R*)-Et-DuPhos](COD)OTf	500	95	[16]
CO_2Me	Et	NHCOPh	Rh[(*R*,*R*)-Et-DuPhos](COD)OTf	500	91	[16]
Ph	Me	NHCOO*t*Bu	$Rh(COD)_2BF_4$/(*R*)-(*S*)-Josiphos	67	86	[17]
Ph	Me	$P(O)Ph_2$	$Rh(NBD)_2BF_4$/(*R*)-(*S*)-Josiphos	500	99	[125]
Ph	Me	$P(O)Ph_2$	$Pd(CF_3CO_2)_2$/(*S*)-SEGPhos	50	96	[114]
Ph	Me	$P(O)Ph_2$	$[Ir(COD)Cl]_2$/(*R*)-SPO ligand	20	70	[105]
Ph	Et	Ts	$Ru(OAc)_2$[(*R*)-BINAP]	20	84	[24]
Ph	Me	Ts	$Pd(CF_3CO_2)_2$/(*S*,*S*,*R*,*R*)-TangPhos	100	99	[112]
Ph	Me	Ts	$Pd(CF_3CO_2)_2$/(*S*)-SYNPhos	50	96	[114]

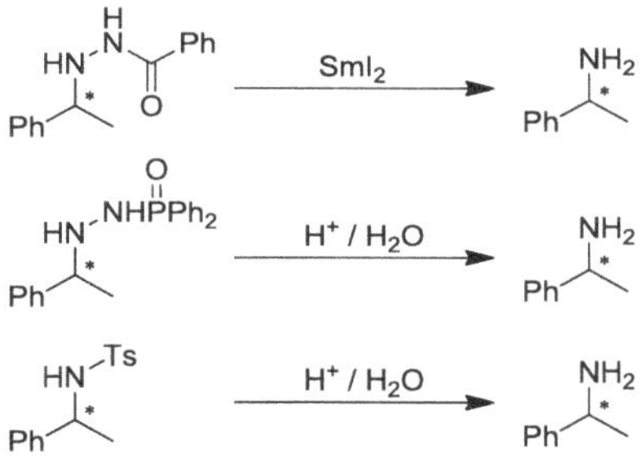

Scheme 41 Deprotection of *N*–X protecting/activating groups to afford chiral primary amines

can be easily deprotected by treatment with benzenesulfonic acid at 60°C to give unprotected hydrazine without affecting enantio purity and the ester group [17]. Similarly, the acid-catalyzed removal of the phosphinyl group is also known to be easy [125]. A reductive cleavage of the tosyl group in *N*-tosyl amines can also be performed with retention of the enantiomeric excess [112].

3.2 Asymmetric Hydrogenation of Non-activated Acyclic Imines

Much effort has been devoted to identifying catalysts that are able to hydrogenate substrates of *N*-arylimines of aryl alkyl ketones and more challenging *N*-arylimines of alkyl alkyl ketones, including the example of the exceptionally successful commercialized process of (*S*)-metolachlor. Numerous research groups have investigated the asymmetric hydrogenation of *N*-arylimines extensively, although other related synthetic methodologies, such as the asymmetric reductive amination [126–128] and the hydrogenation of enamines and enamides [129, 130], have also been demonstrated to afford effectively the same chiral amine products that could serve as key intermediates. Whereas the enantioselective hydrogenation of *N*-aryl imines still confronts major challenges, such as the previously mentioned inhibitory and poisoning effect to the metal catalyst as well as imine and enamine tautomerization, some significant breakthroughs have been achieved in the development of many highly enantioselective and efficient chiral catalysts for asymmetric hydrogenation.

Generally, for *N*-aryl aromatic ketimines (acetophenone type), the relatively simple preparation from the corresponding amine derivatives and carbonyl compounds, stability of isolated substrates, exclusively formed *E* isomers of the imine substrates, and potential of removal of *N*-aryl groups, especially the PMP (*para*-methoxy phenyl) group, are all advantageous properties that have attracted great research interest. Therefore, *N*-aryl aromatic ketimines have typically been chosen as one of the benchmark substrates in evaluating chiral catalysts. Some representative results are summarized in Table 4. Compared to other transition metals, Ir-based catalysts, particularly cationic Ir catalytic species with weakly coordinating counterions such as BARF, have proved remarkably successful in the hydrogenation of this *N*-aryl substrate type.

Whereas the hydrogenation of *N*-aryl alkyl alkyl ketones have been less investigated than the aryl alkyl counterpart, the hydrogenation of the MEA imine from 2-methyl-5-ethylaniline (MEA) and methoxyacetone have enabled the largest scale application of an enantioselective catalytic process by Ir–Xyliphos catalyst (Scheme 11) [8].

Furthermore, there have been few acyclic *N*-alkyl imines, or the corresponding amines have been found to be of practical industrial importance. In the reported studies, *N*-benzyl imine of acetophenone and the analogues thereof have been the benchmark substrate to evaluate the efficacies of different catalyst systems. The advantages of this choice could be the easy preparation of pure crystalline starting material and the readily removable benzyl group by hydrogenolysis after the asymmetric hydrogenation. There have been numerous efficient Ti-, Rh-, Ru-, and Ir-based catalysts, and the representative results are summarized in Table 5. The relatively lower enantioselectivities of *N*-alkylimines may be attributed to the fact that the acyclic *N*-alkylimines exist in the form of interconvertible mixtures of both *E*- and Z- isomers.

Table 4 Representative results of enantioselective hydrogenation of *N*-arylimines

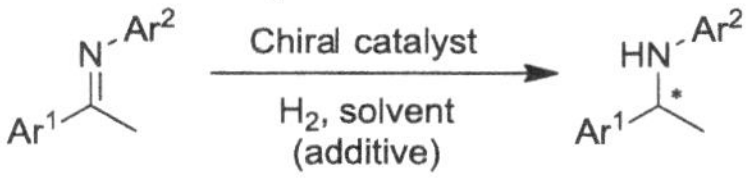

Ar[1]	Ar[2]	Catalyst	S/C	H_2	*ee* (%)	References
Ph	Ph	$[Ir(COD)Cl]_2$/DIOP/I^-	100	20 atm	22	[32]
Ph	Ph	$[Ir(BDPP)HI_2]_2$	1,000	40 atm	40	[103]
Ph	Ph	[Ir(PHOX)(COD)]BARF	1,000	100 atm	89	[63]
Ph	2,6-$(CH_3)_2$-C_6H_3	$[Ir(COD)Cl_2$/f-Binaphane	25	1,000 psi	>99	[42]
Ph	Ph	[Ir(HetPHOX)(COD)] BARF	1,000	50 atm	86	[69]
Ph	Ph	$RuCl_2$(Et-DuPhos) (DACH)	100	15 atm	92	[25]
Ph	Ph	[Ir(COD)(P,N ligand)] BARF	200	20 atm	90	[131]
Ph	Ph	$[Ir(COD)Cl]_2$/phosphine–phosphite ligand	100	30 atm	84	[77]
1-Np	4-MeO-C_6H_4	$[Ir(COD)Cl]_2$/Phosphine-sulfoximine	100	20 atm	98	[73]
Ph	4-CF_3-C_6H_4	[Ir(*t*Bu-BisP*)(COD)] BARF	200	1 atm	99	[51]
Ph	4-Cl-C_6H_4	[Ir(SIPHOX)(COD)] BARF	100	1 atm	97	[71]
Ph	4-MeO-C_6H_4	$[Ir(ddppm)(COD)]PF_6$	200	1 atm	94	[49]
Ph	3,5-$(CH_3)_2$-4-MeO-C_6H_2	[Ir(ferrocene-P,N ligand) (COD)]BARF	100	10 atm	94	[74]
Ph	4-MeO-C_6H_4	Cp*Ir(TsDPEN)(chiral phosphate)	100	20 atm	99	[87]
Ph	Ph	[Ir(DuanPhos)(COD)] BARF	10,000	5 atm	92	[52]
2-Np	Ph	[Ir(DuanPhos)(COD)] BARF	1,000	5 atm	98	[52]
Ph	3,5-$(CH_3)_2$-C_6H_3	$[Ir(COD)_2]$BARF/PipPhos	100	1 atm	>99	[81]
Ph	Ph	[Ir(SpinPHOX)(COD)] BARF	1,000	20 atm	92	[72]
Ph	Ph	[Ir(SimplePHOX)(COD)] BARF	100	5 atm	95.5	[68]
Ph	Ph	[Ir(NeoPHOX)(COD)] BARF	100	5 atm	94	[68]
Ph	Ph	[Ir(PHOX-type ligand) (COD)]BARF	333	50 atm	88	[70]

3.3 *Asymmetric Hydrogenation of N-H Imines*

Unfunctionalized N-H imines are an interesting category of substrates that has been long overlooked. Recently, Zhang and co-workers published the first example of efficient enantioselective hydrogenation of N-H imines [43], in which the typical imine *E*/*Z* stereoisomer problem is completely eliminated. Such conceptually

Table 5 Representative results of enantioselective hydrogenation of *N*-benzylimines

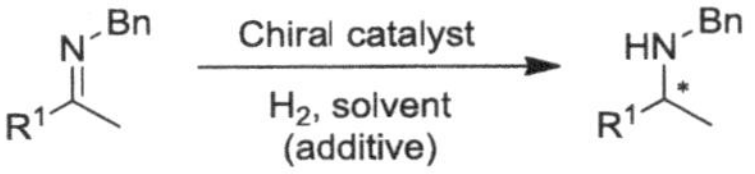

R^1	Catalyst	S/C	H_2	*ee* (%)	References
4-MeO-C_6H_4	$[Rh(NBD)Cl]_2$/Cycphos/KI	100	1,000 psi	91	[12]
Ph	$[Rh(COD)Cl]_2$/sulfonated BDPP	100	70 atm	96	[13]
Ph	Ti–ebthi/*n*BuLi/$PhSiH_3$	20	2,000 psi	85	[20]
Ph	$[Ir(COD)Cl]_2$/Tol-BINAP/$BnNH_2$	100	58.8 atm	70	[106]
4-MeO-C_6H_4	$[Rh(BDPP)(NBD)]ClO_4$/AOT	100	70 atm	92	[14]
Ph	$[Rh(bdpch)(COD)]BF_4$/$BnNH_2$	500	50 atm	72	[15]
Ph	Biphenyl-bridged titanocene/*n*BuLi	1,000	150 atm	76	[22]
Ph	RuHCl(BINAP)(DACH)	500	3 atm	60	[132]
Ph	$RuCl_2$(Tol-BINAP)(DPEN)	100	15 atm	62	[25]
Ph	$[Ir(COD)Cl]_2$/SPO ligand	10	25 atm	76	[105]
Ph	[Ir(COD)(P,N ligand)]BARF	50	50 atm	82	[133]
Ph	$[Ir(COD)Cl]_2$/phosphite-phosphinite ligand/$BnNH_2$	100	70 atm	76	[76]
4-MeO-C_6H_4	$[Ir(COD)Cl]_2$/chiral phosphorus acid diester/PPh_3	100	50 atm	92	[86]
Ph	Cp*RuCl(TsDACH)	100	30 atm	74	[27]
Ph	[Ir(SpinPHOX)(COD)]BARF	100	1 atm	91	[72]
Ph	[Ir(SimplePHOX)(COD)]BARF	200	10 atm	81	[68]
Ph	$[Ru(\mu^6$-cymene)(MsDPEN)]BARF/$(Boc)_2O$	100	50 atm	96	[29]

innovative methodology provides a fundamental step in the development of a most direct and ideally atom-economical approach to access chiral primary amines. Moreover, the hydrogenation of such iminium salt substrates under acidic conditions could also help suppress the inhibitory effects of corresponding amine products. This "acidic activation strategy" has recently encouraged successful examples of asymmetric hydrogenation of imine substrates [120, 121] and heteroaromatics [115, 122, 123]. Furthermore, the preparation of N-H imines is readily accessible via organometallic addition to nitriles followed by quenching with anhydrous MeOH and isolation of the corresponding bench-stable hydrochloride salts.

However, the discovery of an efficient catalyst system is crucial. After extensive screening of metal precursors and diphosphine ligands, it was found that the chiral Ir complex based on ferrocene-based chiral diphosphine f-Binaphane could afford good enantioselectivity but only moderate conversions in CH_2Cl_2, under 100 atm of hydrogen pressure. Then, a mixture of MeOH and CH_2Cl_2 (2:1 v/v) proved to be the best solvent combination, providing enantioselectivity up to 95% and complete conversion. Thus, under optimized conditions (10 atm H_2, room temperature), a variety of aryl alkyl N-H imines were hydrogenated by the Ir–f-Binaphane catalyst in high yields (>90%) with excellent enantioselectivities (up to 95% *ee*, Scheme 42).

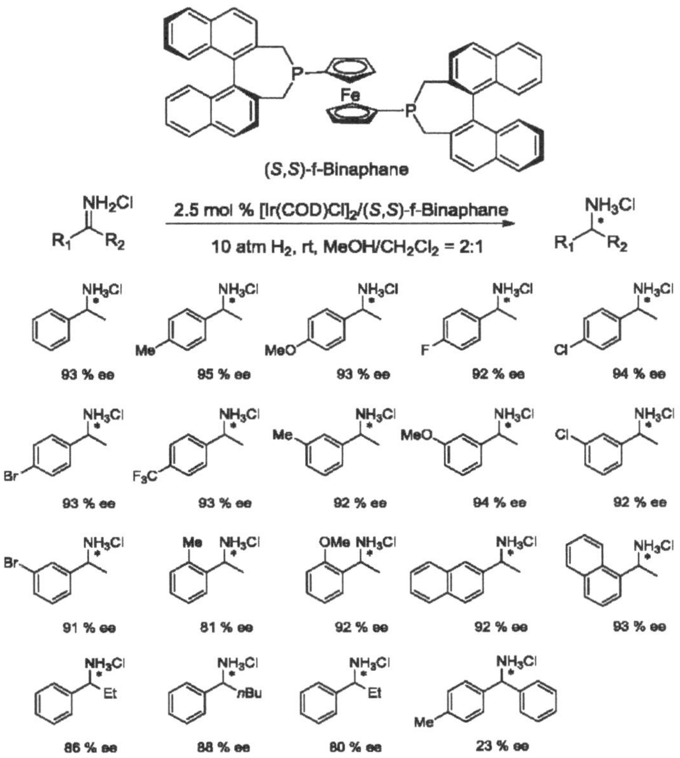

Scheme 42 Enantioselective hydrogenation of N-H imines by Ir-f-Binaphane catalyst

To solve the problem of low enantioselectivity for the hydrogenation of diaryl N-H imine substrate in the above study, Zhang and co-workers further screened a large number of chiral phosphorus ligands, including both bidentate and monodentate classes, to find the solution for the asymmetric hydrogenation of substituted benzophenone imines. Similar to the study of aryl alkyl N-H imines, there is no *E*/*Z* isomer issue or protecting group involved in this straightforward, environmentally sound, atom-economical approach. Surprisingly, the monodentate MonoPhos PE type ligand *N*-Bn-*N*-Me-MonoPhos turned out to be the best choice for this hydrogenation when the reaction was carried out in the solvent mixture of MeOH and CH_2Cl_2 (1:3 v/v) under optimized conditions (500 psi H_2, room temperature). A series of diaryl substituted benzophenone type N-H imines were hydrogenated with good to excellent enantiomeric excesses (up to 98%) [119]. The hydrogenation appears to be sensitive to the steric and electronic nature of the substituent at the *ortho* position of the aromatic ring (Scheme 43).

Enlightened by such an unprotected N-H imines hydrogenation approach, similar "acid activation strategy" has also been successfully applied to tackle other

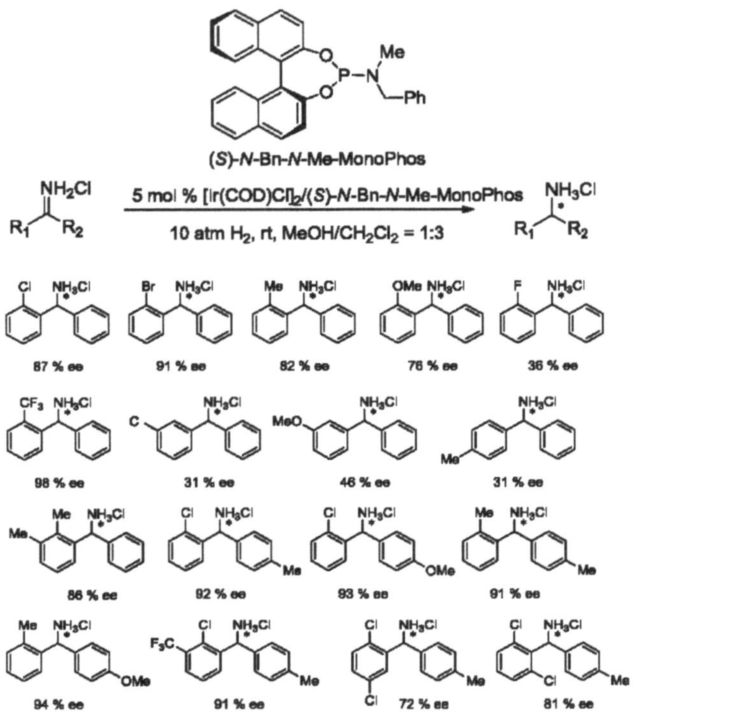

Scheme 43 Enantioselective hydrogenation of benzophenone type N-H imines by Ir-*N*-Bn-*N*-Me-MonoPhos catalyst

challenging substrates, such as unprotected β-enamine ester. Zhang and co-workers demonstrated the excellent efficacy of the same Ir–f-Binaphane complex in the hydrogenation of a wide range of β-enamine hydrochloride esters with excellent enantioselectivities (up to 97% *ee*) and with significantly superior turnover to those of N–H imines [44].

4 Asymmetric Hydrogenation of Cyclic Imines

4.1 Asymmetric Hydrogenation of Activated Cyclic Imines

The asymmetric hydrogenation of cyclic imines provides efficient and straightforward access to the corresponding chiral amines bearing cyclic skeletons, which play important roles as biologically active building blocks and key intermediates in organic synthesis. There are several cyclic analogues of the acyclic activated imines that have been investigated as shown in Scheme 44.

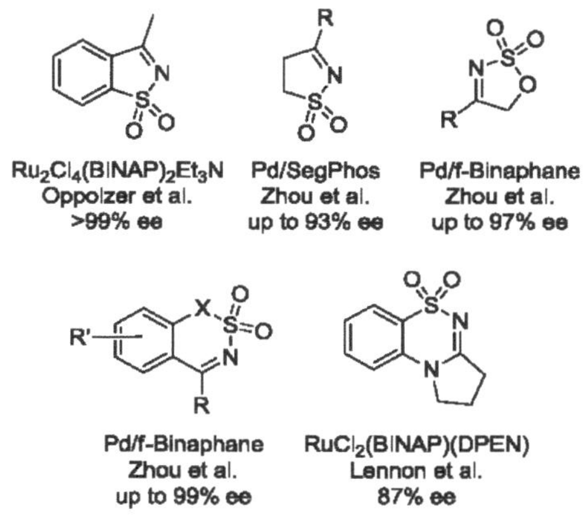

Scheme 44 Activated cyclic imine substrate types [23, 45, 114, 134]

Table 6 Representative results of enantioselective hydrogenation of cyclic imine to afford sultam

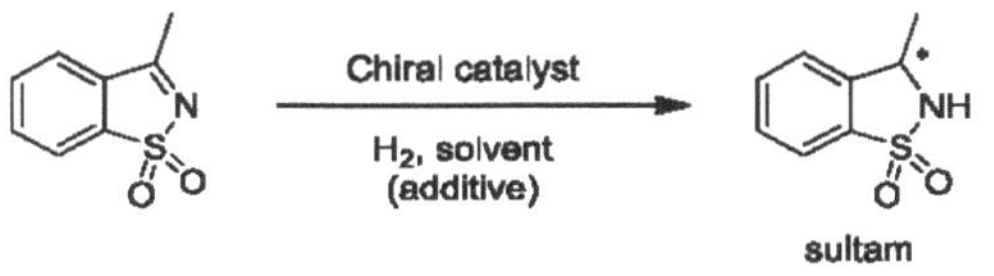

Catalyst	S/C	H_2	*ee* (%)	References
$Ru_2Cl_4(BINAP)_2Et_3N$	90	4 atm	>99	[23]
$Pd(CF_3CO_2)_2$/TangPhos	100	75 atm	94	[112]
$Pd(CF_3CO_2)_2$/SegPhos	50	600 psi	92	[114]
$Pd(CF_3CO_2)_2$/f-Binaphane	50	41 atm	97	[46]

Substituted sultams are important chiral auxiliaries that were first introduced in 1990 by Oppolzer [23] and have since been successfully applied to a number of asymmetric transformations. Asymmetric hydrogenation of the corresponding cyclic *N*-tosylimine is one of the most widely used synthetic transformations to access chiral sultams. Since then, several catalyst systems have been developed to achieve high enantioselectivity for this hydrogenation (Table 6).

4.2 Asymmetric Hydrogenation of Non-activated Cyclic Imine Substrates

In principle, superior enantioselectivities could be expected for cyclic imines because the intrinsic drawback of *E*/*Z* isomerism of acyclic imines is eliminated when the

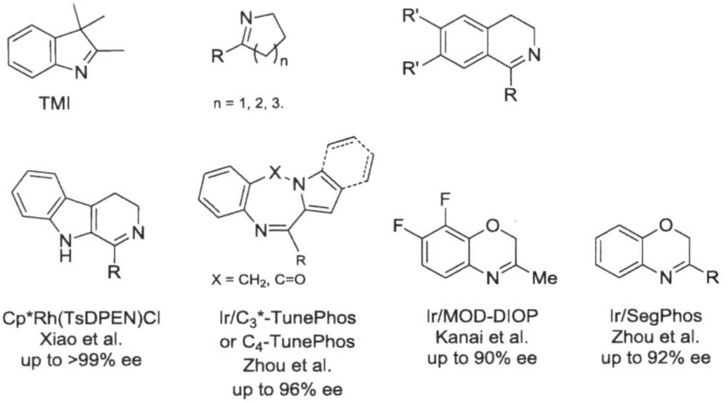

Scheme 45 Non-activated cyclic imine substrate types [19, 101, 137, 138]

C = N double bond is fixed in a ring structure. Nevertheless, compared to the progress that has been made in the hydrogenation of acyclic imines, the solutions for the analogous production of enantiopure amines from cyclic ketimines, while also noteworthy and significant, have been less successful. Until the early 1990s, efficient catalyst systems providing high enantioselectivities in the hydrogenation of non-activated cyclic imines started to emerge. Buchwald et al.'s introduction of the *ansa*-titanocene system developed by Brintzinger [135] is one noted successful example of hydrogenation of cyclic imines and other *N*-benzylimines [20, 21, 136], and the study has been an exceptional example in terms of the high level of enantioselectivity and the transition metal it employs. It should also be noted that virtually all of the other systems employed for imine hydrogenation have been derived from late transition metals. Several common types of cyclic imines such as 2,3,3-trimethylindolenine (TMI) [35], 2-phenyl-1-pyrroline [20], and tetrahydroisoquinolines [20] have been investigated primarily in the past two decades, while some new substrate classes, such as benzodiazepinones [137] and 2*H*-1,4-benzoxazines [138], have recently attracted attention (Scheme 45).

Some representative results of the hydrogenation of these substrate categories are summarized in Tables 7, 8, and 9. The combination of the Ir precursor $[Ir(COD)Cl]_2$ and chiral diphosphine ligands have been the most developed catalyst systems for the hydrogenation of TMI. For the hydrogenation of tetrahydroisoquinoline-type substrates, various catalyst systems based on Rh, Ru, and Ir have proved very efficient.

Table 7 Representative results of enantioselective hydrogenation of 2,3,3-trimethylindolenine (TMI)

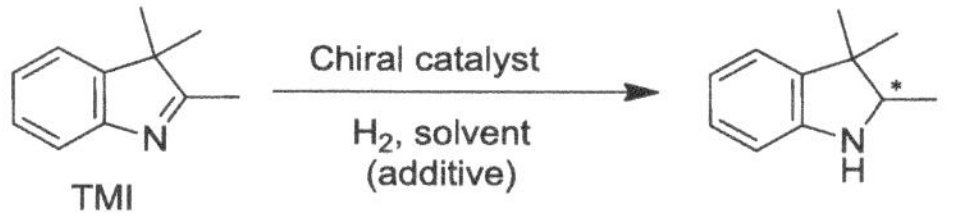

Catalyst	S/C	H_2	*ee* (%)	References
$[Ir(COD)Cl]_2$/MOD-DIOP/Bu_4NI	100	100 atm	81.4	[35]
$[Ir(COD)Cl]_2$/BCPM/BiI_3	100	100 atm	91	[37]
$[Ir(COD)Cl]_2$/MCCPM/BiI_3	100	100 atm	90	[37]
$[Ir(COD)Cl]_2$/BICP/phthalimide	100	1,000 psi	95.1	[38]
$[Ir(COD)Cl]_2$/ferrocene diphosphine ligand	100	65 atm	79	[41]
$[Ir(COD)Cl]_2$/Josiphos/Bu_4NI	250	40 atm	94	[40]
$RuCl_2$(MeO-BIPHEP)(ANDEN)	100	15 atm	88	[25]
$[Ir(COD)Cl]_2$/DIOP*/I_2	100	1,000 psi	85.0	[36]
$[Ir(COD)Cl]_2$/MonoPhos/acridine	100	40 atm	58	[80]

Table 8 Representative results of enantioselective hydrogenation of 2-substituted 1-pyrroline, 2-phenyl-3,4,5,6-tetrahydropyridine and 2-phenyl-4,5,6,7-tetrahydro-3*H*-azepine

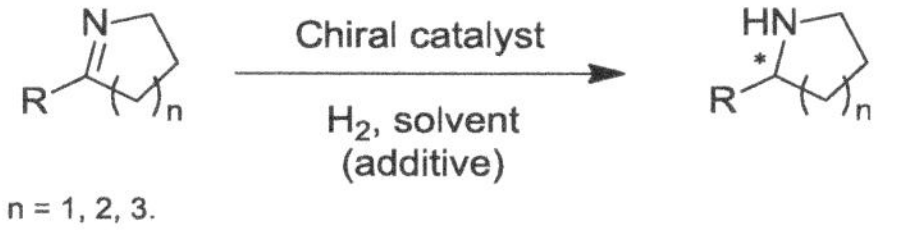

R	*n*	Catalyst	S/C	H_2	*ee* (%)	References
*n*Hex	1	Ti–ebthi/*n*BuLi/$PhSiH_3$	20	2,000 psi	98	[20]
Ph	1	Ti–ebthi/*n*BuLi/$PhSiH_3$	20	80 psi	99	[136]
Ph	2	Ti–ebthi/*n*BuLi/$PhSiH_3$	20	500 psi	98	[136]
Ph	3	Ti–ebthi/*n*BuLi/$PhSiH_3$	20	500 psi	98	[136]
Ph	2	$[Ir(COD)Cl]_2$/Tol-BINAP/$BnNH_2$	100	58.8 atm	90	[106]
Ph	2	$[Ir(COD)Cl]_2$/BICP/phthalimide	100	1,000 psi	64.7	[38]
Ph	1	Biphenyl-bridged titanocene/*n*BuLi	1,000	150 atm	98	[22]
Ph	1	$[[Ir(H)(BINAP)]_2(\mu\text{-}I)_3]^+I^-$	220	60 atm	83	[62]
Ph	2	$[[Ir(H)(BINAP)]_2(\mu\text{-}I)_3]^+I^-$	1,000	60 atm	91	[62]
Ph	3	IrHCl(BINAP)(CH_3CO_2)	100	60 atm	69	[62]
Ph	1	$[Ir(COD)Cl]_2$/f-Binaphane	100	50 atm	85	[47]
Ph	2	$[Ir(COD)Cl]_2$/f-Binaphane	100	50 atm	89	[47]
Ph	3	$[Ir(COD)Cl]_2$/f-Binaphane	100	50 atm	75	[47]
Ph	1	$[Ru(\mu^6$-cymene)(MsDPEN)]BARF/$(Boc)_2O$	100	50 atm	97	[30]
Ph	2	$[Ru(\mu^6$-cymene)(MsDPEN)]BARF/$(Boc)_2O$	100	50 atm	95	[30]
Ph	3	$[Ru(\mu^6$-cymene)(MsDPEN)]BARF/$(Boc)_2O$	100	50 atm	92	[30]

Table 9 Representative results of enantioselective hydrogenation of tetrahydroisoquinolines

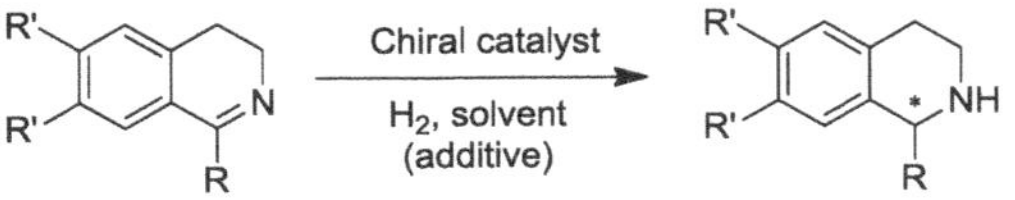

R′	R	Catalyst	S/C	H_2	*ee* (%)	References
MeO	Me	Ti–ebthi/*n*BuLi/PhSiH_3	20	2,000 psi	98	[20]
MeO	Me	[Ir(COD)Cl]$_2$/BCPM/ phthalimide	100	100 atm	93	[88]
MeO	3,4-MeO-C_6H_3-(CH_2)$_2$	[Ir(COD)Cl]$_2$/BINAP/ F_4-phthalimide	100	100 atm	86	[107]
MeO	CH_2OBn	[Ir(COD)Cl]$_2$/BINAP/ phthalimide	200	100 atm	86	[39]
MeO	Me	RuCl_2(Et-DuPhos) (DACH)	100	15 atm	79	[25]
MeO	Me	RuCl_2(Ph-BPM) (DPEN)	100	10 atm	89	[139]
H	Me	Cp*RhCl(TsDPEN)/ AgSbF_6	100	20 atm	99	[19]
MeO	3,4-MeO-C_6H_3-(CH_2)$_2$	Cp*RhCl(TsDPEN)/ AgSbF_6	100	20 atm	99	[19]
MeO	3,4-MeO-C_6H_3	[[Ir(H)(f-Binaphane)]$_2$ (μ-I)$_3$]$^+$I$^-$/I_2	1,000	50 atm	>99	[48]
H	Ph	[[Ir(H)(f-Binaphane)]$_2$ (μ-I)$_3$]$^+$I$^-$/I_2	10,000	50 atm	93	[48]
H	Ph	[Ir(COD)Cl]$_2$/3,5-diMe-Synphos	100	10 atm	94	[140]
H	Me	[Ir(COD)Cl]$_2$/SIPHOS-pe/KI	100	6 atm	99	[84]

5 Conclusion

From the perspectives of efficiency, accessibility, and atom-economy, asymmetric imine hydrogenation, one of the most important applications of homogeneous catalysis in industry, is an ideal synthetic transformation. The significant progress in the asymmetric hydrogenation of olefins and ketones kindled the long-overlooked but challenging field of asymmetric imine hydrogenation. A tremendous amount of effort in the past two decades resulted in a number of chiral metal catalysts that have proved effective for the hydrogenation of various imines achieving high-level enantioselectivities. Although some intrinsic issues of this substrate class, such as *E*/*Z* isomerization and product inhibitory effect, have limited the efficacy of catalysts in terms of enantioselectivity and reactivity, some recent progress has brought new insight and will undoubtedly lead to more groundbreaking advances in this field. Since Ir-based catalysts have proved highly enantioselective for imine hydrogenations but suffer from deactivation due to dimeric or trimeric hydride-bridged complex formation, proper selection of additives or preformation of more reactive catalytic species could help reduce the deactivation of the Ir catalyst. With more variety and fine-tuning of conformationally rigid and electron-donating chiral diphosphine ligands, the hydrogenation of different types of imines is possible. Hydrogenation of imines under acidic conditions can help substrate activation and possibly reduce product inhibition [117]. Such activation strategy has been applied successfully to the enantioselective hydrogenation of N-H imines [43, 119], and in this case the *E*/*Z* isomer issue and necessity of *N*- protecting group are both eliminated. Pd-based catalysts have been the emerging stars in the field of enantioselective hydrogenations of imines, particularly activated imines, and promise to exhibit excellent efficacy in the future.

Although impressive results have been achieved in asymmetric imine hydrogenation, more research advances regarding new catalyst development and equally important mechanistic understanding are still very important in order to address the remaining challenges. The expected progress will enable asymmetric imine hydrogenation to be a more general and viable methodology for synthesizing chiral amines in the near future.

References

1. Shang G, Li W, Zhang X (2010) In: Ojima I (ed) Catalytic asymmetric synthesis, 3rd edn. Wiley, New York, p 343
2. Chi Y, Tang W, Zhang X (2005) In: Evans PA (ed) Modern rhodium-catalyzed organic reactions. Wiley-VCH Verlag GmbH, Weinheim, p 1
3. Blaser H-U, Spindler F (1999) In: Jacobsen EN, Pfaltz A, Yamamoto H (eds) Comprehensive asymmetric catalysis. Springer, Berlin, p 247
4. Spindler F, Blaser H-U (2007) In: De Vries JG, Elsevier CJ (eds) Handbook of homogeneous hydrogenation. Wiley-VCH Verlag GmbH, Weinheim, p 1193
5. Ager D (2007) In: De Vries JG, Elsevier CJ (eds) Handbook of homogeneous hydrogenation. Wiley-VCH Verlag GmbH, Weinheim, p 745
6. Brown JM (1999) In: Jacobsen EN, Pfaltz A, Yamamoto H (eds) Comprehensive asymmetric catalysis. Springer, Berlin, p 121
7. Genet J-P (2008) In: Andersson PG, Munslow IJ (eds) Modern reduction methods. Wiley-VCH Verlag GmbH, Weinheim, p 3
8. Blaser H-U (2002) Adv Synth Catal 344:17
9. Noyori R (2002) Angew Chem. Int Ed 41:2008
10. Roseblade SJ, Pfaltz A (2007) Acc Chem Res 40:1402
11. Levi A, Modena G, Scorrano G (1975) J Chem Soc Chem Commun 6
12. Kang GJ, Cullen WR, Fryzuk MD, James BR, Kutney JP (1988) J Chem Soc. Chem Commun 1466
13. Bakos J, Orosz A, Heil B, Laghmari M, Lhoste P, Sinou D (1991) J Chem Soc. Chem Commun 1684
14. Buriak JM, Osborn JA (1996) Organometallics 15:3161
15. Tararov VI, Kadyrov R, Riermeier TH, Holz J, Borner A (1999) Tetrahedron Asymmetry 10:4009
16. Burk MJ, Feaster JE (1992) J Am Chem Soc 114:6266

17. Yoshikawa N, Tan L, McWilliams JC, Ramasamy D, Sheppard R (2010) Org Lett 12:276
18. Shang G, Yang Q, Zhang X (2006) Angew Chem. Int Ed 45:6360
19. Li C, Xiao J (2008) J Am Chem Soc 130:13208
20. Willoughby CA, Buchwald SL (1992) J Am Chem Soc 114:7562
21. Willoughby CA, Buchwald SL (1994) J Am Chem Soc 116:11703
22. Ringwald M, Stuermer R, Brintzinger HH (1999) J Am Chem Soc 121:1524
23. Oppolzer W, Wills M, Starkemann C, Bernardinelli G (1990) Tetrahedron Lett 31:4117
24. Charette AB, Giroux A (1996) Tetrahedron Lett 37:6669
25. Cobley CJ, Henschke JP (2003) Adv Synth Catal 345:195
26. Arai N, Utsumi N, Matsumoto Y, Murata K, Tsutsumi K, Ohkuma T (2012) Adv Synth Catal 354:2089
27. Shirai S-y, Nara H, Kayaki Y, Ikariya T (2009) Organometallics 28:802
28. Cheruku P, Church TL, Andersson PG (2008) Chem Asian J 3:1390
29. Chen F, Wang T, He Y, Ding Z, Li Z, Xu L, Fan Q-H (2011) Chem Eur J 17:1109
30. Chen F, Ding Z, Qin J, Wang T, He Y, Fan Q-H (2011) Org Lett 13:4348
31. Crabtree RH, Felkin H, Morris GE (1976) J Chem Soc. Chem Commun 716
32. Spindler F, Pugin B, Blaser H-U (1990) Angew Chem. Int Ed Engl 29:558
33. Blaser H-U, Schmidt E (2004) Asymmetric catalysis on industrial scale: challenges, approaches and solutions. Wiley-VCH Verlag GmbH, Weinheim
34. Blaser H-U, Buser H-P, Coers K, Hanreich R, Jalett H-P, Jelsch E, Pugin B, Schneider H-D, Spindler F, Wegmann A (1999) Chimia 53:275
35. Morimoto T, Nakajima N, Achiwa K (1994) Chem Pharm Bull 42:1951
36. Liu D, Li W, Zhang X (2004) Tetrahedron Asymmetry 15:2181
37. Morimoto T, Nakajima N, Achiwa K (1995) Synlett 748
38. Zhu G, Zhang X (1998) Tetrahedron Asymmetry 9:2415
39. Morimoto T, Suzuki N, Achiwa K (1998) Tetrahedron Asymmetry 9:183
40. Blaser HU, Buser HP, Hausel R, Jalett HP, Spindler F (2001) J Organomet Chem 621:34
41. Reetz MT, Beuttenmuller EW, Goddard R, Pasto M (1999) Tetrahedron Lett 40:4977
42. Xiao D, Zhang X (2001) Angew Chem. Int Ed 40:3425
43. Hou G, Gosselin F, Li W, McWilliams JC, Sun Y, Weisel M, O'Shea PD, Chen C-y, Davies IW, Zhang X (2009) J Am Chem Soc 131:9882
44. Hou G, Li W, Ma M, Zhang X, Zhang X (2010) J Am Chem Soc 132:12844
45. Wang Y-Q, Yu C-B, Wang D-W, Wang X-B, Zhou Y-G (2008) Org Lett 10:2071
46. Yu C-B, Wang D-W, Zhou Y-G (2009) J Org Chem 74:5633
47. Chang M, Li W, Hou G, Zhang X (2010) Adv Synth Catal 352:3121
48. Chang M, Li W, Zhang X (2011) Angew Chem. Int Ed 50:10679
49. Dervisi A, Carcedo C, Ooi L-I (2006) Adv Synth Catal 348:175
50. Smidt SP, Pfaltz A, Martinez-Viviente E, Pregosin PS, Albinati A (2003) Organometallics 22:1000
51. Imamoto T, Iwadate N, Yoshida K (2006) Org Lett 8:2289
52. Li W, Hou G, Chang M, Zhang X (2009) Adv Synth Catal 351:3123
53. Lightfoot A, Schnider P, Pfaltz A (1998) Angew Chem. Int Ed 37:2897
54. Kaiser S, Smidt SP, Pfaltz A (2006) Angew Chem Int Ed 45:5194
55. Hedberg C, Källström K, Brandt P, Hansen LK, Andersson PG (2006) J Am Chem Soc 128:2995
56. Liu Q-B, Yu C-B, Zhou Y-G (2006) Tetrahedron Lett 47:4733
57. Barchuk A, Ngai M-Y, Krische MJ (2007) J Am Chem Soc 129:8432
58. Schönleber M, Hilgraf R, Pfaltz A (2008) Adv Synth Catal 350:2033
59. Wang A, Wüstenberg B, Pfaltz A (2008) Angew Chem Int Ed 47:2298
60. Church TL, Andersson PG (2008) Coord Chem Rev 252:513
61. Dorta R, Broggini D, Stoop R, Rüegger H, Spindler F, Togni A (2004) Chem Eur J 10:267
62. Yamagata T, Tadaoka H, Nagata M, Hirao T, Kataoka Y, Ratovelomanana-Vidal V, Genet JP, Mashima K (2006) Organometallics 25:2505

63. Schnider P, Koch G, Pretot R, Wang G, Bohnen FM, Kruger C, Pfaltz A (1997) Chem Eur J 3:887
64. Kainz S, Brinkmann A, Leitner W, Pfaltz A (1999) J Am Chem Soc 121:6421
65. Menges F, Pfaltz A (2002) Adv Synth Catal 344:40
66. Smidt SP, Menges F, Pfaltz A (2004) Org Lett 6:2023
67. Schrems MG, Pfaltz A (2009) Chem Commun 6210
68. Baeza A, Pfaltz A (2010) Chem Eur J 16:4003
69. Cozzi PG, Menges F, Kaiser S (2003) Synlett 833
70. Lu W-J, Chen Y-W, Hou X-L (2010) Adv Synth Catal 352:103
71. Zhu S-F, Xie J-B, Zhang Y-Z, Li S, Zhou Q-L (2006) J Am Chem Soc 128:12886
72. Han Z, Wang Z, Zhang X, Ding K (2009) Angew Chem. Int Ed 48:5345
73. Moessner C, Bolm C (2005) Angew Chem. Int Ed 44:7564
74. Cheemala MN, Knochel P (2007) Org Lett 9:3089
75. Guiu E, Munoz B, Castillon S, Claver C (2003) Adv Synth Catal 345:169
76. Guiu E, Aghmiz M, Diaz Y, Claver C, Meseguer B, Militzer C, Castillon S (2006) Eur J Org Chem 627
77. Vargas S, Rubio M, Suarez A, Pizzano A (2005) Tetrahedron Lett 46:2049
78. van den Berg M, Haak RM, Minnaard AJ, de Vries AHM, de Vries JG, Feringa BL (2002) Adv Synth Catal 344:1003
79. van den Berg M, Minnaard AJ, Haak RM, Leeman M, Schudde EP, Meetsma A, Feringa BL, de Vries AHM, Maljaars CEP, Willans CE, Hyett D, Boogers JAF, Henderickx HJW, de Vries JG (2003) Adv Synth Catal 345:308
80. Faller JW, Milheiro SC, Parr J (2006) J Organomet Chem 691:4945
81. Mrsic N, Minnaard AJ, Feringa BL, de Vries JG (2009) J Am Chem Soc 131:8358
82. Mrsic N, Lefort L, Boogers JAF, Minnaard AJ, Feringa BL, de Vries JG (2008) Adv Synth Catal 350:1081
83. Mrsic N, Jerphagnon T, Minnaard AJ, Feringa BL, de Vries JG (2010) Tetrahedron Asymmetry 21:7
84. Xie J-H, Yan P-C, Zhang Q-Q, Yuan K-X, Zhou Q-L (2012) ACS Catal 2:561
85. Hou C-J, Wang Y-H, Zheng Z, Xu J, Hu X-P (2012) Org Lett 14:3554
86. Reetz MT, Bondarev O (2007) Angew Chem. Int Ed 46:4523
87. Li C, Wang C, Villa-Marcos B, Xiao J (2008) J Am Chem Soc 130:14450
88. Morimoto T, Achiwa K (1995) Tetrahedron Asymmetry 6:2661
89. Hou G-H, Xie J-H, Yan P-C, Zhou Q-L (2009) J Am Chem Soc 131:1366
90. Yan P-C, Xie J-H, Hou G-H, Wang L-X, Zhou Q-L (2009) Adv Synth Catal 351:3243
91. Wang W-B, Lu S-M, Yang P-Y, Han X-W, Zhou Y-G (2003) J Am Chem Soc 125:10536
92. Wang X-B, Zhou Y-G (2008) J Org Chem 73:5640
93. Wang D-W, Wang X-B, Wang D-S, Lu S-M, Zhou Y-G, Li Y-X (2009) J Org Chem 74:2780
94. Tang W-J, Zhu S-F, Xu L-J, Zhou Q-L, Fan Q-H, Zhou H-F, Lam K, Chan ASC (2007) Chem Commun 613
95. Gou F-R, Li W, Zhang X, Liang Y-M (2010) Adv Synth Catal 352:2441
96. Tang W, Xu L, Fan Q-H, Wang J, Fan B, Zhou Z, K-h L, Chan ASC (2009) Angew Chem. Int Ed 48:9135
97. Legault CY, Charette AB (2005) J Am Chem Soc 127:8966
98. Wang X-B, Zeng W, Zhou Y-G (2008) Tetrahedron Lett 49:4922
99. Tang W-J, Tan J, Xu L-J, Lam K-H, Fan Q-H, Chan ASC (2010) Adv Synth Catal 352:1055
100. Tang W, Sun Y, Xu L, Wang T, Fan Q, Lam K-H, Chan ASC (2010) Org Biomol Chem 8:3464
101. Satoh K, Inenaga M, Kanai K (1998) Tetrahedron Asymmetry 9:2657
102. Chan YNC, Meyer D, Osborn JA (1990) J Chem Soc. Chem Commun 869
103. Ng Cheong Chan Y, Osborn JA (1990) J Am Chem Soc 112:9400
104. Blaser H-U, Pugin B, Spindler F, Togni A (2002) C R Chimie 5:379

105. Jiang X-B, Minnaard AJ, Hessen B, Feringa BL, Duchateau ALL, Andrien JGO, Boogers JAF, de VJG (2003) Org Lett 5:1503
106. Tani K, Onouchi J-i, Yamagata T, Kataoka Y (1995) Chem Lett 955
107. Morimoto T, Suzuki N, Achiwa K (1996) Heterocycles 43:2557
108. Tietze LF, Ila H, Bell HP (2004) Chem Rev 104:3453
109. Chen Q-A, Ye Z-S, Duan Y, Zhou Y-G (2013) Chem Soc Rev 42:497
110. Abe H, Amii H, Uneyama K (2001) Org Lett 3:313
111. Shuklov IA, Dubrovina NV, Börner A (2007) Synthesis 2007:2925
112. Yang Q, Shang G, Gao W, Deng J, Zhang X (2006) Angew Chem. Int Ed 45:3832
113. Wang Y-Q, Zhou Y-G (2006) Synlett 1189
114. Wang Y-Q, Lu S-M, Zhou Y-G (2007) J Org Chem 72:3729
115. Wang D-S, Chen Q-A, Li W, Yu C-B, Zhou Y-G, Zhang X (2010) J Am Chem Soc 132:8909
116. Marcazzan P, Patrick BO, James BR (2003) Organometallics 22:1177
117. Yu Z, Jin W, Jiang Q (2012) Angew Chem. Int Ed 51:6060
118. Magee MP, Norton JR (2001) J Am Chem Soc 123:1778
119. Hou G, Tao R, Sun Y, Zhang X, Gosselin F (2010) J Am Chem Soc 132:2124
120. Zhou X-Y, Bao M, Zhou Y-G (2011) Adv Synth Catal 353:84
121. Gao K, Yu C-B, Li W, Zhou Y-G, Zhang X (2011) Chem Commun 47:7845
122. Wang D-S, Tang J, Zhou Y-G, Chen M-W, Yu C-B, Duan Y, Jiang G-F (2011) Chem Sci 2:803
123. Wang D-S, Ye Z-S, Chen Q-A, Zhou Y-G, Yu C-B, Fan H-J, Duan Y (2011) J Am Chem Soc 133:8866
124. Burk MJ, Martinez JP, Feaster JE, Cosford N (1994) Tetrahedron 50:4399
125. Spindler F, Blaser H-U (2001) Adv Synth Catal 343:68
126. Chi Y, Zhou Y-G, Zhang X (2003) J Org Chem 68:4120
127. Hoffmann S, Nicoletti M, List B (2006) J Am Chem Soc 128:13074
128. Storer RI, Carrera DE, Ni Y, MacMillan DWC (2005) J Am Chem Soc 128:84
129. Tang W, Zhang X (2003) Chem Rev 103:3029
130. Xie J-H, Zhu S-F, Zhou Q-L (2011) Chem Rev 111:1713
131. Trifonova A, Diesen JS, Chapman CJ, Andersson PG (2004) Org Lett 6:3825
132. Abdur-Rashid K, Lough AJ, Morris RH (2001) Organometallics 20:1047
133. Blanc C, Agbossou-Niedercorn F, Nowogrocki G (2004) Tetrahedron Asymmetry 15:2159
134. Cobley CJ, Foucher E, Lecouve J-P, Lennon IC, Ramsden JA, Thominot G (2003) Tetrahedron Asymmetry 14:3431
135. Wild FRWP, Zsolnai L, Huttner G, Brintzinger HH (1982) J Organomet Chem 232:233
136. Willoughby CA, Buchwald SL (1994) J Am Chem Soc 116:8952
137. Gao K, Wu B, Yu C-B, Chen Q-A, Ye Z-S, Zhou Y-G (2012) Org Lett 14:3890
138. Gao K, Yu C-B, Wang D-S, Zhou Y-G (2012) Adv Synth Catal 354:483
139. Jackson M, Lennon IC (2007) Tetrahedron Lett 48:1831
140. Berhal F, Wu Z, Zhang Z, Ayad T, Ratovelomanana-Vidal V (2012) Org Lett 14:3308

Top Curr Chem (2014) 343: 145–190
DOI: 10.1007/128_2013_480

Published online: 31 October 2013

Advances in Transition Metal-Catalyzed Asymmetric Hydrogenation of Heteroaromatic Compounds

Yan-Mei He, Feng-Tao Song, and Qing-Hua Fan

Abstract Transition metal-catalyzed asymmetric hydrogenation of heteroaromatic compounds is undoubtedly a straightforward and environmentally friendly method for the synthesis of a wide range of optically active heterocyclic compounds, which are widespread and ubiquitous in naturally occurring and artificial bioactive molecules. Over the past decade, a number of transition metal (Ir, Rh, Ru, and Pd) catalysts bearing chiral phosphorus ligands, amine-tosylamine ligands, and *N*-heterocyclic carbene ligands have been developed for such challenging transformation. This review will describe the significant contributions concerning the transition metal-catalyzed asymmetric hydrogenation of *N*-, *O*-, and *S*-containing heteroaromatic compounds, with emphasis on the evolution of different chiral ligands, related catalyst immobilization, and mechanism investigations.

Keywords Asymmetric hydrogenation · Heteroaromatic · Heterocycles · Homogeneous catalysis · Transition metal

Contents

Y.-M. He, F.-T. Song, and Q.-H. Fan (✉)
Beijing National Laboratory for Molecular Sciences, CAS Key Laboratory of Molecular Recognition and Function, Institute of Chemistry and Graduate School, Chinese Academy of Sciences (CAS), Beijing 100190, P. R. China
e-mail: fanqh@iccas.ac.cn

1 Introduction

Heterocyclic structures, saturated or unsaturated, containing *N*, *O*, *S*, and other hetero atoms in the cyclic rings are widespread and ubiquitous in naturally occurring and artificial bioactive molecules, which are widely used in pharmaceutical and agrochemical production. In recent years the synthesis of chiral heterocycles with chiral centers on the ring atoms has attracted more and more attentions based on the observation of the obvious effects of chiral centers on properties and functions of these heterocyclic compounds. On the other hand, introducing chiral centers into the ring systems will greatly extend the type and increase the amount of heterocyclic compounds, thus offering abundant opportunities for scientists to use these new molecular units in the design of novel functional molecules. Generally, the construction of chiral heterocycles has been realized through various enantioselective cyclization or cycloaddition approaches. Alternatively, catalytic asymmetric reduction of the substituted heteroaromatic compounds, which are usually easy to prepare from readily available starting materials, represents another type of feasible synthetic methods. Among them, transition metal-catalyzed asymmetric hydrogenation with hydrogen gas as the reducing agent has proven to be the most promising straightforward and atom-economic approach. Theoretically, one or more desired chiral centers can be created at the same step through precise control by the suitable match of chiral ligand and metal selection. Considering extensive applications of chiral heterocyclic compounds in organic synthesis, it is of special interest and great significance to develop practical, highly efficient, and highly enantioselective hydrogenation protocols for heteroaromatic substrates.

Although transition metal-catalyzed asymmetric hydrogenation of prochiral olefins, ketones, and imines has been well developed for several decades based on continuing ligand and methodology exploration (For some recent comprehensive reviews, see: [1–7]), the highly enantioselective hydrogenation of heteroaromatic compounds only appeared within the last 10 years (For reviews on reduction of hereroaromatics, see: [8–15]). This delay may be due to the special chemical properties of these heteroaromatic compounds. As we know, these compounds are relatively stable owing to the aromaticity of the conjugated ring systems; and harsh reaction conditions, such as high temperature and/or high hydrogen pressure, are usually required for achieving hydrogenation and thus resulting in low enantioselectivity. In addition, the hetero atoms, such as nitrogen or sulfur atoms in the heterocyclic rings, either in the substrates or in the reduced products, even the ones in the trace impurities, may coordinate to the metal centers and thus poison and/or deactivate the catalysts.

Given the recent discovery of transition metal catalysts together with several novel catalyst or substrate activation strategies, the last 10 years have witnessed significant progress in this field [13]. To date, a number of heteroaromatic compounds, including quinolines, isoquinolines, quinoxalines, pyridines, indoles, pyrroles, furans, thiophenes, imidazoles, and oxazoles, have been successfully applied in the asymmetric hydrogenation with excellent enantioselectivities. The asymmetric hydrogenation of heteroaromatic compounds has become a highly efficient, straightforward, economical, and environmentally friendly method for the synthesis of a wide range of optically active heterocyclic compounds. In this review we would like to describe the most recent advances in homogeneous transition metal-catalyzed asymmetric hydrogenation of different types of heteroaromatic compounds, with emphasis on the evolution of different chiral ligands. Related catalyst immobilization and mechanism investigations will also be highlighted. Considering that several excellent reviews on this topic have recently been published [8–15] (in particular, a comprehensive review was published in 2011 by Zhou and co-workers [13]), it is inevitable that this review will have some overlap with the contents of these previous reviews. Other related excellent work on asymmetric hydrogenation of heteroaromatic compounds using chiral and achiral heterogeneous metal catalysts [15], and asymmetric transfer hydrogenation catalyzed by organocatalysts, will not be included in this review [13, 16, 17].

2 Asymmetric Hydrogenation of Quinolines

Optically active 1,2,3,4-tetrahydroquinoline motif is widely existent in the molecular structures of naturally occurring alkaloids, artificial pharmaceuticals, and agrochemicals [18–20]. Asymmetric hydrogenation of quinoline derivatives, which are easily prepared from readily available starting materials, provides a direct approach for the efficient synthesis of these chiral heterocyclic compounds. To date, a wide variety of 2-substituted and 2,3-disubstituted quinolines have been successfully subjected to the asymmetric hydrogenation, and excellent enantioselectivities have been achieved.

As privileged ligands widely used in the transition metal-catalyzed hydrogenation of prochiral olefins, ketones, and imines, phosphorus-containing ligands are the first choice ligands in the asymmetric hydrogenation of quinolines and also other heteroaromatic compounds. Since the first breakthrough made by Zhou and co-workers in 2003 [21], who applied diphosphine ligand MeO-BIPHEP in the iridium-catalyzed highly enantioselective hydrogenation of 2-substituted quinolines, a large number of phosphorus-containing chiral catalysts have been well developed for this transformation. Notably, the catalytic activity was greatly enhanced by using catalysts bearing electronically deficient phosphorus ligands. Concurrent with phosphorus-containing ligand extension, substrate activation strategy by either chloroformates or Brønsted acids provided other alternatives for the asymmetric hydrogenation of quinolines [22, 23]. Another significant contribution is the application of phosphine-free diamine ligands for this transformation by Fan and co-workers [24, 25]. A broad range of quinoline derivatives were hydrogenated smoothly with diamine-containing ruthenium catalysts, giving the highest *ee* values for most substrates examined.

2.1 Chiral Diphosphine Ligands

Atropisomeric C_2-symmetric biaryl diphosphines are versatile effective ligands for asymmetric catalysis [26]. By tuning stereo and electronic properties of the ligands through rational backbone design and/or substituent modulation, the catalytic activity and stereoselectivity could be well adjusted. After the initial finding of the activation strategy with iodine additive in asymmetric hydrogenation of quinolines, Ir/diphosphine/I_2 became the most popular and widely used catalyst systems for the hydrogenation of heteroaromatic compounds [27–37], and the representative chiral diphosphine ligands used for the asymmetric hydrogenation of quinolines are listed in Fig. 1.

In 2003, Zhou and co-workers reported the first effective and highly enantioselective asymmetric hydrogenation of 2-substituted quinoline derivatives with diphosphine-iridium complex, in situ generated by mixing $[Ir(COD)Cl]_2$ (COD = 1,5-cyclooctadiene) and MeO-BIPHEP [21]. With the addition of iodine into the reaction mixture, the catalytic activity was enhanced enormously. After the optimization of the reaction conditions, a variety of 2-substituted quinoline derivatives were hydrogenated smoothly to chiral 1,2,3,4-tetrahydroquinolines in toluene in yields of 83–95% with *ee* values of 72–96%. It was noticed that the hydrogenation of most 2-alkyl-substituted quinolines gave higher enantioselectivities, while only 72% *ee* was obtained for 2-phenylquinoline. The Ir/MeO-BIPHEP/I_2 catalyst system

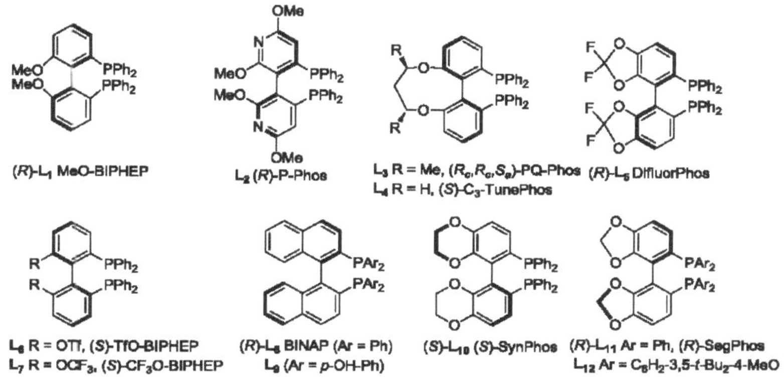

Fig. 1 Chiral diphosphine ligands for asymmetric hydrogenation of quinolines

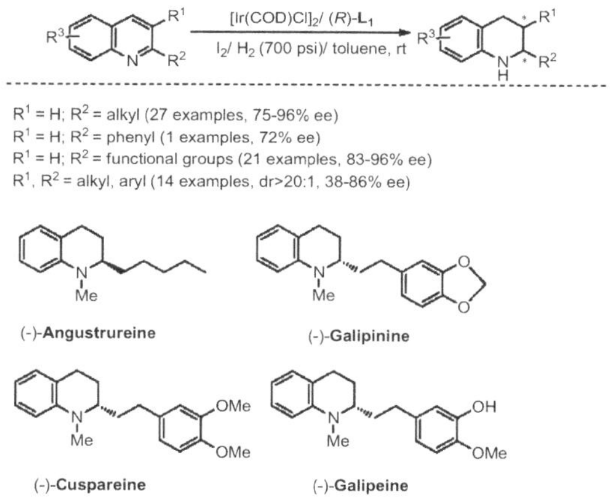

Scheme 1 Asymmetric hydrogenation of quinoline derivatives catalyzed by $[Ir(COD)Cl]_2$/MeO-BIPHEP/I_2

also exhibited good tolerance to hydroxyl and ester groups, but lower enantiomeric excess was found for hydroxymethyl-substituted quinoline. Subsequently, this catalytic system was also applied to the asymmetric synthesis of naturally occurring alkaloids, such as (−)-angustureine, (−)-galipinine, (−)-cuspareine, and (−)-galipeine (Scheme 1) [21, 27].

Later, Zhou and co-workers reported the asymmetric hydrogenation of 2-benzyl-substituted, 2-functionalized, and 2,3-disubstituted quinoline derivatives with the same Ir-catalyst system [28]. The enantioselective hydrogenation of 2-benzylquinolines and 2-functionalized quinolines showed high yields (65–97%) and excellent enantiomeric excesses (80–96%), which were not very sensitive to the electronic and/or steric properties of the substituents. More significantly, the catalyst system could tolerate various functional groups, such as esters, amides, benzenesulfonyl, and TBS (*tert*-butyldimethylsilyl) protected hydroxyl groups. In the investigation of asymmetric hydrogenation with 2,3-disubstituted quinolines as substrates, excellent dr values (>20:1) were achieved, but the *ee* values were under 90%. Similarly, the key intermediates of the gephyrotoxin alkaloid were conveniently synthesized in two steps from the chiral 1,2,3,4-tetrahydroquinoline derivatives in high yields.

Inspired by Zhou's initial work, many other chiral diphosphine ligands were successfully introduced into this transformation. In 2005, Fan, Chan, and co-workers used P-Phos with a 2,2′-bipyridine backbone [38] for the Ir-catalyzed asymmetric hydrogenation of a series of 2-substituted quinolines [29]. The iridium catalyst proved to be air-stable even after exposing the catalyst solution to air for 24 h. The hydrogenation of 2-alkyl-substituted quinolines catalyzed by in situ generated catalyst in THF or liquid poly(ethylene glycol) dimethyl ether without degassing afforded 1,2,3,4-tetrahydroquinolines in high yields (90–99%) with excellent enantiomeric excess values (90–92% *ee*). Most recently, Xu and co-workers found that the catalytic activity was additive-controlled with the same diphosphine ligand [30]. The hydrogenation of 2-methylquinoline could be carried out at much low catalyst loading (up to 4,000 h^{-1} TOF and up to 43,000 TON, TOF = turnover frequency, TON = turnover number) by decreasing the amount of additive I_2.

In addition, the clear effect of the electronic property of diphosphine ligands on catalytic activity in the hydrogenation of quinoline derivatives was noticed by several research groups. It was found that the electron-deficient diphosphine ligands usually displayed enormously improved catalytic reactivity. Xu and co-workers reported that the commercially available and electronically deficient DifluorPhos showed excellent activity (TON up to 43,000) in the Ir-catalyzed hydrogenation of 2-alkyl-substituted quinolines [31] with high enantioselectivities (up to 96% *ee*). The same rule of electronic effect of diphosphine ligand was also demonstrated by Zhou and co-workers [32, 33]. Electron-withdrawing groups (OTf (trifluoromethanesulfonyl) and CF_3O) were introduced into the 6- and 6′-positions of BIPHEP backbone, and the iridium catalysts exhibited high *ee*s (up to 95% and 92%, respectively) and high turnover numbers (up to 14,600 and 25,000, respectively).

In addition to the in situ generated iridium catalysts, Genet, Mashima, Ratovelomanana-Vidal and co-workers synthesized a series of cationic triply halogen-bridged dinuclear Ir (III)-complexes of diphosphines [39, 40], including BINAP, SynPhos, and DifluorPhos (Scheme 2). These iridium complexes were further demonstrated to be highly effective in the hydrogenation of 2-substituted quinoline derivatives. With nearly quantitative conversions, the cationic triply bromo-bridged dinuclear iridium complexes of SynPhos and DifluorPhos showed good enantioselectivities (58–91% *ee*s), although the substrate scope was still limited.

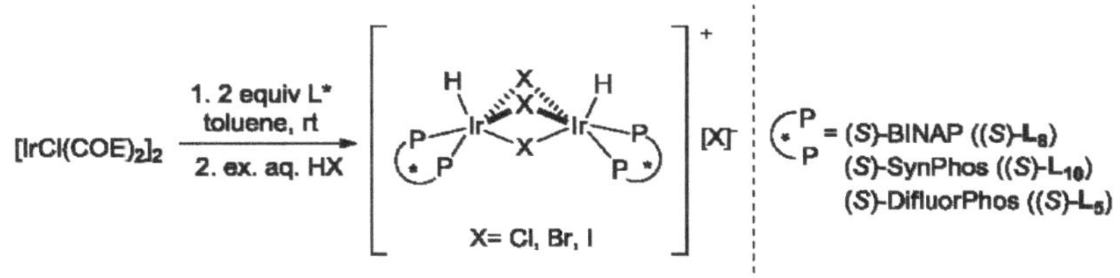

Scheme 2 Preparation of cationic triply halogen-bridged dinuclear iridium(III) complexes

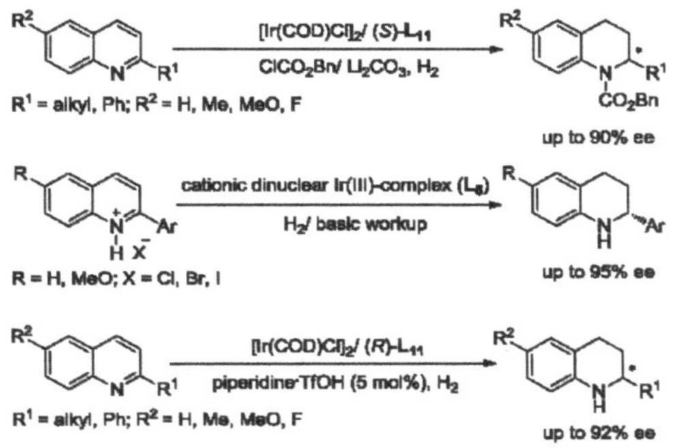

Scheme 3 Asymmetric hydrogenation of quinolines activated by chloroformates and Brønsted acids

The discovery of activation strategies other than catalyst activation by iodine was also explored (Scheme 3). The substrate activation could be achieved by using either chloroformates or Brønsted acids as activating agents, without adding iodine in catalyst systems [22, 41, 42]. In 2006, Zhou and co-workers realized asymmetric hydrogenation of 2-substituted quinolines by the addition of chloroformates [22] in the reaction mixture, opening a new avenue for the hydrogenation of heteroaromatic compounds. The benzyl carboxylate group on *N*-atom of the reduced products could be gently cleaved with H_2/Pd/C in THF.

Recently, Ohshima, Ratovelomanana-Vidal, Mashima, and co-workers developed the first highly enantioselective hydrogenation of 2-arylquinolinium salts by using their cationic dinuclear Ir(III) halide complexes as catalyst precursors [41]. High enantioselectivities (up to 95% *ee*) were achieved. These catalysts were also effective for 2-alkyl quinoline substrates, and all the *ee* values are above 90%.

In 2010, Zhou's group realized Ir-catalyzed hydrogenation of 2-substituted quinolines with catalytic amount of various Brønsted acids as the activator [42]. Full conversions and very good enantioselectivities (84–92% *ee*) were obtained for 2-alkyl-substituted quinolines with catalytic amounts of piperidine · OTf, but only 78% *ee* for 2-phenlyquinoline substrate.

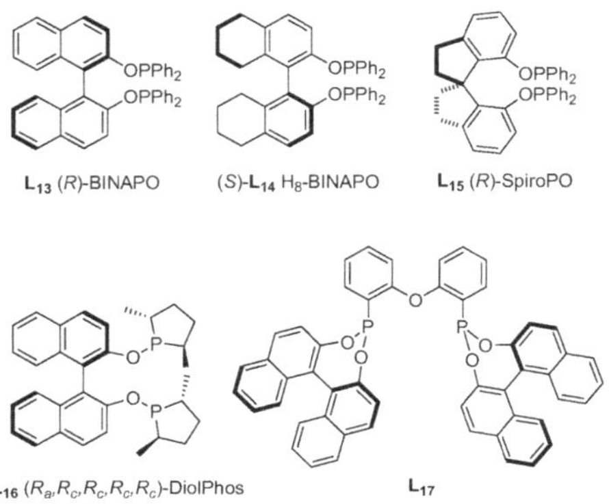

Fig. 2 Chiral diphosphinite and diphosphonite ligands for asymmetric hydrogenation of quinolines

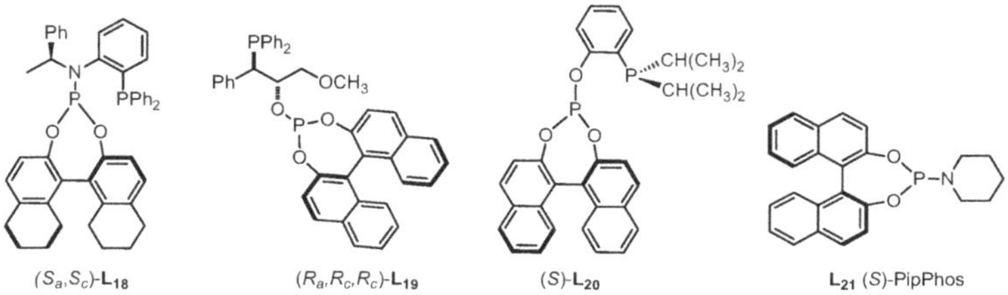

Fig. 3 Chiral phosphine-phosphoramidite, phosphine-phosphite, and monodentate phosphoramidite ligands for asymmetric hydrogenation of quinolines

2.2 *Other Chiral Phosphorus-Containing Ligands*

In addition to chiral diphosphine ligands, other chiral phosphorus-containing ligands (For selected reviews on other phosphorus-containing ligands, see: [43–46]), including diphosphinite, diphosphonite, phosphine-phosphoramidite, phosphine-phosphite, monodentate phosphoramidite and *N,P*-ligands (Figs. 2, 3, and 4) have been also successfully applied to the Ir-catalyzed asymmetric hydrogenation of quinolines [47–58].

In 2005, Fan, Chan, and co-workers first used the electron-deficient chiral diphosphinite H8-BINAPO, which was easily derived from the corresponding diols, for the Ir-catalyzed asymmetric hydrogenation of quinolines [47]. Complete conversions and excellent enantioselectivities (up to 97% *ee*) were obtained. Their further study showed that the chiral diphosphinite ligand derived from the privileged (*R*)-1,1′-spirobiindane-7,7′-diol [59] was highly effective in the Ir-catalyzed asymmetric hydrogenation of quinolines with high substrate/catalyst ratio (up to 5,000) and high enantioselectivity (up to 94% *ee*) [48].

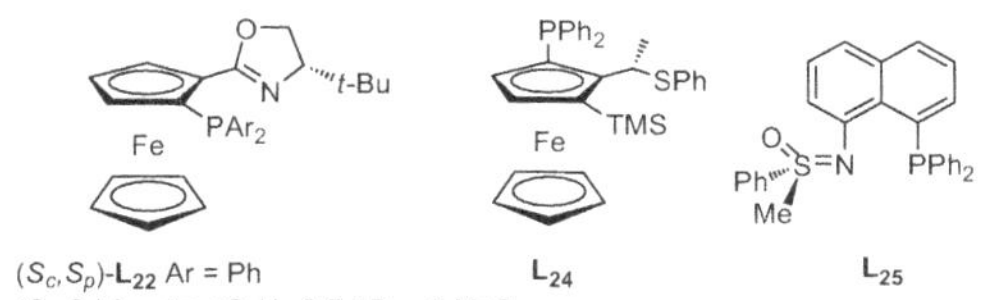

Fig. 4 Chiral *P,N*- and *P,S*-ligands for asymmetric hydrogenation of quinolines

BINOL-derived (BINOL = 1,1′-bi-2-naphthol) chiral diphosphonite ligand bearing an achiral diphenyl ether backbone was also demonstrated to be effective in the iridium-catalyzed hydrogenation of quinolines by Reetz and co-worker [49]. Interestingly, improvement of enantioselectivity was observed by the addition of achiral monodentate phosphine ligand. With this chiral diphosphonite-achiral monophosphine catalyst system, several 2-alkyl-substituted quinolines were efficiently hydrogenated with high *ee* values (up to 96% *ee*).

Most recently, three groups published their research on iridium-catalyzed asymmetric hydrogenations by using electronically dissymmetric *P,P*-ligands independently. A set of highly modular *P-OP* ligands, including various phosphine-phosphoramidites and phosphine-phosphites, were synthesized and evaluated in the iridium-catalyzed asymmetric hydrogenation of 2-substituted quinolines (Fig. 3) [50–52]. The iridium complex with ligand (S_aS_c)-**L$_{18}$** exhibited excellent enantioselectivity (up to 97% *ee*) [52]. However, much lower *ee* (81% *ee*) for 2-methylquinoline was obtained with the mismatched diastereomer.

Besides the bidentate chiral *P,P*-ligands, chiral monodentate phosphines, phosphonites, phosphites and phosphoramidites have recently been reported to present excellent performance in the asymmetric hydrogenation of functionalized olefins [45]. In 2008, the readily accessible and air-stable BINOL-derived phosphoramidite PipPhos proved to be effective in the iridium-catalyzed asymmetric hydrogenation of quinolines by Feringa and co-workers [53]. With addition of both piperidine hydrochloride and achiral monodentate phosphine ligand as additives, great enhancement in enantioselectivity was observed. Under the optimized hydrogenation conditions, a series of 2-substituted quinolines were successfully converted to 1,2,3,4-tetrahydroquinoline derivatives in high conversions with very good enantioselectivities (76–89% *ee*s).

In addition to the bidentate *P,P*-ligands, *P,N*- and *P,S*-bidentate ligands [54, 55] also exhibited good catalytic performance in the enantioselective hydrogenation of quinolines (Fig. 4). Zhou and co-workers first employed the ferrocenyloxazoline-derived *P,N*-ligand to the Ir-catalyzed asymmetric hydrogenation of quinolines [54]. With iodine as the additive, very good enantioselectivities (up to 92% *ee*) were achieved for 2-alkyl-substituted quinolines. In the case of 2-phenylquinoline, however, very poor results (45% yield with 3% *ee*) were obtained. Subsequently, they found that introducing bulky groups on the coordination phosphorus atoms could effectively prevent the formation of inactive dimer species, and thus improved the activity of the iridium catalysts [55]. Up to 93% *ee* was obtained in

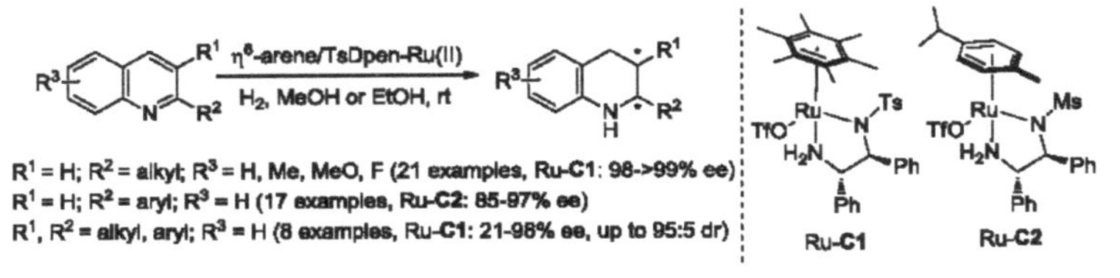

Scheme 4 Asymmetric hydrogenation of quinoline derivatives with Ru-diamine catalysts

the hydrogenation of 2-alkyl-substituted quinolines with the substrate/catalyst ratio up to 25,000.

In 2008, Bolm and co-worker designed and synthesized other types of naphthalene-bridged *P*,*N*-type sulfoximine ligands [57]. Their iridium complexes were found to exhibit good catalytic performance in the hydrogenation of 2-alkyl-substituted quinolines. In contrast to Zhou's catalytic system, the addition of iodine gave negative results in enantioselectivity.

2.3 Chiral Diamine Ligands

In comparison with the chiral phosphorus ligands, chiral bidentate diamine ligands are more readily available, easily tunable, and air-stable [60]. Their transition metal complexes of Ru, Rh, and Ir have been extensively studied in the transfer hydrogenation of aromatic ketones and imines [61, 62].

In 2006, the chiral η^6-arene/Ts-DPEN-Ru(II) complex, which was known as an excellent catalyst only for transfer hydrogenation, proved to be an effective catalyst for asymmetric hydrogenation of simple ketones by Noyori and co-workers [63, 64]. It was found that the hydrogenation with such ruthenium catalyst could be realized simply by switching the conditions from basic to acidic. Inspired by this seminal work, Fan and co-workers first realized the highly effective asymmetric hydrogenation of basic quinoline derivatives with such a cationic ruthenium catalyst (Scheme 4) [24, 65]. Unlike hydrogenation of ketones, which occurred only in methanol or ethanol, quinolines could be hydrogenated smoothly in most common organic solvents, water, ionic liquids, and even under solvent-free/highly concentrated conditions. A comprehensive study revealed that a different catalytic mechanism was involved in the hydrogenation of quinolines as compared to that of ketones, which will be discussed later in Sect. 11.

With this catalytic system, no additive was necessary for achieving high reactivity and/or enantioselectivity. The substrate activation was achieved by the in situ generated H^+ via a ruthenium assisted hetero-split of dihydrogen. Under the optimized reaction conditions, a wide range of quinoline derivatives, including 2-alkylquinolines, 2-arylquinolines, 2-functionalized, and 2,3-disubstituted quinoline derivatives were efficiently hydrogenated to give 1,2,3,4-tetrahydroquinolines with high yields and excellent enantioselectivities (up to >99% *ee* and 5,000 TON). This

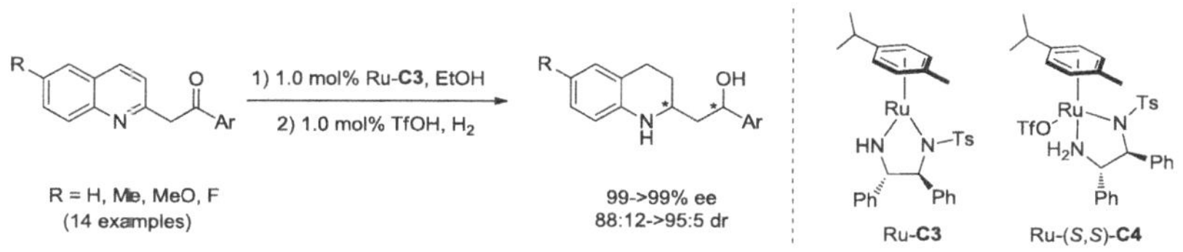

Scheme 5 Asymmetric tandem reduction of 2-(aroylmethyl)quinolines with Ru-diamine catalysts

catalytic protocol was also successfully applied to the gram-scale synthesis of some biologically active tetrahydroquinoline alkaloids, such as (−)-angustureine and 6-fluoro-2-methyl-1,2,3,4-tetrahydroquinoline, a key intermediate for the preparation of antibacterial agent (*S*)-flumequine.

Due to the potential poisoning of catalysts by the substrate and/or reduced product, asymmetric hydrogenation of heteroaromatic compounds under solvent-free conditions is a challenging task [66]. Recently, Fan and co-workers reported the first example of highly enantioselective hydrogenation of quinolines under solvent-free or highly concentrated conditions with the cationic Ru-diamine catalysts [67]. Under optimized conditions, a variety of 2-alkyl-substituted quinolines were hydrogenated at low catalyst loading (low to 0.02 mol%), affording 1,2,3,4-tetrahydroquinolines in high yields with excellent enantioselectivities (up to 97% *ee*). Further application of this solvent-free catalytic system to the gram-scale synthesis of the biologically active alkaloid (−)-angustureine gave 96% overall yield and 94% *ee*.

Subsequently, the same research group disclosed an asymmetric tandem reduction of 2-(aroylmethyl)quinolines (Scheme 5) based on different chemical selectivity of catalyst **Ru-C3** and **Ru-C4** for the reduction of ketone and quinoline motifs [68]. After the C = O bond was reduced under transfer hydrogenation conditions in the presence of 1.0 mol% **Ru-C3**, the asymmetric hydrogenation of quinoline was carried out under 50 atm hydrogen simply by adding 1.0 mol% TfOH (trifluoromethanesulfonic acid), which in situ generated the active hydrogenation catalyst **Ru-C4**. With such catalytic protocol, various 2-(aroylmethyl)quinolines were reduced to give the products bearing two chiral centers with up to 99% *ee* and 95:5 dr.

In 2008, Fan and co-workers extended the application of chiral diamine ligands to Ir-catalyzed asymmetric hydrogenation of quinolines [69]. A series of 2-alkyl-substituted quinolines were hydrogenated in high yields with excellent enantioselectivities (up to 99% *ee*) at a catalyst loading of 0.2 mol% (Scheme 6). More importantly, the reaction was obviously promoted by addition of a catalytic amount of TFA, and could be carried out in undegassed solvent without inert gas protection.

Most recently, Rueping and co-workers reported a Brønsted acid differentiated metal catalyzed hydrogenation of quinolines [70] by kinetic discrimination (Scheme 7). Moderate to good enantioselectivities (up to 82% *ee*) were obtained with the combination of an achiral Ir-diamine complex and a chiral *N*-triflylphosphoramide. In addition, a matched catalyst combination, including

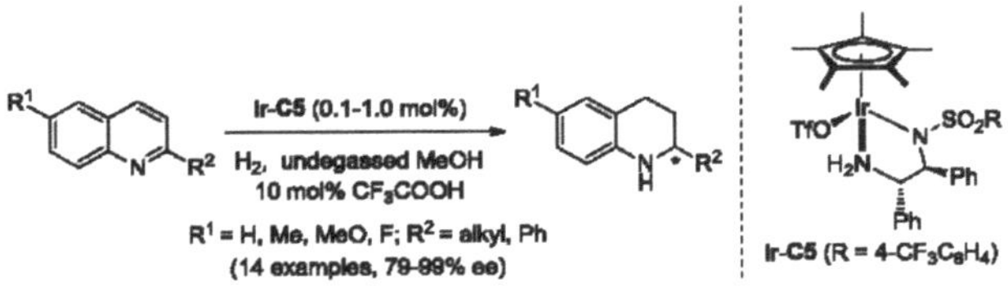

Scheme 6 Asymmetric hydrogenation of quinoline derivatives with Ir-diamine catalyst in undegassed methanol

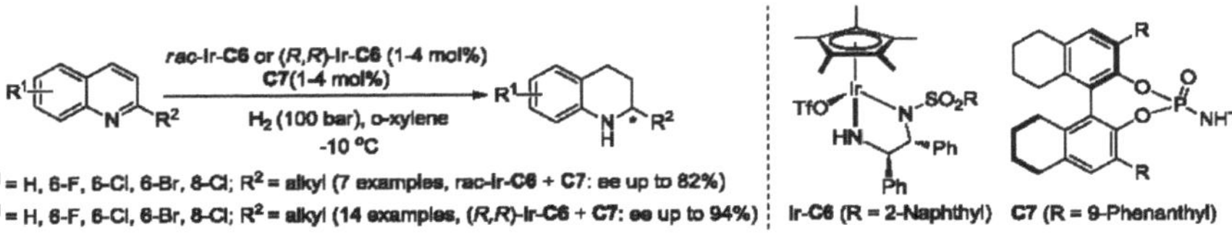

Scheme 7 Brønsted acid differentiated metal catalyzed hydrogenation of quinolines by kinetic discrimination

both chiral iridium-amido complex and chiral Brønsted acid, gave higher enantioselectivities (84–94% *ee*s).

3 Asymmetric Hydrogenation of Isoquinolines

In comparison with quinolines, isoquinolines are more challenging substrates for asymmetric hydrogenation. To date, only two papers on such transformation have been published by Zhou and co-workers [22, 71], which relied on substrate or catalyst activation strategies.

In 2006, Zhou and co-workers reported the partial hydrogenation of 1-substituted isoquinolines catalyzed by Ir/(*S*)-SegPhos complex together with a substrate activation strategy (Scheme 8) [22]. Similar to the asymmetric hydrogenation of quinolines mentioned above, various 1-substituted isoquinoline derivatives were hydrogenated smoothly by using chloroformates as additives, giving the reduced 1,2-dihydroisoquinolines in 46–87% yields and 10–83% *ee*s. Upon further reduction of the products with H_2/Pd/C and $LiAlH_4$, 1,2,3,4-tetrahydroisoquinolines could be achieved, which was exemplified by the synthesis of naturally occurring alkaloid, (*S*)-(−)-carnegine.

The other example was reported recently on the hydrogenation of 3,4-disubstituted isoquinolines by catalyst activation strategy, in which a dynamic kinetic resolution was involved (Scheme 9) [71]. BCDMH (1-bromo-3-chloro-5,5-dimethyl-hydantoin) was used instead of iodine to active the in situ generated catalysts by mixing iridium precursor $[Ir(COD)Cl]_2$ and SynPhos. Lowering the hydrogenation pressure and elevating the reaction temperature led to a further increase in enantioselectivity. A set of 3,4-disubstituted isoquinolines were

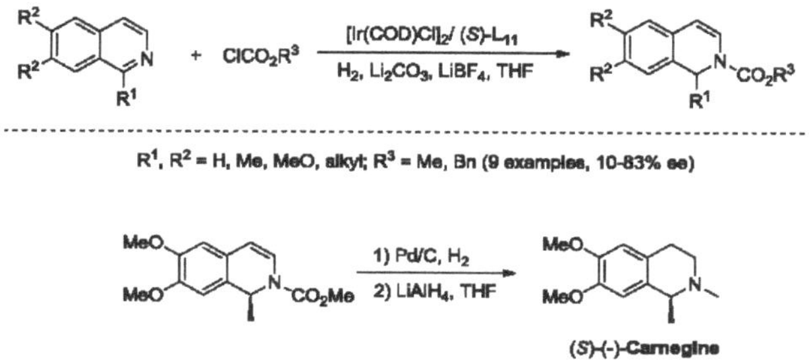

Scheme 8 Asymmetric hydrogenation of activated isoquinolines and its application in the synthesis of naturally occurring alkaloid, (*S*)-(−)-carnegine

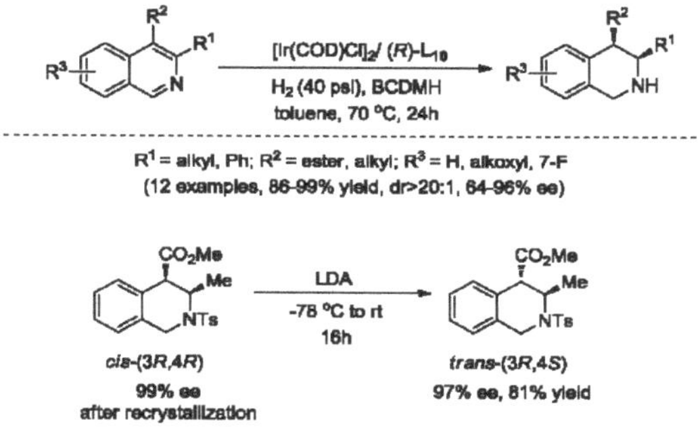

Scheme 9 Iridium-catalyzed enantioselective hydrogenation of 3,4-disubstituted isoquinolines and *cis-trans* transformation using LDA

hydrogenated effectively, giving *cis*-products with 86–99% yields and 64–96% *ee*s. Although it was demonstrated that the ester group at the C-4 position is not necessary for hydrogenation, lower *ee* values were observed for the 4-alkyl-substituted substrates. Subsequent treatment of *cis*-disubstituted products with LDA (lithium diisopropylamide) gave the corresponding *trans*-epimers, which is generally difficult to obtain through direct asymmetric hydrogenation. A plausible hydrogenation mechanism was also proposed, including 1,2-hydrid addition to the C = N bond, acid-catalyzed enamine-imine tautomerization, and the hydrogenation of imine.

4 Asymmetric Hydrogenation of Quinoxalines

The asymmetric hydrogenation of quinoxalines is a potentially cost-efficient and atom-economic method for the preparation of optically pure 1,2,3,4-tetrahydroquinoxaline derivatives, which are of great biological interest. Since the first example of asymmetric hydrogenation catalyzed by rhodium catalyst in 1987 [72], a variety of rhodium, iridium, and ruthenium complexes bearing chiral

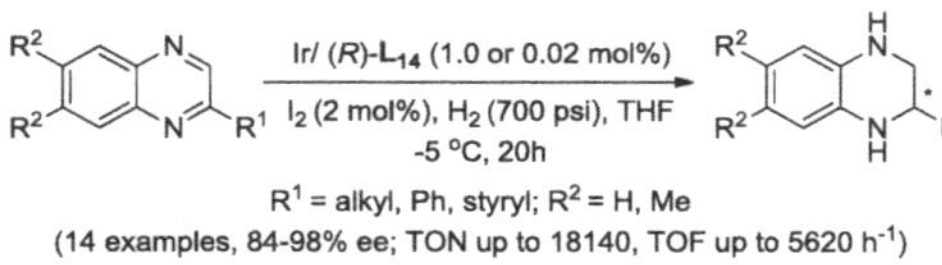

Scheme 10 Asymmetric hydrogenation of substituted quinoxalines catalyzed by Ir-(*R*)-H_8-BINAPO

phosphorus ligands have been applied to such transformation [34, 42, 50, 72–76]. Early reported catalytic systems suffered from low enantioselectivities and/or limited substrate scope [72–76]. On the basis of Zhou's Ir/diphosphine/I_2 catalytic system [21], significant progress has recently been achieved in this reaction by several research groups. In addition, some phosphine-free ligands, including chiral diamines and *N*-heterocyclic carbenes, were found to be highly efficient in the Ru-catalyzed asymmetric hydrogenation of quinoxalines. Notably, the hydrogenation by metal/Brønsted acid relay catalysis has also shown its potential in this research area.

4.1 Chiral Phosphorus-Containing Ligands

Similar to the asymmetric hydrogenation of quinolines, the iridium catalysts bearing diphosphines, diphosphinite, monodentate phosphoramidites, and phosphine-phosphite (*P-OP*) ligands were used for the asymmetric hydrogenation of quinolines [77–81]. In 2009, Fan, Chan, and co-workers reported that chiral diphosphinites derived from H8-BINOL and 1,1′-spiro-biindane-7,7′-diol showed higher enantioselectivities in the iridium catalyzed asymmetric hydrogenation of 2-methylquinoxiline compared to diphosphine-Ir catalysts under the identical reaction conditions (Scheme 10) [77]. It is noticeable that even with 0.005 mol% of Ir catalyst (substrate/catalyst ratio (S/C) = 20,000), the reaction proceeded smoothly in a slightly lowered conversion and with the same enantioselectivity. Under optimal reaction conditions, the hydrogenation of a series of 2-substituted quinoxalines was successfully realized with quantitative conversions and high *ee* values varying from 84% to 96%. It is interesting to note that the enantioselectivities were a little bit higher in some cases as the S/C elevated to 5,000/1, and the enantiomeric excess reached 98%.

At the same time, Feringa and co-worker described the enantioselective hydrogenation of 2-substituted quinoxalines with an Ir-catalyst prepared in situ from $[Ir(COD)Cl]_2$ and monodentate phosphoramidite ligand PipPhos [78]. With 10 mol% piperidine hydrochloride as additive, 71–92% yields and 75–96% *ee*s were obtained at an Ir/ligand mol ratio of 1:2.

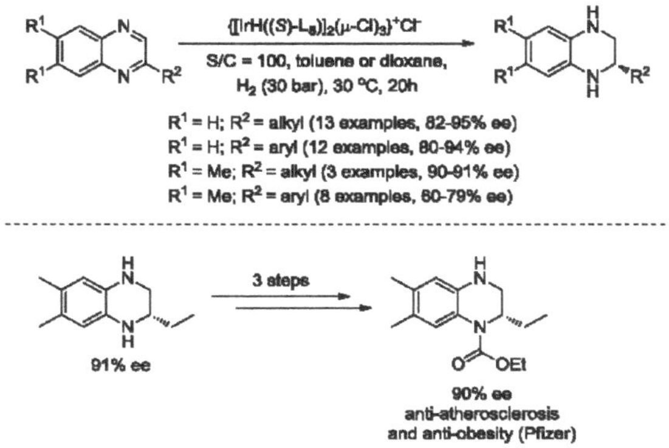

Scheme 11 Asymmetric hydrogenation of quinoxalines catalyzed by cationic dinuclear iridium (III) complex and its application in the synthesis of Pfizer's CETP inhibitor

Oshima, Mashima, and Ratovelomanana-Vidal et al. employed the cationic triply halogen-bridged dinuclear Ir(III)-complexes, mentioned above in quinoline reduction, for the asymmetric hydrogenation of 2-substituted quinoxalines (Scheme 11) [79, 80]. Without iodine as additive, DifluorPhos proved to be optimal for the hydrogenation of a series of 2-alkyl- and 2-aryl-substituted quinoxalines, furnishing 1,2,3,4-tetrahydroquinoxalines with nearly quantitative yields and high enantioselectivities (up to 95% *ee*). Moreover, to illustrate the applicability of this catalyst system, an inhibitor of cholesteryl ester transfer protein (CETP), developed by Pfizer [82], was synthesized with the asymmetric hydrogenation of 2-ethyl-6,7-dimethylquinoxaline as the key step.

Most recently, Mashima and co-workers investigated the additive effects of achiral amines on the asymmetric hydrogenation of 2-aryl-substituted quinoxalines catalyzed by the same chiral cationic dinuclear iridium(III) complexes [81]. In the presence of *N*-methyl-*p*-anisidine (MPA), the catalytic activity was obviously enhanced and the highest enantioselectivity was observed in the hydrogenation of 2-phenylquinoxaline, which is higher than that without amine additive. The following particular mechanistic study indicated the possible existence of dual catalytic cycles in equilibrium. The addition of amine probably shifted the balance, leading to higher activity and enantioselectivity. The detailed description of the experimental processes will be included in the mechanistic study session.

4.2 *Chiral Phosphine-Free Ligands*

After the amazing discovery of excellent performance of chiral diamines in the ruthenium catalyzed asymmetric hydrogenation of quinolines, Fan and co-workers recently reported the extended application of such Ru-diamine catalysts in the hydrogenation of 2- and 2,3-substituted quinoxalines (Scheme 12) [83]. It was found that the weakly coordinating counteranions of the half-sandwich Ru (II) complexes showed an evident influence on the enantioselectivity and/or

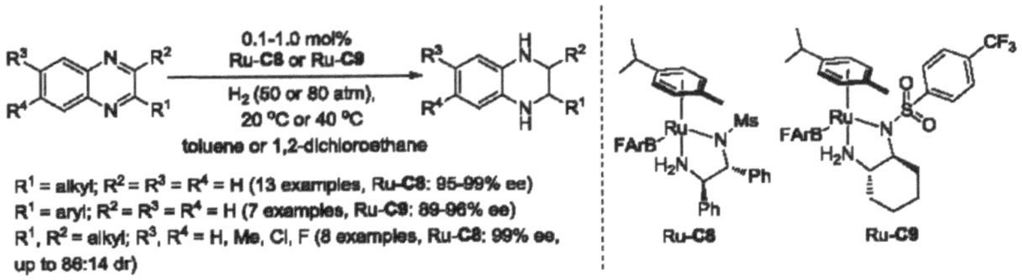

Scheme 12 Asymmetric hydrogenation of 2- and 2,3-substituted quinoxalines with chiral cationic ruthenium diamine catalysts

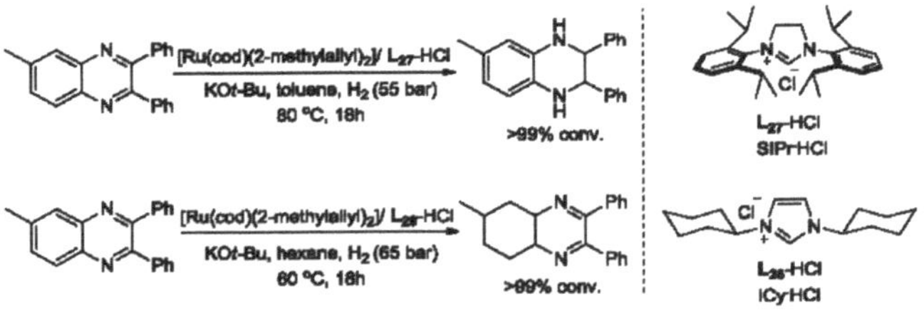

Scheme 13 Ligand-controlled regioselective hydrogenation of 6-methyl-2,3-diphenylquinoxaline

diastereoselectivity of the reaction. The Ru complexes with a bulky $BArF^-$ [tetrakis (3,5-bistrifluoromethylphenyl)borate] counteranion were selected as optimal catalyst, and were applied in the asymmetric hydrogenation of 2-alkyl-substituted, 2-aryl-substituted, and 2,3-dialkyl-substituted quinoxaline derivatives with high yields, high *ee*s, and moderate dr values. Excellent enantioselectivities ranging from 95% to 99% *ee* were achieved for 2-alkyl-substituted quinoxalines, which are the best results reported to date for this type of heteroaromatic compound.

In 2011, Glorius and co-workers reported a ligand-controlled highly regioselective and asymmetric hydrogenation of quinoxaline derivatives for the first time by using monodentate *N*-heterocyclic carbenes (NHC) as ligand [84]. With different achiral NHC ligands, the hydrogenation of 2,3-diphenyl-6-methylquinoxaline catalyzed by the in situ formed ruthenium catalysts proceeded smoothly, affording partially reduced products, 1,2,3,4-tetrahedroquinoxaline and 5,6,7,8-tetrahydroquinoxaline, with full conversions but different regioselectivities, respectively (Scheme 13). Based on this finding, the Ru-catalyzed asymmetric hydrogenation of 2,3-diphenyl-substituted quinoxalines was realized by using chiral NHC ligands (Scheme 14). Under optimized reaction conditions, only the aromatic carbocyclic ring was selectively hydrogenated with moderate to very good enantioselectivity (up to 88% *ee*), which is totally different from the reported regioselectivity of asymmetric hydrogenation of bicyclic benzo-heteroaromatics with other types of catalyst.

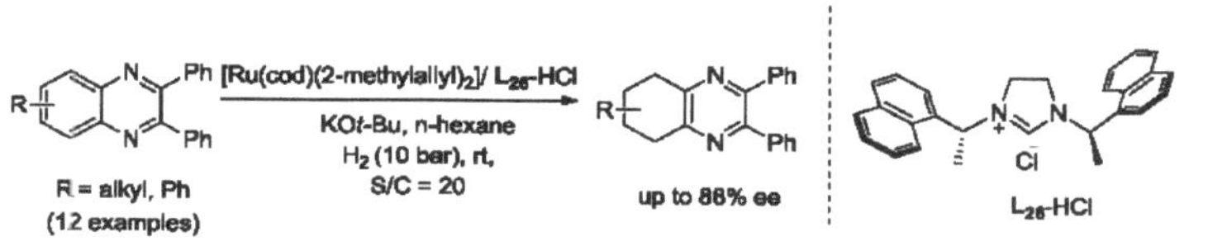

Scheme 14 Asymmetric hydrogenation of 2,3-diphenyl-substituted quinoxalines with chiral ruthenium-NHC catalysts

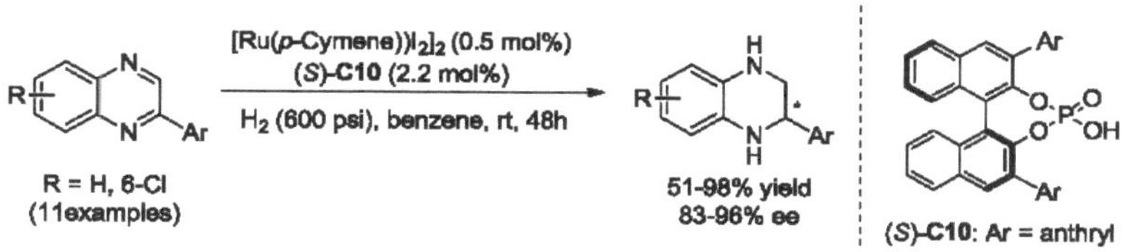

Scheme 15 Asymmetric reduction of quinoxalines through convergent disproportionation of dihydroquinoxalines using metal/Brønsted acid relay catalysis system

4.3 *Metal/Brønsted Acid Catalytic System*

In 2011, Zhou and co-workers reported a relay catalysis to prepare chiral tetrahydroquinoxalines by using a combination of achiral ruthenium complex ([Ru(*p*-cymene)$I_2]_2$) and sterically demanding chiral Brønsted acid **(*S*)-C10** (Scheme 15) [85]. After the partial hydrogenation of 2-aryl quinoxalines, the dihydroquinoxaline intermediates were enantioselectively reduced through convergent asymmetric disproportionation, in which a self-transfer hydrogenation process was involved. With this metal/Brønsted acid relay catalyst system, a variety of quinoxalines were hydrogenated to 1,2,3,4-tetrahydroquinoxalines in high yields with good to excellent enantioselectivities (up to 96% *ee*). The proposed mechanism will be discussed later in Sect. 11.

5 Asymmetric Hydrogenation of Pyridines

As a wide-existent structural motif in natural alkaloids and many biologically active compounds, the enantioselective synthesis of chiral piperidine derivatives is of great interest [86]. Although asymmetric hydrogenation of pyridine derivatives is the most straightforward and convenient route to such chiral *N*-containing heterocyclic compounds, few successful examples were reported. Early study on this reaction focused on heterogeneous catalytic systems, which suffered from low enantioselectivities and/or reactivities, and narrow substrate scope [15]. In 2000, Studer and co-workers reported the first homogeneous transition metal-catalyzed asymmetric hydrogenation of the activated 2- and 3-pyridine carboxylic acid and their corresponding ethyl esters [87]. However, only 17–27% *ee*s were observed for

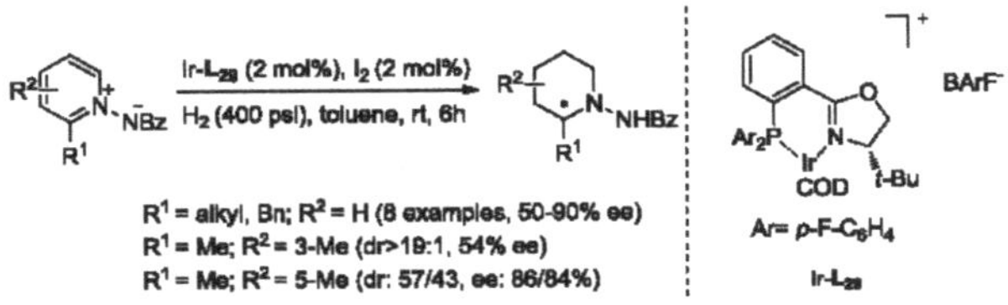

Scheme 16 Asymmetric hydrogenation of N-benzoyliminopyridinium ylides catalyzed by Ir-PHOX complex

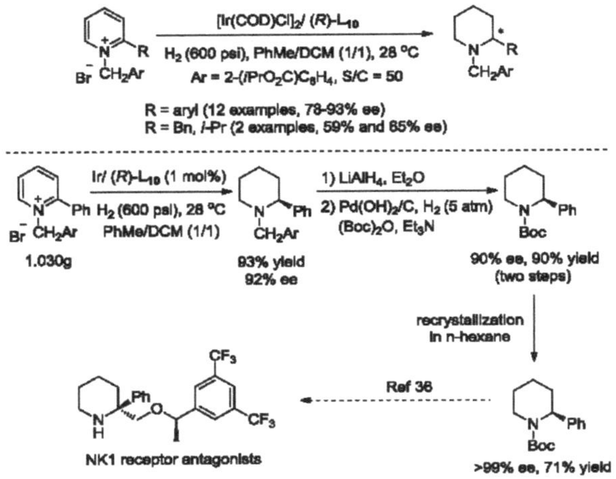

Scheme 17 Iridium-catalyzed asymmetric hydrogenation of 2-substituted pyridinium salts and its application for the formal synthesis of NK1 receptor antagonist

only four substrates tested. Recently, a breakthrough was made in the asymmetric hydrogenation of monocyclic pyridine derivatives via substrate activation strategy by Charette's and Zhou's groups, respectively.

In 2005, Charette and co-workers investigated the asymmetric hydrogenation of a special kind of activated pyridine, 2-alkyl-substituted N-iminopyridinium ylides (Scheme 16) [88]. A series of chiral P,N-ligands were screened in the Ir/I_2 catalyzed hydrogenations. Under the optimized reaction conditions, a set of 2-alkyl-substituted N-benzoyliminopyridinium ylides were successfully hydrogenated to afford chiral piperidines with modest to high yields and enantioselectivities (50–90% *ee*s). In some cases, partial reduced byproducts were observed. The hydrogenation of 2,3- and 2,5-dimethyl N-benzoyliminopyridinium ylides also proceeded smoothly, giving products with good stereoselectivities. In addition, the role of iodine was to oxidize the initial Ir(I) complex to the Ir(III) pre-catalyst [89].

Most recently, Zhou and co-workers described the asymmetric hydrogenation of pyridinium salts [90], another type of activated pyridine, affording 2-substituted N-benzyl piperidines in 60–99% yields and 59–93% *ee*s with Ir/diphosphine catalytic system (Scheme 17). Upon screening of various diphosphine ligands, the

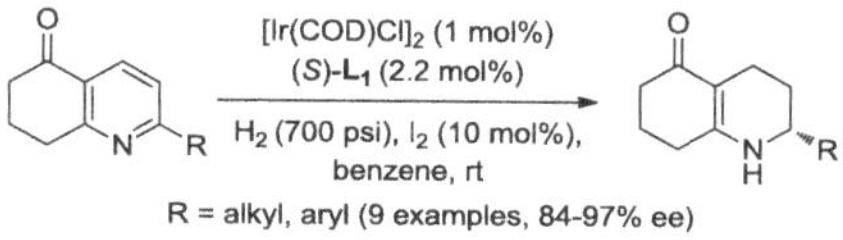

Scheme 18 Ir-catalyzed asymmetric hydrogenation of 7,8-dihydro-quinolin-5(6*H*)-ones

electron-rich (*R*)-SegPhos and (*R*)-SynPhos displayed the highest enantioselectivity. It is noticeable that introducing an electron-withdrawing ester group at the *ortho* position of the activating benzyl group could significantly improve the enantioselectivity. Interestingly, the 2-aryl-substituted pyridines offered better reactivities and enantioselectivities than 2-alkyl-substituted pyridines. To demonstrate the practicability of this methodology, they developed a gram-scale synthesis of *N*-benzyl-2-phenyl piperidine, which was further used for the formal synthesis of an orally active NK1 receptor antagonist (Scheme 17) [91, 92].

In contrast to monocyclic pyridines, the activated bicyclic trisubstituted pyridines, 7,8-dihydro-quinolin-5(6*H*)-ones, could be hydrogenated in a relatively easy way. In 2008, Zhou and co-workers found that their Ir/diphosphine/I_2 catalytic system was also effective for the asymmetric hydrogenation of such pyridine derivatives (Scheme 18) [93]. Under optimal reaction conditions, a wide range of 2-alkyl and 2-phenyl substituted pyridines were hydrogenated smoothly, providing chiral hexahydroquinolines in moderate to high yields (up to 98%) with high enantiomeric excesses of 84–97%.

Electron deficient diphosphines, such as DifluorPhos and P-Phos, based on chiral biphenyl backbone were also employed in the Ir/diphosphine/I_2 catalyzed asymmetric hydrogenation of 7,8-dihydro-quinolin-5(6*H*)-ones by Xu and co-workers [30, 31]. Excellent enantioselectivities (up to 98% *ee*) and nearly quantitative yields were achieved for both catalysts. In contrast to the hydrogenation of quinolines with electron deficient diphosphines, no enhancement of catalytic activity was noted with pyridine substrates.

6 Asymmetric Hydrogenation of Indoles and Pyrroles

Indoline and pyrroline motifs are widely existent in the structures of naturally occurring alkaloids and other biologically active molecules [94]. The asymmetric hydrogenation of indole and pyrrole derivatives provides a straightforward method to prepare chiral compounds. Different from the basic six-membered ring pyridine and quinoline, the mono-nitrogen containing five-membered pyrroles and indoles exhibit weak acidic properties and opposite electron distribution. In most cases, the indole substrates only bearing an electron-withdrawing protecting group, such as Ac, Ts, or Boc, on the nitrogen atom could be hydrogenated. Chiral phosphorus ligands, particularly chiral diphosphines, were usually used for this transformation [95–101] and some of them are listed in Fig. 5. In addition, a breakthrough in the

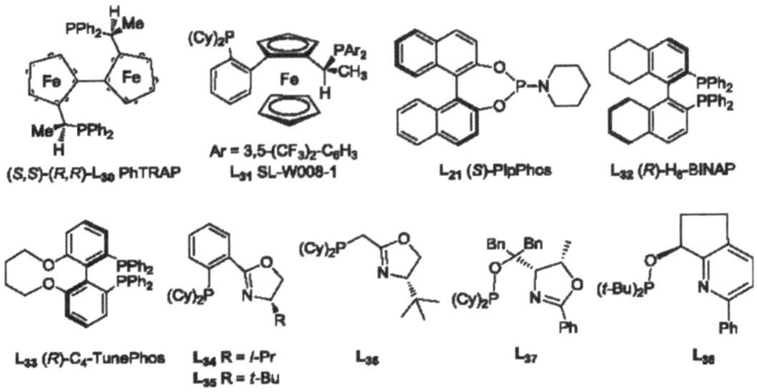

Fig. 5 Representative diphosphines and *P,N*-ligands for the asymmetric hydrogenation of indoles and pyrroles

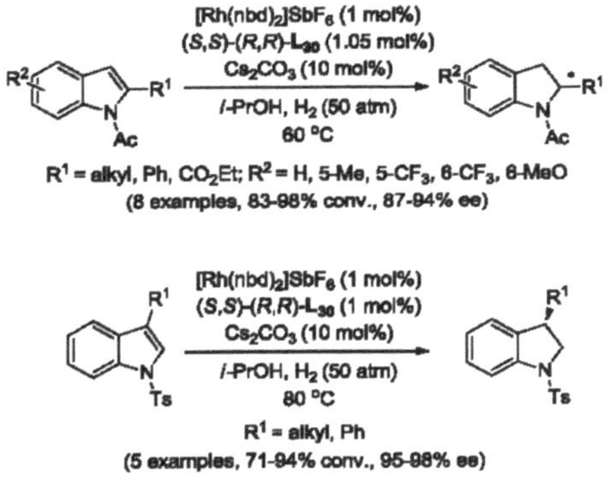

Scheme 19 Asymmetric hydrogenation of 2- and 3-substituted indoles catalyzed by $[Rh(nbd)_2]SbF_6$/PhTRAP

asymmetric hydrogenation of unprotected simple indoles was recently made by Zhou with homogeneous palladium catalytic system together with substrate activation strategy.

Early in 2000, Kuwano and co-workers reported the first efficient asymmetric hydrogenation of indoles with rhodium catalysts [95]. With a *trans*-chelating chiral diphosphine, (*S*,*S*)-(*R*,*R*)-PhTRAP, a series of 2-substituted *N*-acetyl indoles were tested, and high yields (83–98%) and excellent enantioselectivities (87–94% *ee*s) were achieved (Scheme 19). Later, they extended the substrate scope to a wide range of 2- and 3-substituted indoles with different *N*-protecting groups [96, 97]. Similarly, the combination of PhTRAP and $[Rh(nbd)_2]SbF_6$ (nbd = 2,5-norbornadiene) was demonstrated to be optimal, and excellent *ee* values were achieved for 3-substituted *N*-tosyl-protected indoles. This method was also applied to the synthesis of a Wierenga's synthetic intermediate for the left-hand segment of the antitumor agent (+)-CC-1065 [102].

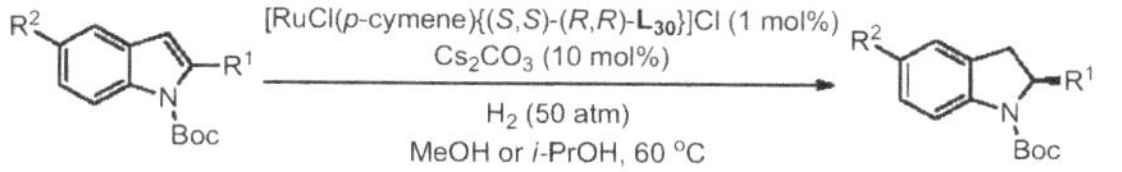

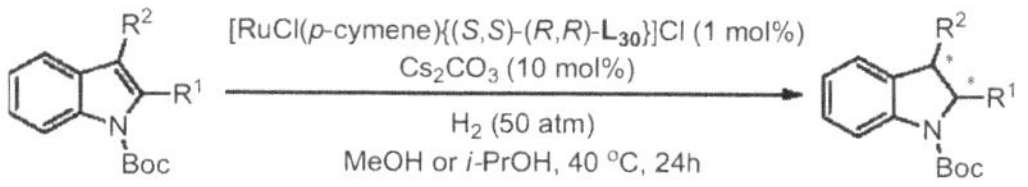

Scheme 20 Asymmetric hydrogenation of *N*-Boc-protected 2-substituted indoles catalyzed by Ru-PhTRAP complex

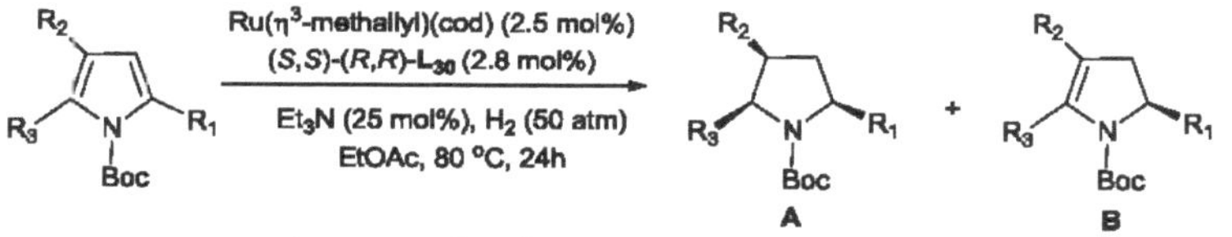

Scheme 21 Asymmetric hydrogenation of trisubstituted pyrroles catalyzed by Ru-PhTRAP complex

Subsequently, the same research group applied the same (*S*,*S*)-(*R*,*R*)-PhTRAP ligand in the ruthenium catalyzed hydrogenation of *N*-Boc-protected indoles [100], which could not gain satisfying stereo induction with rhodium catalysis (Scheme 20). With [RuCl(*p*-cymeme){(*S*,*S*)-(*R*,*R*)-PhTRAP}]Cl/$CsCO_3$ combination, a set of 2- and 3-substituted *N*-Boc-protected indoles were hydrogenated effectively in high yields (up to 99%) with high enantioselectivities (up to 95% *ee*). This catalytic system was also used for the hydrogenation of 2,3-dimethylindole, giving *cis*-2,3-dimethylindoline in moderate yield with good enantioselectivity.

On the basis of the above success, in 2008 Kuwano and co-workers developed the first highly enantioselective hydrogenation of *N*-Boc protected 2,3,5-trisubstituted pyrroles with the previously reported ruthenium catalytic system [101]. Under optimized reaction conditions, a series of pyrrole derivatives were hydrogenated smoothly with their chiral Ru-PhTRAP catalyst in the presence of a suitable base, giving chiral pyrrolidines (**A**) and/or 4,5-dihydropyrroles (**B**) with full conversions and excellent *ee* values (98–99.7%) (Scheme 21).

In addition to the diphosphine-containing metal (Rh and Ru) catalysts, most recently, Pfaltz and co-worker found the cationic iridium complexes derived from

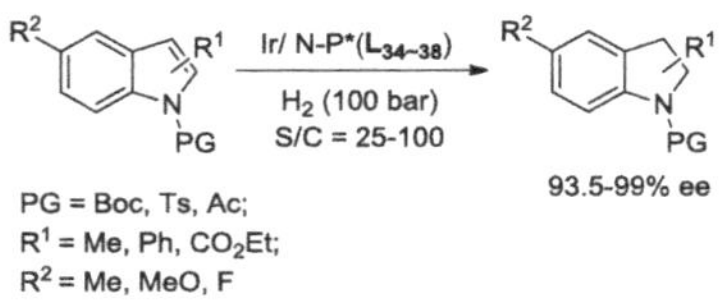

Scheme 22 Asymmetric hydrogenation of 2- and 3- substituted indoles catalyzed by iridium-*P,N* complex

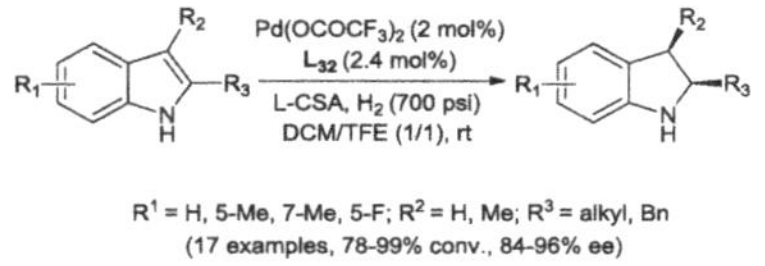

Scheme 23 Asymmetric hydrogenation of unprotected indoles using $Pd(OCOCF_3)_2$/(*R*)-H8-BINAP with a Brønsted acid as an activator

PHOX or other chiral *P,N*-ligands to be efficient catalysts for the asymmetric hydrogenation of *N*-protected indoles (Scheme 22) [103]. In contrast to the findings reported by Kuwano, adding a base additive to the reaction mixture led to negative results. With proper combination of the protecting group and the chiral iridium catalyst, hydrogenation of series of 2- and 3-substituted indoles proceeded smoothly with full conversions and good to excellent enantioselectivities (up to 99% *ee*).

Recently, chiral diphosphine-containing palladium complexes have proven to be effective catalysts for the asymmetric hydrogenation of imines, olefins, and ketones [104–110]. However, in contrast to the Rh, Ir, and Ru catalysts, the Pd catalysts are rarely utilized for the asymmetric hydrogenation of heteroaromatic compounds. In 2010, the first example of the Pd-catalyzed asymmetric hydrogenation of heteroaromatics was reported by Zhou and co-workers [111]. They found that the palladium complex bearing (*R*)-H8-BINAP was an efficient catalyst for the highly enantioselective hydrogenation of *N*-unprotected indoles by using a stoichiometric amount of strong Brønsted acid as activator (Scheme 23). Under optimized reaction conditions, a variety of 2-substituted indoles were hydrogenated in high yields (78–99%) with good to excellent enantioselectivities (84–96% *ee*s). The existence of a reversible process of protonation and deprotonation (enamine/imine isomerization) was demonstrated by isotopic labeling experiments; and the equilibrium proved to be faster than hydrogenation. For the hydrogenation of 2,3-disubstituted indoles, a dynamic kinetic resolution was involved; and *cis*-indolines were obtained with excellent enantioselectivities. In addition, a special mixed solvent system containing TFE (TFE = trifluoroethanol) was crucial for attaining high enantioselectivity and reactivity.

Subsequently, Zhou and co-workers found that 3-(α-hydroxy-alkyl)indoles, which were easily synthesized by nucleophilic additions, could undergo dehydration to form vinylogous iminium salts in the presence of Brønsted acid. Subsequent asymmetric

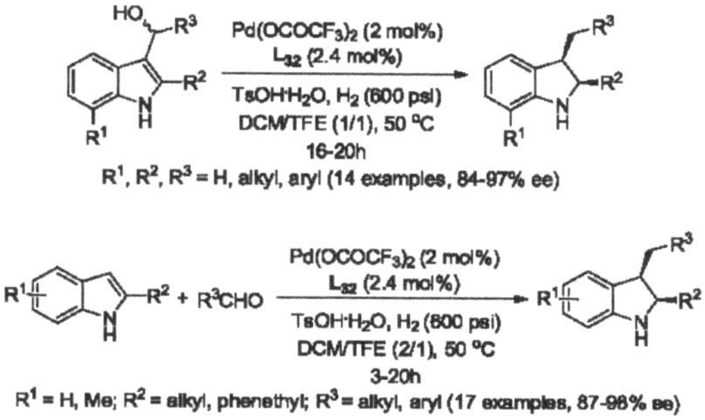

Scheme 24 Enantioselective synthesis of 2,3-disubstituted indolines *via* Pd-catalyzed asymmetric hydrogenation of 3-(α-hydroxy-alkyl)indoles and one-pot Brønsted acid/Pd-promoted tandem reactions

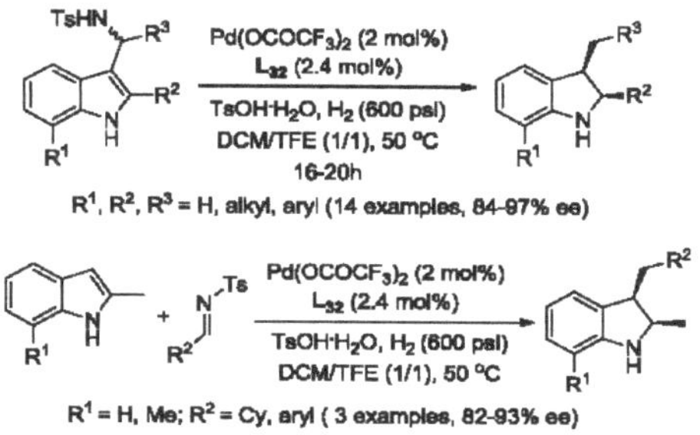

Scheme 25 Enantioselective synthesis of 2,3-disubstituted indolines *via* Pd-catalyzed asymmetric hydrogenation of 3-(toluenesulfonamidoalkyl)indoles and one-pot Brønsted acid/Pd-promoted tandem reactions

hydrogenation of the in situ generated vinylogous iminium intermediate [112] in the presence of Pd/H8-BINAP provided 2,3-disubstituted indolines with excellent enantioselectivities (up to 97% *ee*) (Scheme 24). Based on this finding, they further realized one-pot Brønsted acid/Pd-H8-BINAP promoted tandem Friedel-Crafts/dehydration/hydrogenation reactions with 2-substituted indoles and aldehydes as the starting materials (Scheme 24), giving 2,3-disubstituted indolines with excellent enantioselectivities (87–98% *ee*s) [113].

Very recently a series of 2-substituted 3-(toluenesulfonamidoalkyl)indoles were also synthesized by the same group and applied to the asymmetric hydrogenation [114]. Similar Brønsted acid/Pd-H8-BINAP promoted $TsNH_2$-elimination/hydrogenation processes provided 2,3-disubstituted indolines with comparable *ee* values to those obtained from 3-(α-hydroxy-alkyl)indoles as substrates. In addition, one-pot tandem Friedel-Crafts/hydrogenation reactions with *N*-tosyl imines and 2-substituted indoles as starting materials were also realized through relay catalysis by Brønsted acid/H8-BINAP/Pd (Scheme 25), giving the reduced products in high yields with excellent enantioselectivities.

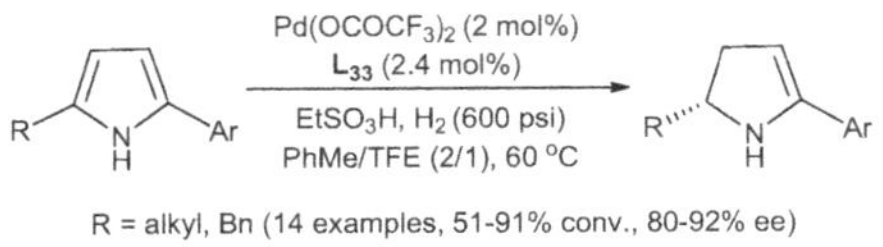

Scheme 26 Asymmetric hydrogenation of 2,5-disubstituted pyrroles catalyzed by Pd-(*R*)-C_4-TunePhos complex

In 2011, Zhou and co-workers used a palladium/diphosphine catalytic system for the asymmetric hydrogenation of simple 2-alkyl-5-arylpyrroles [115]. With a Brønsted acid as activator and Pd/C_4-TunePhos as catalyst, a series of partially reduced chiral 2,5-disubstituted 1-pyrrolines were obtained in moderate to high yields with very good enantioselectivities (Scheme 26). It was found that the length and steric property of alkyl side chain and the electronic and steric properties of aryl substituent showed a clear influence on the enantioselectivity. Substrates with electron-withdrawing and/or steric hindered aryl substituents displayed better results.

7 Asymmetric Hydrogenation of Furans and Benzofurans

Compared to the enormous effort put into the research of asymmetric hydrogenation of *N*-containing heteroaromatic compounds, other hetero atom-containing heteroaromatics, such as furans, thiophenes, benzofurans, and benzothiophenes, are much less explored; and their highly enantioselective hydrogenations are still a big challenge in the field of asymmetric hydrogenation.

7.1 Chiral Phosphorus-Containing Ligands

Furan is among the limited compounds which were investigated in the pioneering stage of the research area of asymmetric hydrogenation of heteroaromatics (For examples of heterogeneous asymmetric hydrogenation of furans, see: [116–118]). In 1995, Takaya and co-workers published their initial results on the first homogeneous enantioselective hydrogenation of 2-methylfuran with Ru-BINAP catalyst, but only 50% *ee* of enantioselectivity was obtained under relatively harsh reaction conditions [118]. Five years later, Studer and co-workers reported the first rhodium-catalyzed asymmetric hydrogenation of 2-functionalized furans [87], including 2-furan carboxylic acid and 2-hydroxylmethylfuran. After screening of a series of diphosphine ligands, only less than 30% *ee* values were obtained, whereas the catalytic activity was pretty high. Recently, Albert and co-workers reported asymmetric hydrogenation of 2,5-disubstituted furans with cationic rhodium/

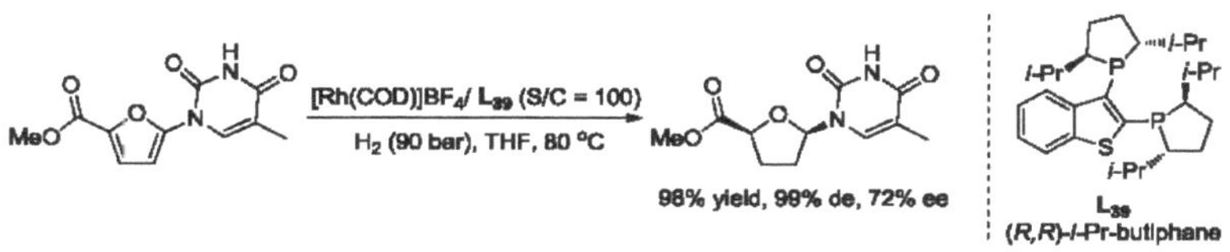

Scheme 27 Rhodium-catalyzed asymmetric hydrogenation of 2,5-disubstituted furan

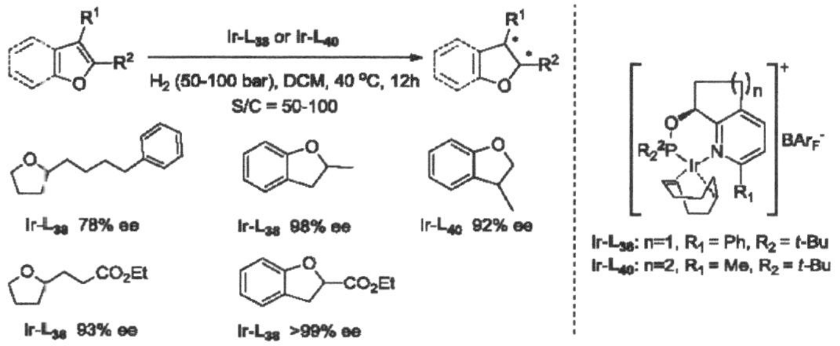

Scheme 28 Asymmetric hydrogenation of furan and benzofuran derivatives catalyzed by Ir-*P,N* complexes

diphosphine complex [119], giving 2′,3′-dideoxynucleoside analogue with high diastereoselectivity and moderate enantioselectivity (Scheme 27).

A breakthrough in the asymmetric hydrogenation of simple furan and benzofuran derivatives was reported by Pfaltz and co-workers in 2006 (Scheme 28) [120]. They found that iridium complexes containing pyridine-phosphinite ligand, which have proven to be excellent catalysts for the hydrogenation of unfunctionalized olefins [121], were effective for the asymmetric hydrogenation of simple furans and benzofurans. With the cationic iridium catalysts bearing a bulky electron-rich $(t\text{-Bu})_2P$ group on the *P,N*-ligand, several 2-alkyl-substituted furans and 2- or 3-substituted benzofurans were successfully hydrogenated, providing tetrahydrofurans and benzodihydrofurans with high *ee* values (78% to >99% *ee*).

7.2 *Chiral* N-*Heterocyclic Carbene Ligands*

After the successful application and unique performance in the asymmetric hydrogenation of quinoxaline derivatives, the chiral *N*-heterocyclic carbene ligands were employed in the ruthenium-catalyzed enantioselective hydrogenation of benzofurans in 2012 by Glorius and co-workers (Scheme 29) [122]. The regioselective reduction of the heterocyclic ring was observed, offering 2,3-dihydrobenzofurans as the only products. Under the optimized reaction conditions, a series of 2-aryl and 2-alkyl substituted benzofuran derivatives were smoothly hydrogenated in the

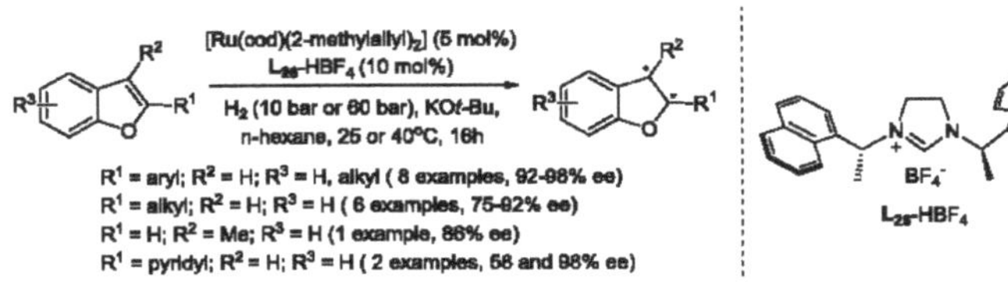

Scheme 29 Asymmetric hydrogenation of 2,3-disubstituted and 2,3,6-trisubstituted benzofuran catalyzed by Ru-NHC complex

presence of 10 mol% ruthenium catalyst under 10 or 60 bar of H_2 and mild reaction temperature. In most cases, full conversions and excellent enantioselectivities (up to 98% *ee*) were achieved. It was found that the 2-aryl substituted benzofurans bearing electron-withdrawing substituents exhibited higher reactivity than those containing electron-donating groups. In addition, higher reaction temperature and hydrogen pressure were needed for sterically demanding substrates of 2-(*o*-tolyl)-benzofuran and 2-(*tert*-butyl)-benzofuran, and lower conversions (73% and 38%) and/or enantioselectivities (92% and 75% *ee*) were observed, respectively. Further kinetic study revealed that full conversion of 2-methyl benzofuran was achieved at a catalyst loading of 0.5 mol%, providing a TON of 200 and TOF of 1,092 h^{-1}; and an induction period of 1 h was required to form the catalytically active species, and the reason was unknown.

8 Asymmetric Hydrogenation of Thiophenes and Benzothiophenes

Optically active dihydro- and tetrahydrothiophene structures are widely existent in naturally occurring and artificial bioactive molecules [123]. Asymmetric hydrogenation of the corresponding aromatic organosulfur compounds is undoubtedly a straightforward and efficient method for the synthesis of such chiral heterocyclic compounds. However, thiophene, particularly substituted thiophene derivatives, are challenging substrates for homogeneous catalysis, probably due to its aromaticity and the strong coordination/poisoning ability of the substrate and/or the reduced products. To date, the only successful example of asymmetric hydrogenation of thiophenes and benzothiophenes was reported in 2012 by Glorius and co-workers [124].

It was found that the Ru/NHC complex generated in situ from Ru(COD)(2-methylally)$_2$ and monodentate NHC **L26** was a highly efficient catalyst for the hydrogenation of thiophenes and benzothiophenes. Under optimized reaction conditions, hydrogenation of a series of 2-alkyl-substituted benzothiophenes proceeded smoothly, providing chiral 2,3-dihydrobenzothiophenes in moderate to high yields with excellent enantioselectivities of 96–98% *ee*s (Scheme 30). Similarly, 3-methylbenzothiophene could also be reduced in 79% yield with 98% *ee*. However,

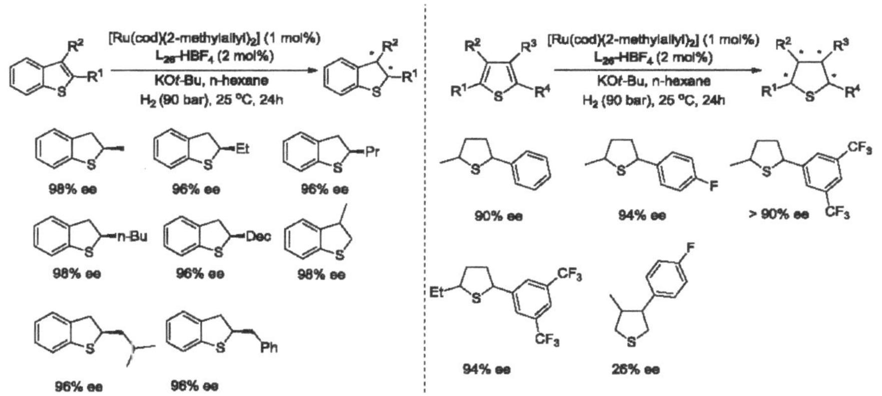

Scheme 30 Asymmetric hydrogenation of substituted benzothiophenes and thiophenes catalyzed by Ru-NHC complex

no conversion of arylated substrates was observed. In the case of thiophenes, it was found that the hydrogenation of 2- and 3-alkyl, and 2-aryl substituted thiophenes proceeded smoothly, but giving racemic tetrahydrothiophenes. Remarkably, 2,5-disubstituted thiophene derivatives could be hydrogenated to tetrahydrothiophenes with moderate to high yields and perfect diastereoselectivities (only *cis*-products were observed) and excellent enantioselectivities (up to 94% *ee*). In addition, a 3,4-disubstituted thiophene was also hydrogenated to give the reduced product in good yield but with lower enantioselectivity (26% *ee*).

9 Asymmetric Hydrogenation of Imidazoles and Oxazoles

Like thiophenes and benzothiophenes, the catalytic asymmetric hydrogenation of five-membered heteroaromatic rings containing two or more heteroatoms are less studied. The only example of highly enantioselective hydrogenation of such monocyclic heteroaromatics was reported by Kuwano and co-workers in 2011 [125]. With chiral Ru-PhTRAP complex as the catalyst, which has been successfully applied for the asymmetric hydrogenation of indoles and pyrroles, both imidazoles and oxazoles were hydrogenated smoothly, giving partially reduced chiral imidazolines and oxazolines, respectively (Scheme 31). It was found that excellent *ee* values (96–99% *ee*) were achieved for the hydrogenation of 2-phenyl-4-alkyl-substituted *N*-Boc-imidazoles. In the case of oxazoles, both 4- and 5-substituted 2-phenyloxazoles were smoothly hydrogenated to chiral oxazolines with good to excellent enantioselectivities (72–98% *ee*s). In most cases, adding TMG (*N*,*N*,*N*′,*N*′-tetramethylguanidine) as a base was necessary for fast conversions. In addition, subsequent hydrolysis of the obtained chiral imidazolines and oxazolines provided acyclic chiral 1,2-diamine and β-amino alcohols, respectively, which are widely used as chiral auxiliaries in organic synthesis.

Scheme 31 Asymmetric hydrogenation of *N*-Boc-protected imidazoles and oxazoles catalyzed by Ru-PhTRAP complex

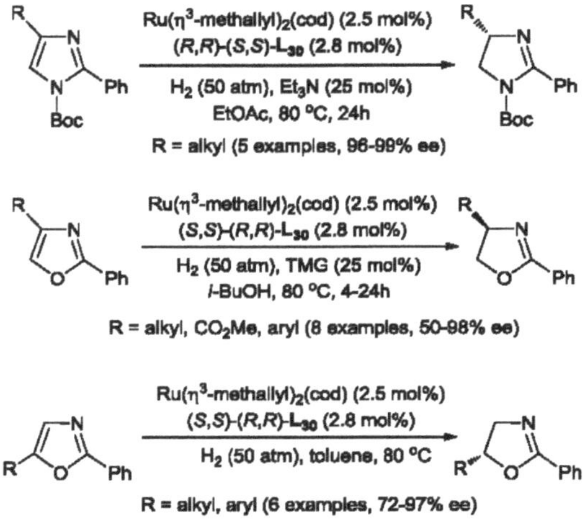

10 Catalyst Immobilization

As described above, various transition metal catalysts involving rhodium, iridium, ruthenium, and palladium complexes bearing different types of chiral ligands have proven to be effective in the asymmetric hydrogenation of *N*-, *O*-, and *S*-containing heteroaromatic compounds. However, most catalytic systems suffered from low catalyst efficiency as evidenced by the fact that good results could only be obtained at a low substrate-catalyst ratio of 100. Considering that all these catalysts bearing noble metals and chiral ligands are very expensive and also toxic, immobilization of these catalysts is highly desirable. To date, a number of methods for recycling homogeneous chiral catalysts have been developed over the past several decades [126–131]. However, only a few examples have been reported on the immobilization of transition metal catalysts in the asymmetric hydrogenation of quinoline derivatives.

10.1 Biphasic Catalytic Systems

Recently, liquid polyethylene glycol (PEG) has been adopted as a new approach for catalyst recycling due to their benign characteristics [132, 133]. In 2005, Fan, Chan, and co-workers developed the first immobilized iridium catalytic system for the asymmetric hydrogenation of heteroaromatics (Scheme 32) [29, 134]. At the beginning, ionic liquids and PEG instead of organic solvents were used in the Ir/P-Phos catalyzed asymmetric hydrogenation of quinolines. However, poor conversion and/or enantioselectivity were observed. When the less polar

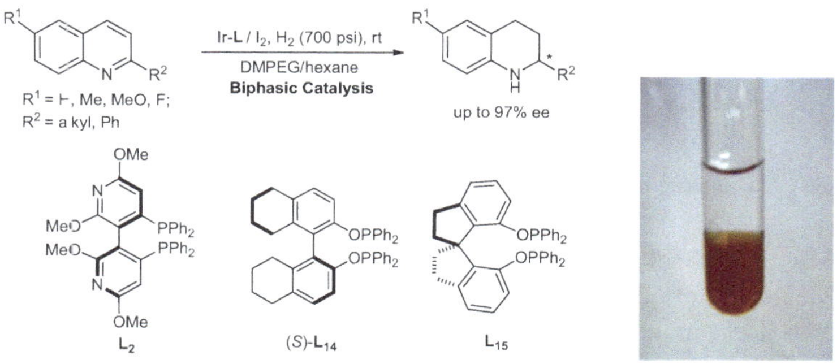

Scheme 32 Biphasic catalytic asymmetric hydrogenation of quinolines with iridium catalysts in DMPEG/hexane

polyethylene glycol dimethyl ether (DMPEG) was used as the solvent, complete conversion and very good enantioselectivity were obtained. To facilitate the separation of the product and the recycling of the catalyst, the DMPEG/hexane biphasic catalytic system was applied in this reaction. A series of 2-substituted quinolines were hydrogenated to 1,2,3,4-tetrahydroquinolines in high yields with excellent enantioselectivities, which were comparable to those obtained in THF. More importantly, the reduced product could easily be separated via simple decantation and the DMPEG layer containing the iridium catalyst was used in the next catalytic run. The catalyst was reused at least eight times without obvious loss of reactivity and enantioselectivity [29].

Subsequently, this immobilization approach was extended to the iridium complexes bearing chiral diphosphinites H8-BINAPO and SpiroPO [47, 48]. When the reactions were carried out in a biphasic DMPEG/hexane system, comparable and even higher enantioselectivities were observed as compared to those obtained in aprotic organic solvents [47]. Similarly, the catalyst could be reused for several times, but the recovered catalyst showed decreased reactivity due to the possible decomposition of the catalyst in the course of recycling.

10.2 Catalyst Immobilization with Soluble Linear Polymer

Unlike the cross-linked polymer-supported homogeneous chiral catalysts, the linear polymeric chiral catalysts bearing catalytically active units on the main chain catalyze organic reactions in a completely homogeneous manner, similar to the conventional homogeneous catalytic reaction. When the reaction is complete, the catalyst can be separated by either solvent precipitation or membrane filtration. Therefore, soluble polymers as alternative catalyst supports have recently attracted much attention [135, 136].

Recently, Zhou and co-worker developed a series of tunable axial chiral diphosphine ligands by attaching MeO-BIPHEP onto soluble polyethylene glycol

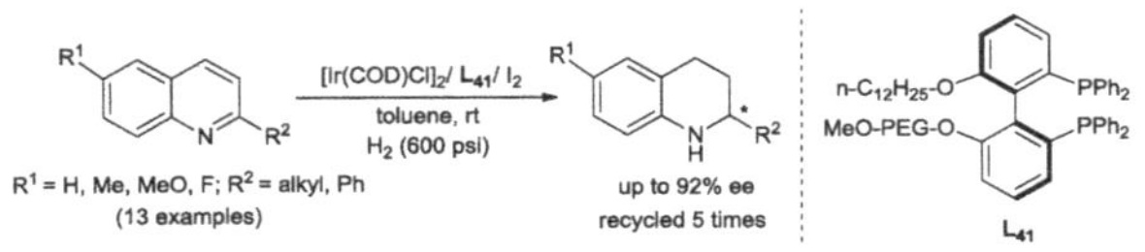

Scheme 33 Asymmetric hydrogenation of quinolines with PEG-supported iridium catalysts

(PEG, Mw = 1,000, 1,600, and 5,000) support via covalent bonding (Scheme 33) [137]. MeO-PEG (1,600)-based chiral ligands bearing different alkyl substituents gave 80–91% *ee*s in the Ir-catalyzed asymmetric hydrogenation of 2-methyl quinoline. The different enantioselectivities might be due to the varying of the dihedral angle of the chiral backbone. The attachment of PEG support with different molecular weight onto the ligand had no obvious effect on the *ee* values. With MeO-PEG (1,600)-supported MeO-BIPHEP as the ligand, a series of 2-substituted quinolines were hydrogenated to 1,2,3,4-tetrahydroquinolines in high yields (85–97%) with very good *ee* values (89–92% *ee*s). Interestingly, the MeO-PEG-supported iridium catalyst was found to be more air-stable and could be recovered by solvent precipitation. The catalyst was reused at least five times with the maintenance of the reactivity, but the enantioselectivity gradually decreased from 91% *ee* to 84% *ee* during the process.

10.3 Chiral Dendrimeric Catalyst

In recent years, attachment of homogeneous catalysts to dendrimers has been attracting considerable attention owing to its well-defined molecular architecture (For recent reviews on organometallic dendrimer catalysts, see [138–143]). Unlike traditional polymer-supported catalysts, dendrimeric catalysts offer opportunities for the study of support-catalyst interactions to fine-tune both catalytic activity and stereoselectivity through systematic adjusting of their structure, size, shape, and solubility. In the case of the core-functionalized dendrimers, it is expected that the steric shielding or blocking effect of the specific microenvironment created by the branched shell could modulate the catalytic behavior of the core.

In 2007, Fan and co-workers applied their BINAP-cored poly(benzyl ether) dendrimers to the iridium catalyzed asymmetric hydrogenation of 2-alkyl-substituted quinolines (Scheme 34) [144, 145]. All four generations of dendrimer catalysts, generated in situ from BINAP-cored dendrimers and $[Ir(COD)Cl_2]$, were found to be effective even at an extremely high substrate/catalyst ratio in the asymmetric hydrogenation of 2-methyl quinoline with I_2 as the additive. It was found that the catalytic activity gradually increased with increasing dendrimer generation, and the maximum initial TOF from the third-generation catalyst could reach 3,450 h^{-1}, which represents the highest TOF obtained so far for the asymmetric hydrogenation of quinolines. In addition, the reaction proceeded smoothly

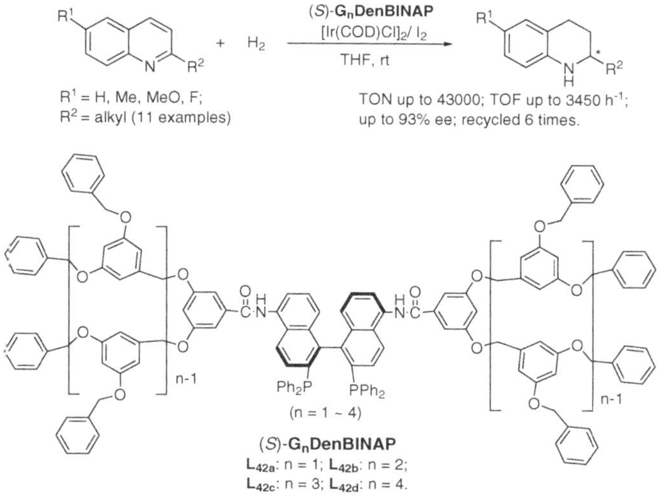

Scheme 34 Asymmetric hydrogenation of quinolines with dendrimeric iridium catalysts

under rather low catalyst loading in a large scale reaction (~18 g substrate) to give a TON of 43,000, which is also the highest TON reported to date for such a reaction.

This rate enhancement of the dendrimer catalysts was further demonstrated by the time-conversion curves (Fig. 6). This was probably due to the encapsulation of such an iridium complex into a dendrimer framework, which would reduce the formation of inactive iridium dimer and therefore enhance the productivity of the catalyst. In addition, the third-generation dendrimer catalyst could be quantitatively recovered by precipitation with methanol, and reused at least six times with similar enantioselectivities but at the expense of relatively low catalytic activities.

10.4 Catalyst Immobilization in Ionic Liquid

The use of room temperature ionic liquids as alternative reaction media is particularly good in enhancing catalyst stability and reaction rates, and improving stereoselectivity in transition metal-catalyzed asymmetric hydrogenation. In addition, ionic liquids have served as a promising means to immobilize catalysts, therefore facilitating product isolation and offering an opportunity to reuse the catalyst [146–150].

In 2008, Fan and co-workers first used phosphine-free cationic Ru-TsDPEN catalyst for the asymmetric hydrogenation of quinolines [24] and found that the

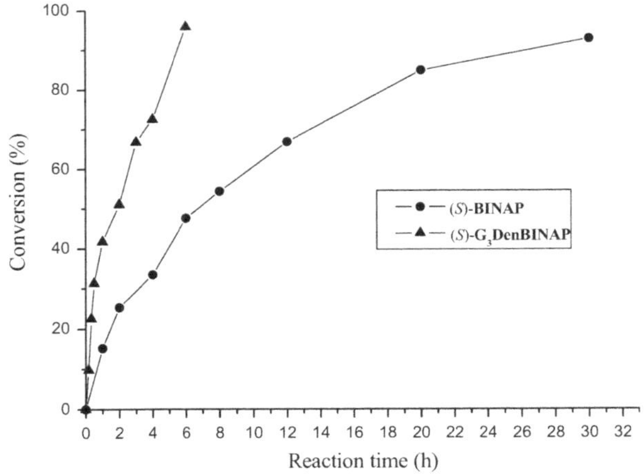

Fig. 6 Time course curves for the Ir-catalyzed hydrogenation of 2-methyl quinoline using the third-generation dendrimer and BINAP

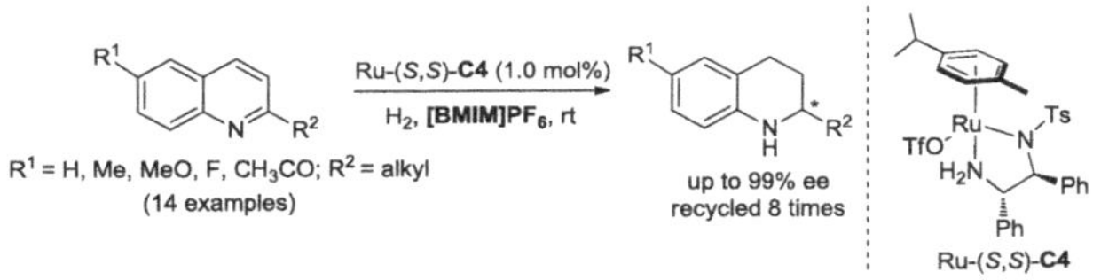

Scheme 35 Asymmetric hydrogenation of quinolines with ruthenium catalyst in ionic liquid

reaction proceeded smoothly in neat ionic liquid [BMIM]PF_6 (Scheme 35, BMIM = 1-*n*-butyl-3-methylimidazolium). Unlike the Ir/diphosphine/I_2 catalytic system, which gave unsatisfactory results in ionic liquid [29], unprecedented reactivities and excellent enantioselectivities were achieved, which are better than those in methanol. Under the optimized reaction conditions, a variety of 2-alkyl quinolines were hydrogenated to tetrahydroquinolines in high yields (87–97%) with excellent enantioselectivities (96–99% *ee*). More interestingly, it was found that the hydrogenation in ionic liquid was selective for the C = N (quinoline) over the C = O bond, which were both reduced in methanol.

In addition, the catalyst stability was obviously enhanced in ionic liquid. Even after exposing the catalyst solution to air for 30 days, almost the same activity and enantioselectivity were observed. In contrast, the catalyst in methanol was found to be decomposed within 1 week under otherwise identical conditions. The dramatic enhancement of the catalyst stability by an ionic liquid was probably a result of the solvation effect of [BMIM]PF_6 on the cationic Ru catalyst and the very low solubility of oxygen in the ionic liquid. On the basis of the high stability of the catalyst in ionic liquid, manipulation of catalyst recycling was easy and reliable. Upon completion of the reaction, the reduced products were separated by extraction

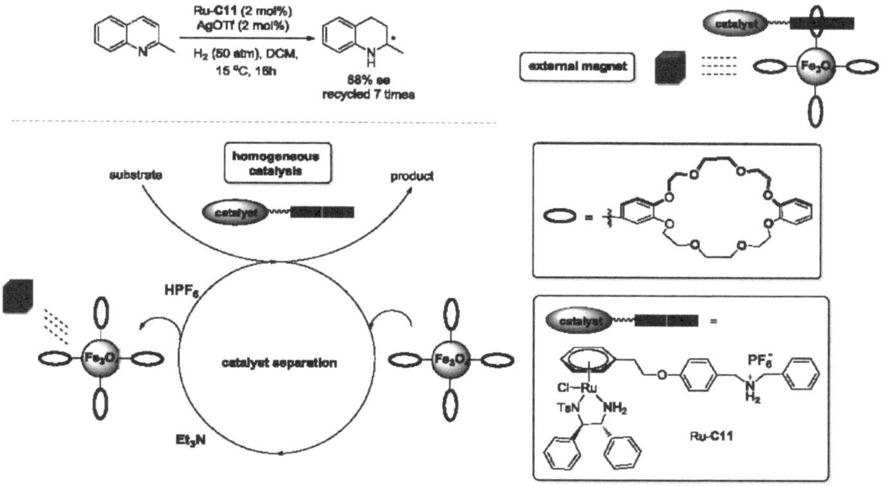

Scheme 36 Homogeneous asymmetric hydrogenation of 2-methylquinoline and catalyst recycling by heterogeneous magnetic nanoparticles through supramolecular protocol

with *n*-hexane. The ionic liquid layer was recharged with quinoline and subjected to the next catalytic run. The ruthenium catalyst was reused at least eight times without obvious decrease in reactivity and enantioselectivity.

10.5 Catalyst Immobilization with Magnetic Nanoparticles

Nanoparticles are considered to be robust and readily available heterogeneous catalysts as well as catalyst supports for catalytic transformations due to their high surface area [151–155]. Based on its heterogeneous nature, covalently bonded nanoparticle-immobilized catalysts are still suffering from decreased catalytic activity and selectivity. By introducing non-covalent interactions between the nanoparticle support and homogeneous catalyst, both homogeneous catalysis and heterogeneous catalyst separation could be realized in the same reaction. In recent years, magnetic nanoparticles as solid support for catalysis have attracted more and more attentions due to its simple separation from liquid reaction mixtures with the aid of an external magnet [154, 155].

In 2011, a homogeneous catalyst/heterogeneous catalyst separation protocol combining the use of magnetic nanoparticles and host-guest assembly was developed by Fan and co-workers [156]. This novel approach was applied in the Ru-catalyzed asymmetric hydrogenation of 2-methylquinoline (Scheme 36). After the homogeneous hydrogenation catalyzed by the chiral Ru-diamine catalyst bearing a dialkylammonium salt tag, crown ether functionalized magnetic nanoparticles were added to facilitate host-guest assembly. The assembled heterogeneous catalyst was readily recovered by magnet-assisted decantation, and was further basified to release the tagged catalyst. The recycled catalyst was then acidified and reused in the hydrogenation again by recharging 2-methylquinoline. The catalyst was reused seven times with similar *ee* values (88–89%) and nearly quantitative conversions.

11 Mechanistic Aspects

As one of the most straightforward and powerful approaches for the creation of optically active heterocyclic compounds, asymmetric hydrogenation has been successfully used for different types of heteroaromatics, including quinolines, isoquinolines, quinoxalines, pyridines, indoles, pyrroles, furans, thiophenes, imidazoles and oxazoles, with excellent enantioselectivities. However, only a few detailed studies of the reaction mechanism have been published to date. This might be, on one hand, due to the short history of research in this topic. On the other hand, such transformations are more complicated relative to hydrogenation of alkenes, ketones, and imines, which generally involve reduction of different types of double bonds, together with isomerization and dehydrogenation processes. Most recently, efforts have been devoted to the mechanism studies of ruthenium and iridium catalyzed asymmetric hydrogenation of quinolines and quinoxalines. Both stepwise inner-sphere and stepwise outer-sphere pathways were proposed, depending on the catalytic system, addition of additives, and substrate used.

11.1 Mechanism for Asymmetric Hydrogenation of Quinolines

In general, the reduction of quinolines includes a hydrogenation sequence of C = C and C = N bonds. In 2008, Fan and co-workers proposed an ionic and cascade reaction pathway, including 1,4-hydride addition, isomerization, and 1,2-hydride addition for the asymmetric hydrogenation of quinolines in ionic liquid [24]. Subsequently, Fan, Yu, and co-workers further investigated the catalytic pathway of this reaction systematically by using a combination of stoichiometric reaction, intermediate characterization and isotope labeling patterns together with DFT (density functional theory) calculations [65].

In the stoichiometric reaction between 2-substituted quinolines and ruthenium hydride complex, it was found that the addition of 2 equiv. of Brønsted acid was crucial for full conversion. Unlike the hydrogenation of aromatic ketones with a similar ruthenium catalyst, quinolines should be activated by the Brønsted acid before hydrogenation (Scheme 37). In addition, the existence of an imine intermediate was confirmed by the evidence from ^{1}H NMR (nuclear magnetic resonance)

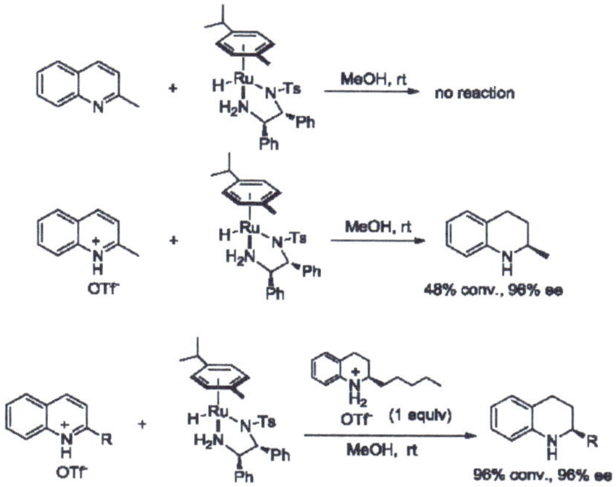

Scheme 37 Stoichiometric reduction of quinoline with isolated Ru-H hydride

Scheme 38 The possible hydrogenation pathway of quinoline with Ru-H hydride

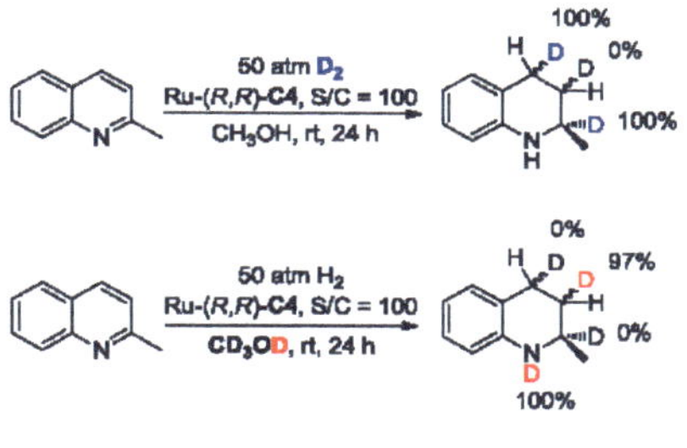

Scheme 39 Deuterium-labeling studies

and ESI-MS (electrospray ionization-mass spectrometry) characterization. These observations suggested a possible ionic catalytic pathway including 1,4-hydride addition, isomerization, and then 1,2-hydride addition process (Scheme 38). The deuteration study further provided positive evidence, such as that 100% deuterium at both C-2 and C-4 positions were observed when D_2 was employed, and 97% deuterium at the C-3 position was observed when CD_3OD was used as the solvent (Scheme 39). Furthermore, the 1,4-hydride addition step proved to be reversible by deuterium scrambling into recovered 2-phenyl quinoline at the C4-position, and 1,2,3,4-tetrahydroquinoline at the C2- and C4-positions.

To further understand the mechanism and the origins of enantioselectivity, theoretical calculations were carried out to study the hydrogenation process of the

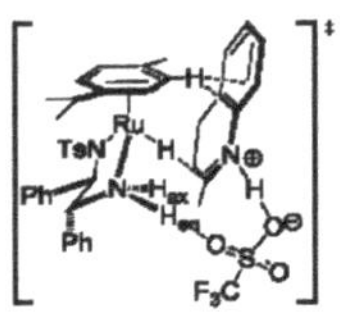

Fig. 7 Calculated transition structures for the hydride transfer to imine

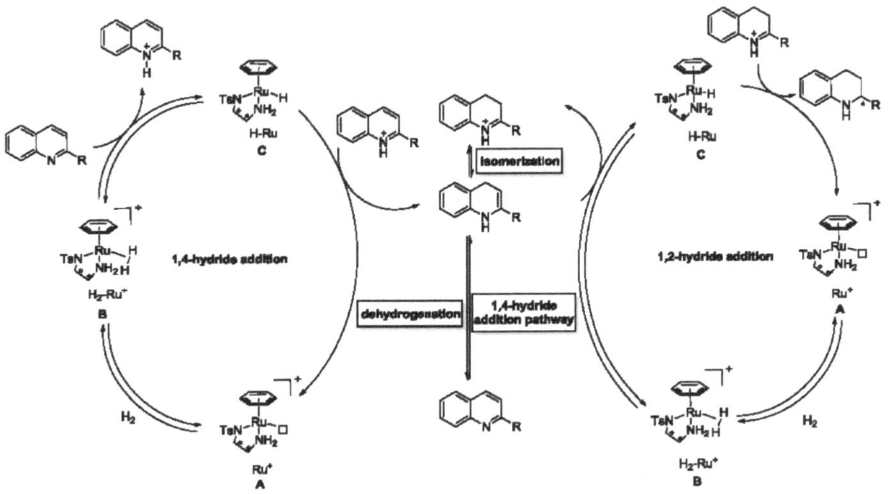

Scheme 40 Catalytic cycle proposed for the asymmetric hydrogenation of quinoline using Ru(II)-TsDPEN catalyst (ethylenediamine ligand and OTf^- are omitted for clarity)

protonated imine intermediate in situ generated by the 1,4-hydride addition and isomerization. The computational results indicated that the final 1,2-hydride addition should proceed through an ionic pathway involving the following transition state (Fig. 7). Similar to the ketone reduction reported by Noyori and co-workers [157, 158], the enantioselectivity originated from the CH/π attraction between the η^6-arene ligand in the Ru-complex and the fused phenyl ring of dihydroquinoline via a 10-membered ring transition state with the participation of TfO^- anion. This proposed transition state also offered new insights into the mechanism of other imine reduction involved reactions with similar ruthenium catalysts [159].

Based on these experimental and theoretical results, a plausible mechanism for this process was thus proposed (Scheme 40). First, dihydrogen is reversibly coordinated to the ionized ruthenium complex **A**. The formed dihydrogen complex **B** is then deprotonated by quinoline to generate both the active ruthenium hydride species **C** and the activated substrate. A subsequent 1,4-hydride transfer affords the enamine intermediate and the regenerated **A**. Similarly, the enamine serves as a base to deprotonate the dihydrogen ligand, resulting in species **C** and the iminium cation.

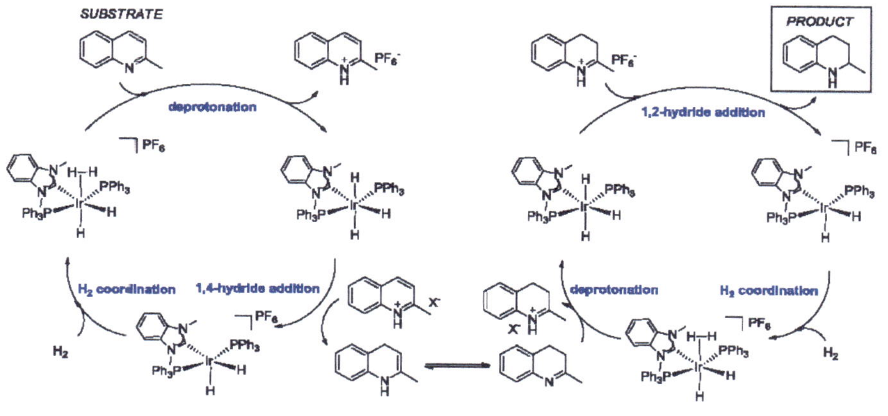

Scheme 41 Catalytic cycle proposed for the hydrogenation of 2-methylquinoline using Ir(I)-NHC catalyst

Finally, 1,2-hydride transfer via a ten-membered ring transition state as described above gives 1,2,3,4-tetrahydroquinoline enantioselectively, and regenerates ruthenium complex **A**.

In 2011, Crabtree and co-workers developed a new homogeneous achiral iridium catalyst for the hydrogenation of quinolines under mild conditions [160]. Based on experimental and theoretical results, a similar ionic and cascade reaction pathway and an unusual stepwise outer-sphere mechanism were proposed (Scheme 41).

As described in Sect. 2, the first highly enantioselective hydrogenation of quinolines was realized with iridium/diphosphine/I_2 catalytic system. In 2009, Zhou and co-workers proposed two possible hydrogenation pathways depending on the initial 1,2- or 1,4-hydrogen addition [28]. A combination of experimental and theoretical studies indicated that hydrogenation of quinolines favored an initial 1,4-hydride transfer tandem reaction pathway. A conventional inner-sphere mechanism was proposed through a cascade reaction of 1,4-hydride addition, enamine-imine isomerization, and 1,2-hydride addition (Scheme 42). First, the oxidative addition of I_2 to the Ir(I) catalyst precursor **A** results in Ir(III) species, which was involved in the subsequent heterolytic cleavage of the H_2 to form the catalytic active Ir(III)-H species **B**. Then quinoline (**R**) coordinates with **B** to form **C**, followed by 1,4-hydride transfer to give the intermediate **D**. Subsequent heterolytic cleavage of H_2 with **D** affords the enamine intermediate and the regenerated **B**. The enamine isomerizes to imine catalyzed by the in situ generated HI. Similarly, coordination of the formed imine with **B** gives the intermediate **H**. Subsequently, the enantioselective 1,2-insertion of C = N bond into the Ir-H bond gives the intermediate **I**. Finally, the reduced product 1,2,3,4-tetrahydroquinoline (**P**) is released via hydrogenolysis.

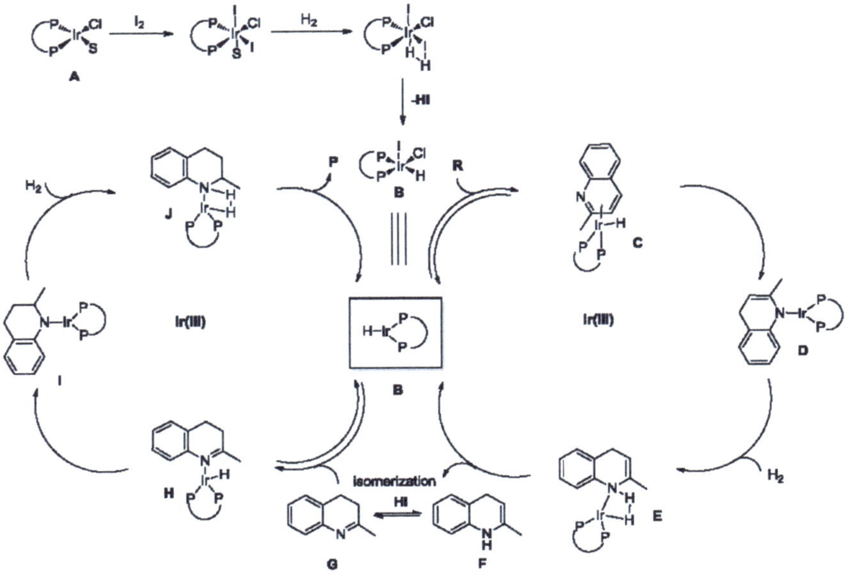

Scheme 42 Catalytic cycle proposed for the asymmetric hydrogenation of quinoline using Ir/MeO-BIPHEP/I_2 catalytic system

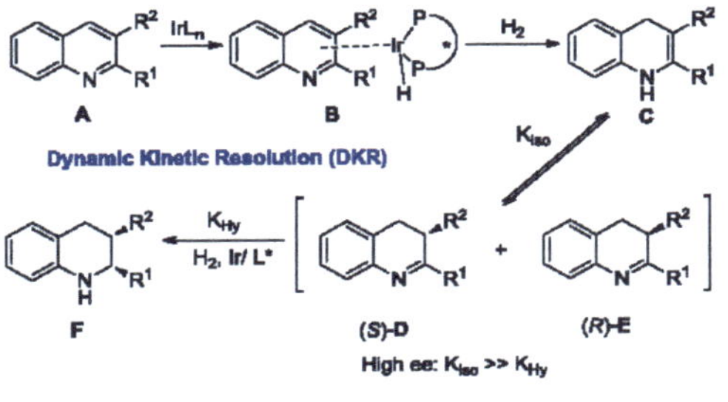

Scheme 43 Plausible mechanism for Ir-catalyzed asymmetric hydrogenation of 2,3-disubstituted quinolines via dynamic kinetic resolution

In addition, the hydrogenation process of 2,3-disubstituted quinolines is more complicated than that of 2-substituted quinolines. The enantio-determining steps include the isomerization of enamine to imine and subsequent hydrogenation of the C = N bond (Scheme 43). The first step is in fact a dynamic kinetic resolution process. It was found that high enantioselectivities were observed under high temperature and low hydrogen pressure, which accelerate the isomerization and slow down the hydrogenation. In both cases, the origin of enantioselectivity was still untouched and awaits further exploration.

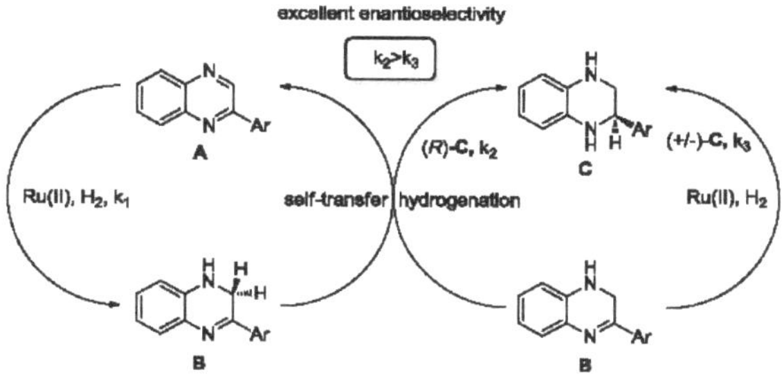

Scheme 44 Catalytic cycle proposed for the metal/Brønsted acid induced asymmetric hydrogenation of 2-aryl quinoxalines

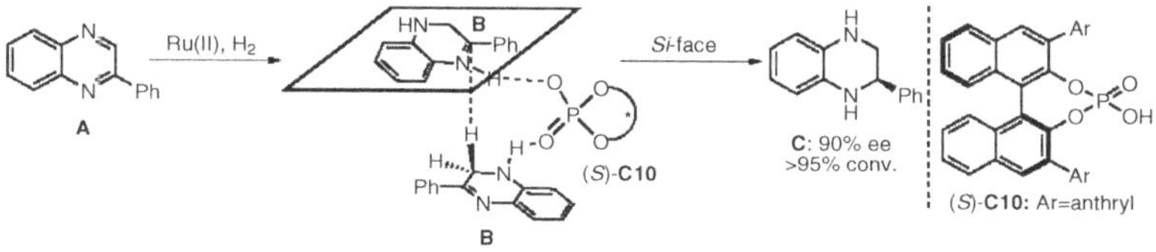

Scheme 45 The proposed "three-point contact model" for the self-transfer hydrogenation

11.2 *Mechanism for Asymmetric Hydrogenation of Quinoxalines*

Over the past few decades, several homogeneous catalytic systems have been developed for the highly enantioselective hydrogenation of quinoxalines. However, fewer mechanistic studies of this reaction have been reported than those of quinolines. In 2011, Zhou and co-workers described the first mechanistic study [85] related to the convergent asymmetric disproportionation of dihydroquinoxalines induced by chiral phosphoric acid in a transition metal/Brønsted acid catalyzed asymmetric hydrogenation of 2-aryl quinoxalines. In their proposed mechanism (Scheme 44), hydrogenation of quinoxaline **A** first generates the intermediate dihydroqinoxaline **B** by using achiral ruthenium(II) as the catalyst. Subsequently, the intermediate **B** undergoes self-transfer hydrogenation catalyzed by chiral phosphoric acid, giving the starting material **A** and the final product 1,2,3,4-tetrahydroquinoxaline **C**. The intermediate **B** was observed in the hydrogenation of **A** in the absence of phosphoric acid, suggesting that the first hydrogenation process catalyzed by ruthenium complex is the rate-determining step. The observed high enantioselectivities were thus due to the fact that the reaction rate of the chiral phosphoric acid-catalyzed self-transfer hydrogenation is higher than that of the achiral ruthenium-catalyzed hydrogenation of **B**. To understand the origin of enantioselectivity, a "three-point contact model" was also proposed on the basis of experimental results and DFT calculations (Scheme 45).

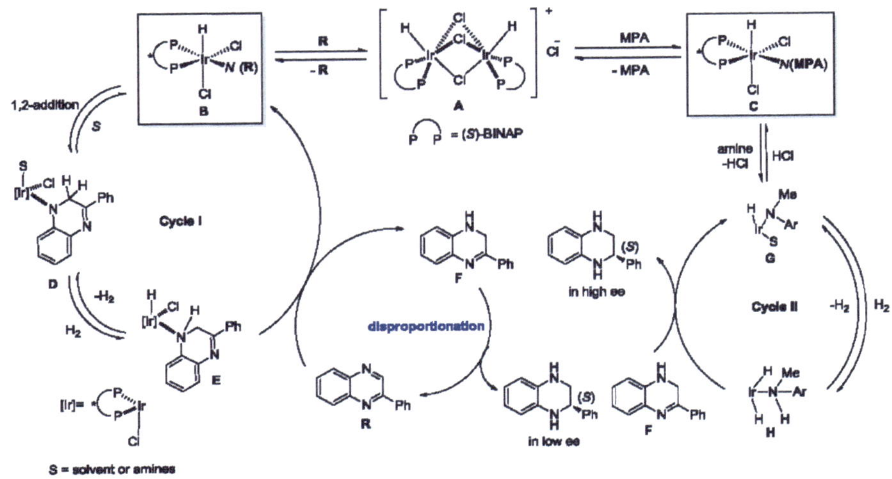

Scheme 46 Catalytic cycles proposed for the asymmetric hydrogenation of quinoxalines using chiral cationic dinuclear iridium complexes

For asymmetric hydrogenation of *N*-heteroaromatic compounds with diphosphine ligands, addition of additives is crucial for achieving high reactivity and/or enantioselectivity. Most recently, Mashima and co-workers discovered the additive effects of amine in the asymmetric hydrogenation of 2-aryl quinoxalines with their chiral cationic dinuclear triple-halide-bridged iridium complexes [81]. The pK_a value and amount of amine additive were found to be crucial for the enhancement of catalytic activity. With the addition of 1 equiv. of the optimal amine *N*-methyl-*p*-anisidine (MPA), a series of 2-aryl quinoxalines were hydrogenated in nearly quantitative conversions with high *ee* values. The remarkable additive effects were then investigated in detail by solution dynamics studies of iridium complexes in the presence of MPA and by labeling experiments, which revealed a plausible dual mechanism comprised of two individual catalytic cycles in equilibrium (Scheme 46). Each of the proposed intermediates in these cycles was either isolated or spectroscopically characterized.

At the beginning, the reversible and competing coordination of substrate (cycle I) and amine additive MPA (cycle II) to the dinuclear iridium precursor **A** generates two active catalytic species **B** and **C**, respectively. For cycle I, reversible 1,2-insertion of the C = N bond of quinoxaline (**R**) into the Ir–H bond of **B** gives the intermediate **D**, followed by the addition of H_2 to the Ir–N bond to produce the intermediate **E**. Subsequent replacement reaction of **E** with the substrate gives the partially reduced product **F** and the regenerated **B**. Then acidic protons derived from catalyst **A** induce rapid racemic disproportionation of the partially reduced product **F** to give quinoxaline and racemic tetrahydroquinoxaline; at the same time, the iridium catalyst induces slow asymmetric hydrogenation to give the chiral product, thus resulting in low enantioselectivity at the early stage of hydrogenation. As the reaction proceeds, accumulation of amine product may suppress the acid-catalyzed racemic disproportionation.

For cycle II, addition of excess amounts of MPA shift the equilibrium to the formation of catalytic species **C**. Subsequent elimination of HCl from **C** affords the amide-hydride species **G**, which reacts with H_2 to give an amine-dihydride species **H**. Finally, asymmetric hydrogenation of the $C = N$ bond of dihydroquinoxaline and quinoxaline with the catalytic species **H** to give tetrahydroquinoxaline with high enantioselectivity through the bifunctional outer-sphere mechanism, similar to that described by Noyori and co-workers [157, 158]. The amine additive MPA serves not only as a ligand to generate the highly reactive and enantioselective Ir-amide species but also as a Brønsted base to bypass the acid-catalyzed racemic disproportionation of half-reduced compounds. Similarly, the amine product also exhibits additive effects as well as positive feedback enhancement [161], which was observed for the first time in asymmetric hydrogenation.

12 Summary and Perspectives

As reviewed in this chapter, transition metal-catalyzed asymmetric hydrogenation of *N*-, *O*-, and *S*-containing heteroaromatic compounds, which are generally regarded as more challenging substrates, has received increasing attention in recent years. A number of transition metal (Ir, Rh, Ru, and Pd) catalysts bearing chiral phosphorus ligands, amine-tosylamine ligands, and *N*-heterocyclic carbene ligands have been examined. Different types of heteroaromatics, including quinolines, isoquinolines, quinoxalines, pyridines, indoles, pyrroles, furans, thiophenes, imidazoles and oxazoles, have been successfully applied to such transformations, providing a facile and economic access to a variety of optically active heterocycle compounds. The significant progress made in this challenging hydrogenation reaction was mainly attributed to the discovery of new catalytic systems, and the development of new catalyst and substrate activation strategies. Furthermore, several recoverable catalytic systems have also been developed and found to be highly efficient in the asymmetric hydrogenation of quinolines. Mechanistic studies on the hydrogenation of quinolines and quinoxalines have revealed the catalytic pathway and/or offered some insight into the origin of enantioselectivity. However, despite significant progress having been made, this field is still far from being mature as compared to the asymmetric hydrogenation of prochiral olefins and ketones, and there are still many challenges which limit its application. The focus for further research in this area is expected to be the development of more efficient homogeneous and heterogeneous chiral catalysts and the extension of substrate scope. In addition, detailed mechanistic investigations for different heteroaromatic compounds will form part of future studies, which are likely to be beneficial for the discovery of new generation catalysts and new substrates of heteroaromatics for asymmetric hydrogenation.

Acknowledgements We are grateful for the financial support from the National Basic Research Program of China (973 Program, Nos. 2010CB833300 and 2011CB808600) and the National Natural Science Foundation of China (No 21232008).

References

1. Noyori R (1994) Asymmetric catalysis in organic synthesis. Wiley, New York
2. Ojima I (ed) (2000) Catalytic asymmetric synthesis, 2nd edn. Wiley, New York
3. Jacobson EN, Pfaltz A, Yamamoto H (eds) (1999) Comprehensive asymmetric catalysis, vol 2. Springer, Berlin
4. Lin G, Li Y, Chan ASC (2001) Principles and applications of asymmetric synthesis. Wiley-Interscience, New York
5. Tang W-J, Zhang X-M (2003) Chem Rev 103:3029–3070
6. Nugent TC, El-Shazly M (2010) Adv Synth Catal 352:753–819
7. Xie J-H, Zhu S-F, Zhou Q-L (2011) Chem Rev 111:1713–1760
8. Dyson PJ (2003) Dalton Trans 2964–2974
9. Glorius F (2005) Org Biomol Chem 3:4171–4175
10. Lu S-M, Han X-W, Zhou Y-G (2005) Chin J Org Chem 25:634–636
11. Zhou Y-G (2007) Acc Chem Res 40:1357–1366
12. Kuwano R (2008) Heterocycles 76:909–922
13. Wang D-S, Chen Q-A, Lu S-M, Zhou Y-G (2012) Chem Rev 112:2557–2590
14. Church TL, Andersson PG (2010) In: Nugent TC (ed) Chiral amine synthesis. New York, Wiley, Chapter 6, 179–223
15. Besson M, Pinel C (2003) Top Catal 25:43–61
16. Rueping M, Sugiono E, Schoepke FR (2010) Synlett 6:852–865
17. You S-L (2007) Chem Asian J 2:820–827
18. Katritzky AR, Rachwal S, Rachwal B (1996) Tetrahedron 52:15031–15070
19. Scott JD, Williams RM (2002) Chem Rev 102:1669–1730
20. Sridharan V, Suryavanshi PA, Menéndez JC (2010) Chem Rev 111:7157–7259
21. Wang W-B, Lu S-M, Yang P-Y, Han X-W, Zhou Y-G (2003) J Am Chem Soc 125:10536–10537
22. Lu S-M, Wang Y-Q, Han X-W, Zhou Y-G (2006) Angew Chem Int Ed 45:2260–2263
23. Yu Z, Jin W, Jiang Q (2012) Angew Chem Int Ed 51:6060–6072
24. Zhou H-F, Li Z-W, Wang Z-J, Wang T-L, Xu L-J, He Y-M, Fan Q-H, Pan J, Gu L-Q, Chan ASC (2008) Angew Chem Int Ed 47:8464–8467
25. He Y-M, Fan Q-H (2010) Org Biomol Chem 8:2497–2504
26. Shimizu H, Nagasaki I, Saito T (2005) Tetrahedron 61:5405–5432
27. Yang P-Y, Zhou Y-G (2004) Tetrahedron Asymmetry 15:1145–1149
28. Wang D-W, Wang X-B, Wang D-S, Lu S-M, Zhou Y-G, Li Y-X (2009) J Org Chem 74:2780–2787
29. Xu L-J, Lam KH, Ji J-X, Wu J, Fan Q-H, Lo WH, Chan ASC (2005) Chem Commun 1390–1392
30. Tang W-J, Tan J, Xu L-J, Lam KH, Fan Q-H, Chan ASC (2010) Adv Synth Catal 352:1055–1062
31. Tang W-J, Sun Y-W, Xu L-J, Wang T-L, Fan Q-H, Lam KH, Chan ASC (2010) Org Biomol Chem 8:3464–3471
32. Zhang D-Y, Wang D-S, Wang M-C, Yu C-B, Gao K, Zhou Y-G (2011) Synthesis 2796–2802
33. Zhang D-Y, Yu C-B, Wang M-C, Gao K, Zhou Y-G (2012) Tetrahedron Lett 53:2556–2559
34. Qiu L-Q, Kwong FY, Wu J, Lam WH, Chan S-S, Yu W-Y, Li Y-M, Guo R-W, Zhou Z-Y, Chan ASC (2006) J Am Chem Soc 128:5955–5965
35. Jahjah M, Alame M, Pellet-Rostaing S, Lemaire M (2007) Tetrahedron Asymmetry 18:2305–2312
36. Gou F-R, Li W, Zhang X-M, Liang Y-M (2010) Adv Synth Catal 352:2441–2444
37. Maj AM, Suisse I, Méliet C, Hardouin C, Niedercorn FA (2012) Tetrahedron Lett 53:4747–4750
38. Wu J, Chan ASC (2006) Acc Chem Res 39:711–720

39. Yamagata T, Tadaoka H, Nagata M, Hirao T, Kataoka Y, Vidal VR, Genet JP, Mashima K (2006) Organometallics 25:2505–2513
40. Deport C, Buchotte M, Abecassis K, Tadaoka H, Ayad T, Ohshima T, Genet JP, Mashima K, Vidal VR (2007) Synlett 2743–2747
41. Tadaoka H, Cartigny D, Nagano T, Gosavi T, Ayad T, Genêt JP, Ohshima T, Vidal VR, Mashima K (2009) Chem Eur J 15:9990–9994
42. Wang D-S, Zhou Y-G (2010) Tetrahedron Lett 51:3014–3017
43. van Leeuwen PWNM, Kamer PCJ, Claver C, Pàmies O, Diéguez M (2011) Chem Rev 111:2077–2118
44. Fernández-Pérez H, Etayo P, Panossian A, Vidal-Ferran A (2011) Chem Rev 111:2119–2176
45. Teichert JF, Feringa BL (2010) Angew Chem Int Ed 49:2486–2528
46. Roseblade SJ, Pfaltz A (2007) Acc Chem Res 40:1402–1411
47. Lam KH, Xu L-J, Feng L-C, Fan Q-H, Lam FL, Lo WH, Chan ASC (2005) Adv Synth Catal 347:1755–1758
48. Tang W-J, Zhu S-F, Xu L-J, Zhou Q-L, Fan Q-H, Zhou H-F, Lam K, Chan ASC (2007) Chem Commun 613–615
49. Reetz MT, Li X-G (2006) Chem Commun 2159–2160
50. Rico JLN, Pérez HF, Buchholz JB, Ferran AV (2010) Organometallics 29:6627–6631
51. Rubio M, Pizzano A (2010) Molecules 15:7732–7741
52. Eggenstein M, Thomas A, Theuerkauf J, Franciò G, Leitner W (2009) Adv Synth Catal 351:725–732
53. Mršić N, Lefort L, Boogers JAF, Minnaard AJ, Feringa BL, de Vries JG (2008) Adv Synth Catal 350:1081–1089
54. Lu S-M, Han X-W, Zhou Y-G (2004) Adv Synth Catal 346:909–912
55. Zhao Y-J, Wang Y-Q, Zhou Y-G (2005) Chin J Catal 26:737–739
56. Wang D-S, Zhou J, Wang D-W, Guo Y-L, Zhou Y-G (2010) Tetrahedron Lett 51:525–528
57. Lu S-M, Bolm C (2008) Adv Synth Catal 350:1101–1105
58. Hammerer T, Weisgerber L, Schenk S, Stelzer O, Englert U, Leitner W, Franciò G (2012) Tetrahedron Asymmetry 23:53–59
59. Xie J-H, Zhou Q-L (2008) Acc Chem Res 41:581–593
60. Fache F, Schulz E, Tommasino ML, Lemaire M (2000) Chem Rev 100:2159–2232
61. Noyori R, Hashigushi S (1997) Acc Chem Res 30:97–102
62. Gladiali S, Alberico E (2006) Chem Soc Rev 35:226–236
63. Ohkuma T, Utsumi N, Tsutsumi K, Murata K, Sandoval CA, Noyori R (2006) J Am Chem Soc 128:8724–8725
64. Sandoval CA, Ohkuma T, Utsumi N, Tsutsumi K, Murata K, Noyori R (2006) Chem Asian J 1:102–110
65. Wang T-L, Zhuo L-G, Li Z-W, Chen F, Ding Z-Y, He Y-M, Fan Q-H, Xiang J-F, Yu Z-X, Chan ASC (2011) J Am Chem Soc 133:9878–9891
66. Walsh PJ, Li H, de Parrodi CA (2007) Chem Rev 107:2503–2545
67. Wang Z-J, Zhou H-F, Wang T-L, He Y-M, Fan Q-H (2009) Green Chem 11:767–769
68. Wang T-L, Ouyang G-H, He Y-M, Fan Q-H (2011) Synlett 939–942
69. Li Z-W, Wang T-L, He Y-M, Wang Z-J, Fan Q-H, Pan J, Xu L-J (2008) Org Lett 10:5265–5268
70. Rueping M, Koenigs R (2011) Chem Commun 304–306
71. Shi L, Ye Z-S, Cao L-L, Guo R-N, Hu Y, Zhou Y-G (2012) Angew Chem Int Ed 51:8286–8289
72. Murata S, Sugimoto T, Matsuura S (1987) Heterocycles 26:763–766
73. Bianchini C, Barbaro P, Scapacci G, Farnetti E, Grazian M (1998) Organometallics 17:3308–3310
74. Bianchini C, Barbaro P, Scapacci G (2001) J Organomet Chem 621:26–33
75. Cobley CJ, Henschke JP (2003) Adv Synth Catal 345:195–201
76. Henschke JP, Burk MJ, Malan CG, Herzberg D, Peterson JA, Wildsmith AJ, Cobley CJ, Casy G (2003) Adv Synth Catal 345:300–307

77. Tang W-J, Xu L-J, Fan Q-H, Wang J, Fan B-M, Zhou Z-Y, Lam KH, Chan ASC (2009) Angew Chem Int Ed 48:9135–9138
78. Mršić N, Jerphagnon T, Minnaard AJ, Feringa BL, de Vries JG (2009) Adv Synth Catal 351:2549–2552
79. Cartigny D, Nagano T, Ayad T, Genêt JP, Oshima T, Mashima K, Vidal VR (2010) Adv Synth Catal 352:1886–1891
80. Cartigny D, Berhal F, Nagano T, Phansavath P, Ayad T, Genêt JP, Oshima T, Mashima K, Vidal VR (2012) J Org Chem 77:4544–4556
81. Nagano T, Iimuro A, Schwenk R, Oshima T, Kita Y, Togni A, Mashima K (2012) Chem Eur J 18:11578–11592
82. Chang G, Didiuk MT, Finneman JI, Garigipati RS, Kelley RM, Perry DA, Ruggeri RB, Bechle BM, Pollastri MP (2004) 1,2,4-Substituted 1,2,3,4-tetrahydro- and 1,2-dihydroquinoline and 1,2,3,4-tetrahydro-quinoxaline derivatives as CETP inhibitors for the treatment of atherosclerosis and obesity. WO/2004/085,401; PCT/IB2004/000,836. Pfizer Products Inc
83. Qin J, Chen F, Ding Z-Y, He Y-M, Xu L-J, Fan Q-H (2011) Org Lett 13:6568–6571
84. Urban S, Ortega N, Glorius F (2011) Angew Chem Int Ed 50:3803–3806
85. Chen Q-A, Wang D-S, Zhou Y-G, Duan Y, Fan HJ, Yang Y, Zhang Z (2011) J Am Chem Soc 133:6126–6129
86. Rubiralta M, Giralt E, Diez A (1991) Piperidines: structure, preparation, reactivity and synthetic applications of piperidine and its derivatives. Elsevier, Amsterdam
87. Studer M, Exl CW, Spindler F, Blaser HU (2000) Monatshefte für Chemie 131:1335–1343
88. Legault CY, Charette AB (2005) J Am Chem Soc 127:8966–8967
89. Legault CY, Charette AB, Cozzi PG (2008) Heterocycles 76:1271–1283
90. Ye Z-S, Chen M-W, Chen Q-A, Shi L, Duan Y, Zhou Y-G (2012) Angew Chem Int Ed 51:10181–10184
91. Xiao D, Lavey BJ, Palani A, Wang C, Aslanian RG, Kozlowski JA, Shih N-Y, McPhail AT, Randolph GP, Lachowicz JE, Duffy RA (2005) Tetrahedron Lett 46:7653–7656
92. Sheikh NS, Leonori D, Barker G, Firth JD, Campos KR, Meijer AJHM, Brien PO, Coldham I (2012) J Am Chem Soc 134:5300–5308
93. Wang X-B, Zeng W, Zhou Y-G (2008) Tetrahedron Lett 49:4922–4924
94. Fattorusso E, Taglialatela-Scafati O (2008) Modern alkaloids: structure, isolation: synthesis and biology. Wiley-VCH, Weinheim
95. Kuwano R, Sato K, Kurokawa T, Karube D, Ito Y (2000) J Am Chem Soc 122:7614–7615
96. Kuwano R, Kaneda K, Ito T, Sato K, Kurokawa T, Ito Y (2004) Org Lett 6:2213–2215
97. Kuwano R, Kashiwabara M, Sato K, Ito T, Kaneda K, Ito Y (2006) Tetrahedron Asymmetry 17:521–535
98. Maj AM, Suisse I, Méliet C, Niedercorn FA (2010) Tetrahedron Asymmetry 21:2010–2014
99. Mršić N, Jerphagnon T, Minnaard AJ, Feringa BL, de Vries JG (2010) Tetrahedron Asymmetry 21:7–10
100. Kuwano R, Kashiwabara M (2006) Org Lett 8:2653–2655
101. Kuwano R, Kashiwabara M, Ohsumi M, Kusano H (2008) J Am Chem Soc 130:808–809
102. Wierenga W (1981) J Am Chem Soc 103:5621–5623
103. Baeza A, Pfaltz A (2010) Chem Eur J 16:2036–2039
104. Abe H, Amii H, Uneyama K (2001) Org Lett 3:313–315
105. Yang Q, Shang G, Gao W, Deng J, Zhang X (2006) Angew Chem Int Ed 45:3832–3835
106. Wang Y-Q, Lu S-M, Zhou Y-G (2007) J Org Chem 72:3729–3734
107. Goulioukina NS, Shergold IA, Bondarenko GN, Ilyin MM, Davankov VA, Beletskaya IP (2012) Adv Synth Catal 354:2727–2733
108. Yu C-B, Gao K, Wang D-S, Shi L, Zhou Y-G (2011) Chem Commun 5052–5054
109. Yu C-B, Gao K, Chen Q-A, Chen M-W, Zhou Y-G (2012) Tetrahedron Lett 53:2560–2563
110. Wang Y-Q, Lu S-M, Zhou Y-G (2005) Org Lett 7:3235–3238
111. Wang D-S, Chen Q-A, Li W, Yu C-B, Zhou Y-G, Zhang X-M (2010) J Am Chem Soc 132:8909–8911

112. Wang D-S, Tang J, Zhou Y-G, Chen M-W, Yu C-B, Duan Y, Jiang G-F (2011) Chem Sci 2:803–806
113. Duan Y, Chen M-W, Ye Z-S, Wang D-S, Chen Q-A, Zhou Y-G (2011) Chem Eur J 17:7193–7197
114. Duan Y, Chen M-W, Chen Q-A, Yu C-B, Zhou Y-G (2012) Org Biomol Chem 10:1235–1238
115. Wang D-S, Ye Z-S, Chen Q-A, Zhou Y-G, Yu C-B, Fan H-J, Duan Y (2011) J Am Chem Soc 133:8866–8869
116. Polyak F, Dorofeeva T, Zelchan G (1995) Synth Commun 25:2895–2900
117. Sebek M, Holz J, Börner A, Jähnisch K (2009) Synlett 3:461–465
118. Ohta T, Miyake T, Seido N, Kumobayashi H, Takaya H (1995) J Org Chem 60:357–363
119. Feiertag P, Albert M, Nettekoven U, Spindler F (2006) Org Lett 8:4133–4135
120. Kaiser S, Smidt SP, Pfaltz A (2006) Angew Chem Int Ed 45:5194–5197
121. Bell S, Wüstenberg B, Kaiser S, Menges F, Netscher T, Pfaltz A (2006) Science 311:642–644
122. Ortega N, Urban S, Beiring B, Glorius F (2012) Angew Chem Int Ed 51:1710–1713
123. Benetti S, Risi CD, Pollini GP, Zanirato V (2012) Chem Rev 112:2129–2163
124. Urban S, Beiring B, Ortega N, Paul D, Glorius F (2012) J Am Chem Soc 134:15241–15244
125. Kuwano R, Kameyama N, Ikeda R (2011) J Am Chem Soc 133:7312–7315
126. Fan Q-H, Li Y-M, Chan ASC (2002) Chem Rev 102:3385–3466
127. Trindade AF, Gois PMP, Afonso CAM (2009) Chem Rev 109:418–514
128. Wang Z, Chen G, Ding K (2009) Chem Rev 109:322–359
129. Heitbaum M, Glorius F, Escher I (2006) Angew Chem Int Ed 45:4732–4762
130. de Vos DE, Vankelekom IFJ, Jacobs PA (eds) (2000) Chiral catalyst immobilization and recycling. Wiley-VCH, Weinheim
131. Ding K, Uozumi Y (eds) (2008) Handbook of asymmetric heterogeneous catalysis. Wiley-VCH, Weinheim
132. Chen J, Spear SK, Huddleston JG, Rogers RD (2005) Green Chem 7:64–82
133. Zhou H-F, Fan Q-H, He Y-M, Gu L-Q, Chan ASC (2007) Prog Chem 19:1517–1528
134. Chan SH, Lam KH, Li Y-M, Xu L-J, Tang W-J, Lam FL, Lo WH, Yu W-Y, Fan Q-H, Chan ASC (2007) Tetrahedron Asymmetry 18:2625–2631
135. Bergbreiter DE (2002) Chem Rev 102:3345–3384
136. Dickerson TJ, Reed NN, Janda KD (2002) Chem Rev 102:3325–3344
137. Wang X-B, Zhou Y-G (2008) J Org Chem 73:5640–5642
138. Oosterom GE, Reek JNH, Kamer PCJ, van Leeuwen PWNM (2001) Angew Chem Int Ed 40:1828–1849
139. Astruc D, Chardac F (2001) Chem Rev 101:2991–3024
140. Kassube JK, Gade LH (2006) Top Organomet Chem 20:61–96
141. Fan Q-H, Deng G-J, Feng Y, He Y-M (2008) In: Ding K, Uozumi Y (eds) Handbook of asymmetric heterogeneous catalysis. Wiley-VCH, Weinheim, pp 131–180
142. Ma B, Fan Q-H (2010) Sci Sinica Chim 40:827–836
143. Fan Q-H, Ding K (2011) Top Organomet Chem 36:207–246
144. Fan Q-H, Chen Y-M, Chen X-M, Jiang D-Z, Xi F, Chan ASC (2000) Chem Commun 789–790
145. Wang Z-J, Deng G-J, Li Y, He Y-M, Tang W-J, Fan Q-H (2007) Org Lett 9:1243–1246
146. Dupont J, de Souza RF, Suarez PAZ (2002) Chem Rev 102:3667–3692
147. Hallett JP, Welton T (2011) Chem Rev 111:3508–3576
148. Song CE (2004) Chem Commun 1033–1043
149. Durand J, Teuma E, Gómez M (2007) C R Chimie 10:152–177
150. Luo S, Zhang L, Cheng J-P (2009) Chem Asian J 4:1184–1195
151. Astruc D, Lu F, Aranzaes JR (2005) Angew Chem Int Ed 44:7852–7872
152. Mallat T, Orglmeister E, Baiker A (2007) Chem Rev 107:4863–4890
153. Schätz A, Reiser O, Stark WJ (2010) Chem Eur J 16:8950–8967
154. Polshettiwar V, Luque R, Fihri A, Zhu H, Bouhrara M, Basset J-M (2011) Chem Rev 111:3036–3075

155. Zhu Y, Stubbs LP, Ho F, Liu R, Ship CP, Maguire JA, Hosmane NS (2010) ChemCatChem 2:365–374
156. Wu L, He Y-M, Fan Q-H (2011) Adv Synth Catal 353:2915–2919
157. Yamakawa M, Ito H, Noyori R (2000) J Am Chem Soc 122:1466–1478
158. Yamakawa M, Yamada I, Noyori R (2001) Angew Chem Int Ed 40:2818–2821
159. Ding Z-Y, Chen F, Qin J, He Y-M, Fan Q-H (2012) Angew Chem Int Ed 51:5706–5710
160. Dobereiner GE, Nova A, Schley ND, Hazari N, Miller SJ, Eisenstein O, Crabtree RH (2011) J Am Chem Soc 133:7547–7562
161. Soai K, Shibata T, Sato I (2000) Acc Chem Res 33:382–390

Top Curr Chem (2014) 343: 191–260
DOI: 10.1007/128_2013_500

Published online: 26 November 2013

Asymmetric Hydroamination

Alexander L. Reznichenko, Agnieszka J. Nawara-Hultzsch, and Kai C. Hultzsch

Abstract The direct addition of amine N–H bonds across an unsaturated carbon–carbon linkage gives fast and highly atom-economical access to amines. This review provides an overview of the most effective stereoselective methods using various catalyst systems based on alkali, alkaline earth, and transition metals, as well as enzymatic and organocatalytic approaches. Except for a few sporadic examples in the last century, this field has evolved primarily over the last decade. Catalyst systems have reached enantioselectivities exceeding 90% *ee* for many substrate classes, but significant challenges remain, in particular in the stereoselective intermolecular hydroamination of unactivated alkenes.

Keywords Alkenes · Amines · Asymmetric catalysis · Brønsted acid catalysis · Heterofunctionalization · Hydroamination · Nitrogen heterocycles · Transition metal catalysis

Contents

A.L. Reznichenko (✉), A.J. Nawara-Hultzsch (✉), and K.C. Hultzsch (✉)
Department of Chemistry and Chemical Biology, Rutgers, The State University of New Jersey, 610 Taylor Road, Piscataway, NJ 08854-8087, USA
e-mail: hultzsch@rci.rutgers.edu

Abbreviations

Ar	Aryl
BDPP	2,4-Bis(diphenylphosphino)pentane
BINAP	2,2′-Bis(diphenylphosphino)-1,1′-binaphthyl
Bn	Benzyl
Boc	*tert*-Butoxycarbonyl
Bu	Butyl
cat	Catalyst
Cbz	Benzyloxycarbonyl
COD	Cyclooctadiene
COE	Cyclooctene
Cp*	Pentamethylcyclopentadienyl
Cy	Cyclohexyl
d	Day(s)
DCE	1,2-Dichloroethane
DCM	Dichloromethane
DIOP	Bis(diphenylphosphino-methyl)-2,2-dimethyl-1,3-dioxolane
DMF	Dimethylformamide
dr	Diastereomer ratio
ebthi	Ethylenebis(tetrahydroindenyl)
ee	Enantiomer excess
equiv	Equivalent(s)
Et	Ethyl
Fmoc	Fluorenylmethoxycarbonyl
Grubbs-I	Benzylidene-bis(tricyclohexylphosphine)dichlororuthenium
Grubbs-II	Benzylidene[1,3-bis(2,4,6-trimethylphenyl)-2-imidazolidinylidene]dichloro(tricyclohexylphosphine)ruthenium
h	Hour(s)
i-Pr	Isopropyl
KHMDS	Potassium hexamethyldisilazide potassium bis(trimethylsilyl) amide
LDA	Lithium diisopropylamide
Mbs	*p*-Methoxy benzenesulfonyl
Me	Methyl
MeO-BIPHEP	6,6-Dimethoxy-2,2-bis(diphenylphosphino)-1,1-biphenyl
Mes	Mesityl 2,4,6-trimethylphenyl

mol	Mole(s)
MS	Molecular sieves
Mts	Mesitylenesulfonyl
MW	Microwave
Nf	Nonafluorobutanesulfonyl
NMO	*N*-Methylmorpholine *N*-oxide
N_t	Turnover frequency
PAL	Phenylalanine ammonia lyase
PG	Protecting group
Ph	Phenyl
PMB	4-Methoxyphenyl
PNB	4-Nitrobenzoyl
rt	Room temperature
SEGPHOS	4,4′-Bi-1,3-benzodioxole-5,5′-diylbis(diphenylphosphane)
TBDPS	*tert*-Butyldiphenylsilyl
TBS	*tert*-Butyldimethylsilyl
t-Bu	*tert*-Butyl
Tf	Trifluoromethanesulfonyl (triflyl)
TFAA	Trifluoroacetic anhydride
THF	Tetrahydrofuran
T*i*PP	2,4,6-Triisopropyphenyl
TIPS	Triisopropylsilyl
Tol	4-Methylphenyl
Troc	2,2,2-Trichloroethoxycarbonyl
Ts	Tosyl 4-toluenesulfonyl
Xylyl-BINAP	2,2′-Bis[di(3,5-xylyl)phosphino]-1,1′-binaphthyl

1 Introduction

The addition of an amine N–H bond across an unsaturated carbon–carbon linkage (Scheme 1), the so-called hydroamination, allows a facile and highly atom-economical access to industrially relevant nitrogen containing basic and fine chemicals as well as naturally occurring alkaloid skeletons [1–6]. In general, the addition of amines to non-activated simple alkenes represents the greatest challenge, while reactions involving alkynes, allenes, and dienes proceed more readily. However, the addition of an amine to a polyene moiety may not necessary result in the formation of a new stereocenter, if an imine is generated.

Significant research efforts have led to the development of efficient catalyst systems for hydroamination reactions. Many catalytic systems based on transition metals (groups 3–5 and 8–10, as well as lanthanides and actinides) and main group metals (alkali and alkaline earth metals), as well as Brønsted acids, have been developed over the last 6 decades, especially the last 2 decades have seen significant progress. However, the majority of these catalyst systems are confined to a

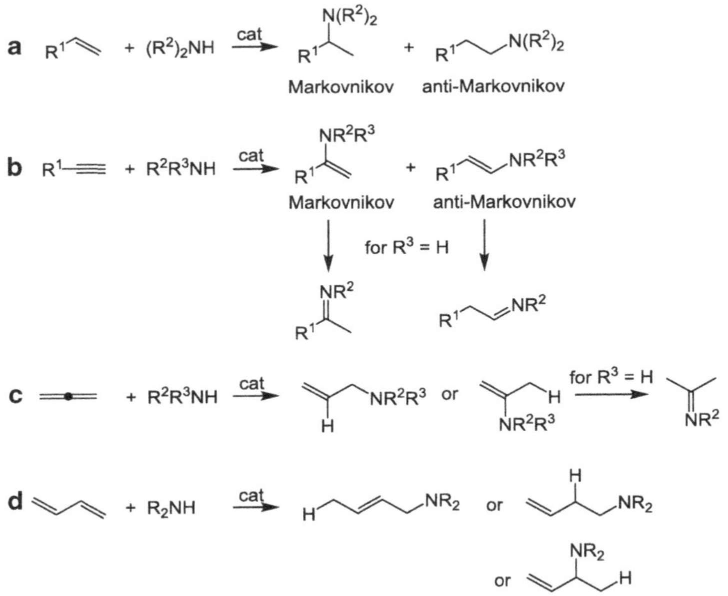

Scheme 1 The catalytic hydroamination of (**a**) alkenes, (**b**) alkynes, (**c**) allenes, and (**d**) dienes leads to amines, imines, and enamines. The reactions may also be performed in an intramolecular fashion (not shown)

limited set of substrates, e.g., requiring activated multiple C–C bonds, such as 1,3-dienes, vinyl arenes, allenes, or ring-strained olefins. Because of its potential relevance for the synthesis of pharmaceuticals and biological-active molecules, most of which are chiral, the development of chiral catalysts for stereoselective hydroamination (including kinetic resolution of chiral amines) has drawn increasing attention over the last decade [7–16].

The direct addition of amines to alkenes is thermodynamically feasible ($\Delta G^{\theta} \approx -14.7$ kJ mol^{-1} for the addition of ammonia to ethylene) [17, 18] with a slightly exothermic to thermoneutral reaction enthalpy. However, this seemingly simple reaction is hampered by a high activation barrier caused by an electrostatic repulsion between the nitrogen lone pair of the approaching amine and the π-bond of the electron rich olefin. The activation barrier cannot be overcome by increasing the reaction temperature because the negative reaction entropy shifts the equilibrium towards the starting materials. Therefore, amines add commonly only to activated, electron deficient alkenes (e.g., vinyl ethers or Michael acceptors) in the absence of a catalyst. For the scope of this overview, such *aza*-Michael reactions will not be considered and the reader is referred to pertinent reviews in the literature [19, 20].

2 Hydroamination of Alkenes

2.1 *Metal-Catalyzed Intermolecular Hydroamination of Alkenes*

Although several significant contributions to the field of intermolecular hydroamination of simple alkenes have been made within the last decade, the asymmetric hydroamination of *non-activated* alkenes with *unprotected* amines remains one of the most challenging problems in the field. Unfavorable factors for the intermolecular hydroamination are not only the negative reaction entropy, but also the competition between strongly coordinating amines and – in particular for early transition metal catalysts – weakly binding alkenes. Many catalyst systems are deactivated by strongly basic amines that block binding of the alkene to the catalyst metal center.

An asymmetric intermolecular hydroamination of terminal, non-activated alkenes with cyclic ureas has been reported only recently (Scheme 2) [21]. The Markovnikov addition proceeds with up to 78% *ee* utilizing the axial chiral MeOBIPHEP-ligated bis(gold(I))-catalyst system (*S*)-**1**. Unfortunately, the reaction requires a large excess of the alkene substrate and lower alkene loadings lead to diminished enantioselectivities.

An iridium-catalyzed addition of amides to unreactive α-olefins has been reported to generate a mixture of the hydroamination product **3a** and oxidative amination product **3b** (Eq. 1) [22]. The amide **3a** is formed in low enantioselectivities of up to 19% *ee*. Again, a large excess of alkene is required to achieve high conversion. The enamide by-product **3b** can be reduced to the amide **3a** via transfer hydrogenation using *i*-PrOH.

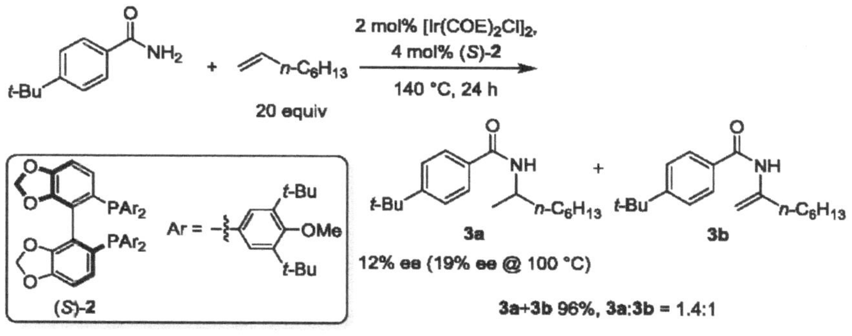

Equation 1. Intermolecular iridium-catalyzed addition of amides to unactivated α-olefins [22]

The asymmetric hydroamination of non-activated alkenes with *N*-unprotected amines remains very elusive as well. Primary alkylamines react with terminal unactivated olefins with excellent Markovnikov selectivity in the presence of the yttrium binaphtholate catalyst (*R*)-**4**-Y (Scheme 3) [23]. The reactions require

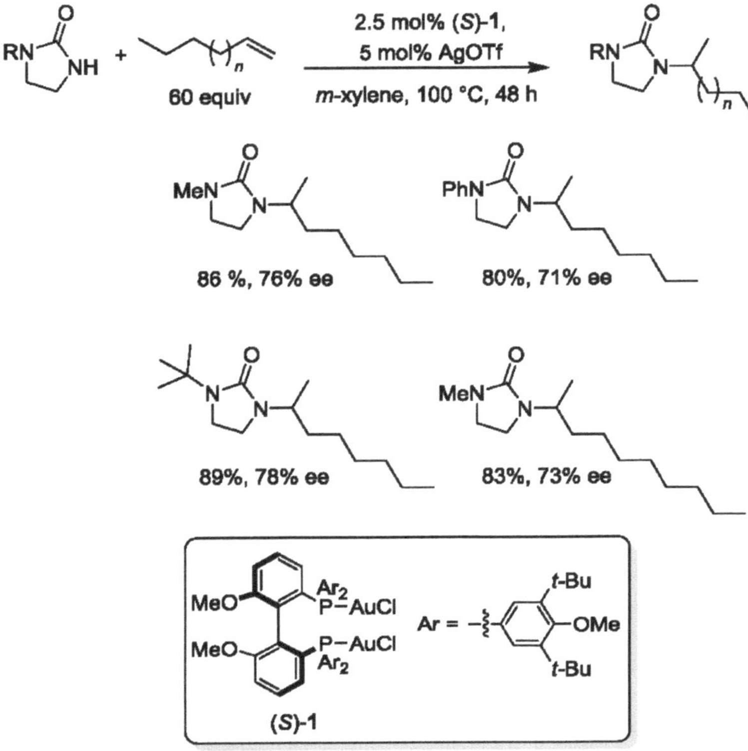

Scheme 2 Gold-catalyzed asymmetric hydroamination of alkenes with cyclic ureas [21]

a 9- to 15-fold excess of the alkene and high reaction temperatures. Despite the harsh reaction conditions, moderate enantioselectivities of up to 61% *ee* are observed. Secondary amines and internal alkenes are unreactive; moreover, sterically hindered terminal olefins, such as vinyl cyclohexane, also display significantly diminished reactivity.

The catalytic activity of rare earth metal complexes in intermolecular hydroaminations is controlled by the electron deficiency of the metal center (even more so than in intramolecular reactions; see Sect. 2.4.1). Therefore, catalysts comprised of a C_1-symmetric NOBIN-based aminodiolate ligand exhibit lower catalytic activity and lower enantioselectivities (up to 40% *ee*) compared to the binaphtholate complex (*R*)-**4**-Y [24]. The decrease in catalytic activity can be rationalized by the presence of an additional amine donor in the aminodiolate ligand framework. While the catalytic activity generally increases with increasing size of the rare earth metal in intramolecular hydroaminations, this does not seem to be the case for intermolecular hydroaminations of alkenes where yttrium and lanthanum catalysts possess comparable activity [24].

Since more reactive alkenes, such as vinyl arenes or sterically strained polycycles, react more readily in the hydroamination reaction, several asymmetric

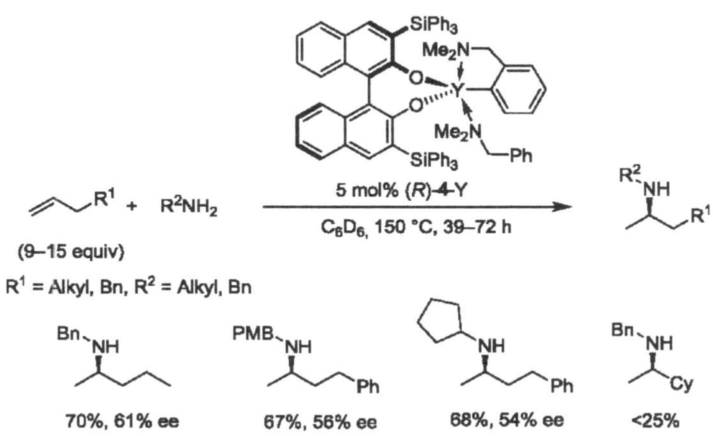

Scheme 3 Asymmetric intermolecular hydroamination of unactivated alkenes [23]

hydroamination reactions utilizing these substrates have been reported. Weakly basic anilines react with vinyl arenes to give the Markovnikov addition products with good yields and moderate to high enantioselectivities in the presence of chiral palladium complexes utilizing BINAP (Eq. 2) or the 4,4′-disubstituted SEGPHOS **5** (Eq. 3) [25–27].

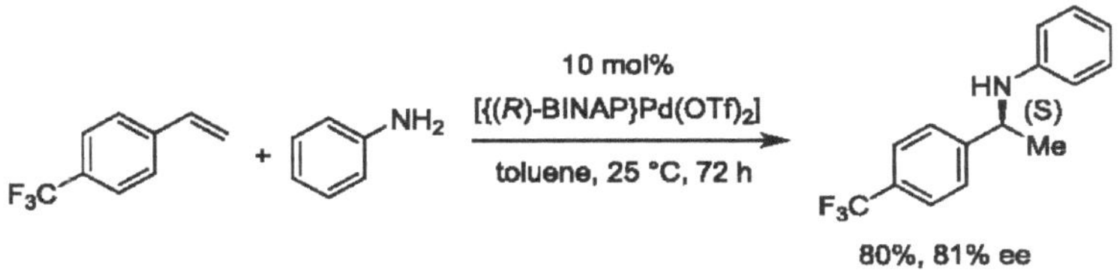

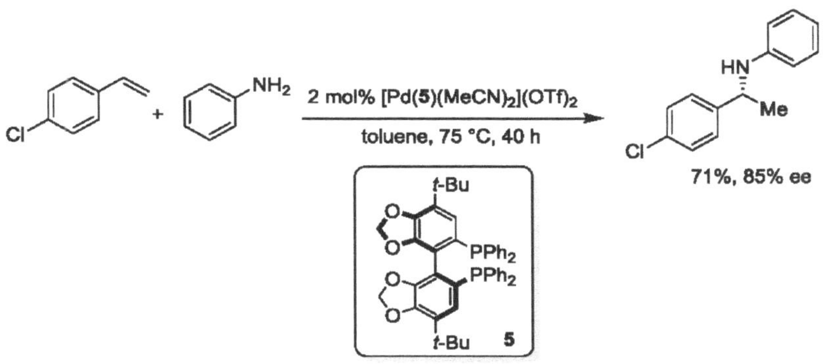

Equations 2 and 3. Palladium-catalyzed asymmetric Markovnikov addition of aniline to vinyl arenes [25, 27]

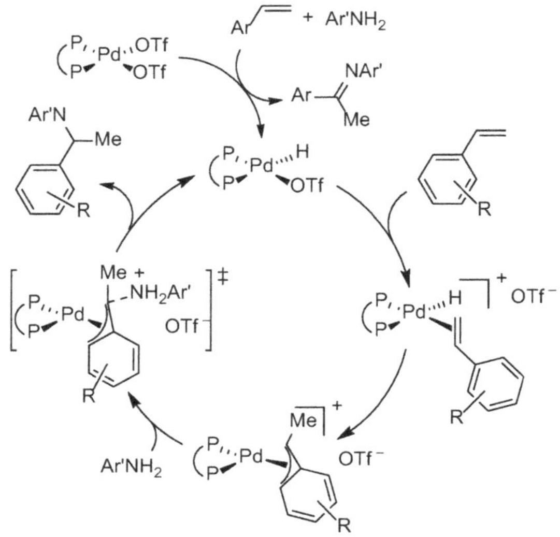

Scheme 4 Proposed mechanism for the palladium-catalyzed hydroamination of vinyl arenes

The proposed mechanism for this process involves an insertion of the vinyl arene into a palladium hydride species, followed by a nucleophilic attack of the amine on the resulting η^3-benzylic palladium intermediate (Scheme 4) [28–30]. The high Markovnikov regioselectivity results from the electronically favored secondary insertion of the vinyl arene in the palladium hydride bond. Similar mechanistic models have been suggested for the late transition metal-catalyzed hydroamination of allenes [31–33], dienes [30, 34–36], and trienes [37]. Generally, the η^3-allyl species are identified as the resting state of the catalyst during the hydroamination reaction, indicating that nucleophilic attack is rate determining [28, 36].

Two mechanistically plausible scenarios for nucleophilic attack on the η^3-benzyl palladium species seem feasible. Formation of the C–N bond could occur either via external attack of the amine via inversion of configuration at the carbon stereocenter, or alternatively, the amine could coordinate to palladium followed by an internal attack on the η^3-benzyl ligand. Mechanistic investigations [28] using stoichiometric amounts of the enantio- and diastereomerically pure η^3-benzyl palladium complex [{(*R*)-Tol-BINAP}{η^3-1-(2-naphthyl)ethyl}Pd](OTf) (**6**) reveal that the reaction with aniline produces predominantly (*R*)-*N*-1-(2-naphthyl)ethylaniline ((*R*)-**7**), consistent with external nucleophilic attack (Scheme 5) [28]. However, the catalytic reaction of [{(*R*)-Tol-BINAP}Pd(OTf)$_2$] with vinyl arenes and amines produces preferentially the opposite enantiomeric (*S*)-amine hydroamination product, indicating, in analogy to asymmetric rhodium-catalyzed hydrogenation [38], that the minor diastereomer of the catalytic active benzyl palladium species is responsible for the majority of product formed in the catalytic process.

The scope of amine substrates can be expanded to more basic alkylamines (Eq. 4) [29]. The arylethanamine **8** is obtained in moderate yield and enantioselectivity utilizing the (*R*,*R*)-Et-FerroTANE ligand. Note that almost stoichiometric

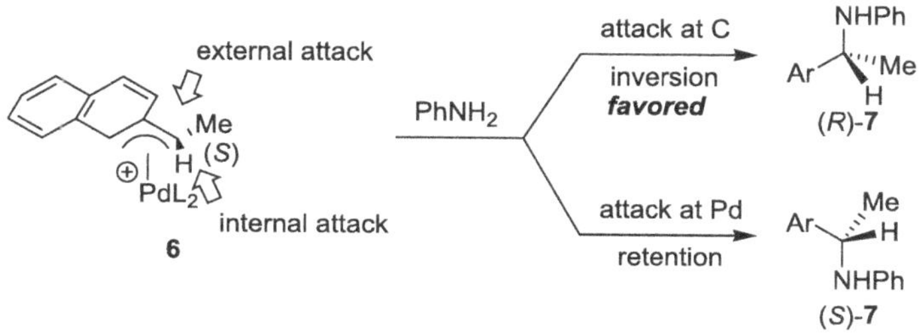

Scheme 5 Preferential external attack of the aniline nucleophile leads to inversion of stereochemistry at the benzyl palladium intermediate [28]

amounts of a strong Brønsted acid are required to afford the Markovnikov hydroamination product of the vinyl arene.

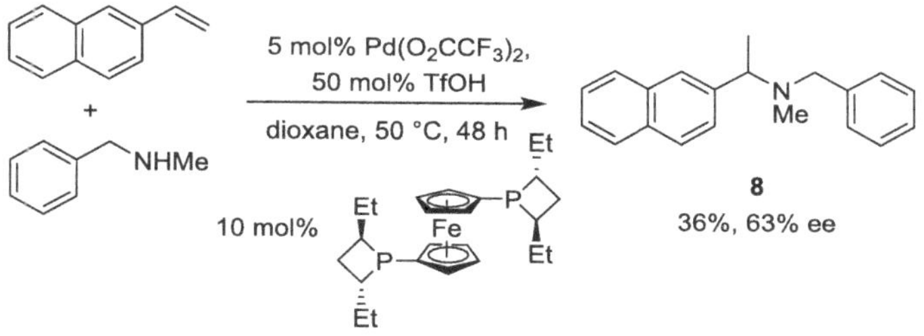

Equation 4. Palladium-catalyzed asymmetric Markovnikov addition of an alkylamine to a vinyl arene [29]

The enhanced reactivity of strained polycyclic olefins, such as norbornene, has made them attractive model compounds for intermolecular hydroamination. In most cases the reaction employs a late transition metal-based catalyst and requires elevated temperatures and extended reaction times. The first chiral iridium-based catalyst system was reported by Togni in 1997 (Eq. 5) [39]. The activity and enantioselectivity of this catalyst system is significantly enhanced by addition of Schwesinger's "naked" fluoride $[N\{P(NMe_2)_3\}_2]F$. In some cases the addition of fluoride also results in a reversal of absolute product configuration. However, the reason for the strong fluoride effect remains unclear, but it has been suggested that hydrogen bridging of the fluoride could play a role.

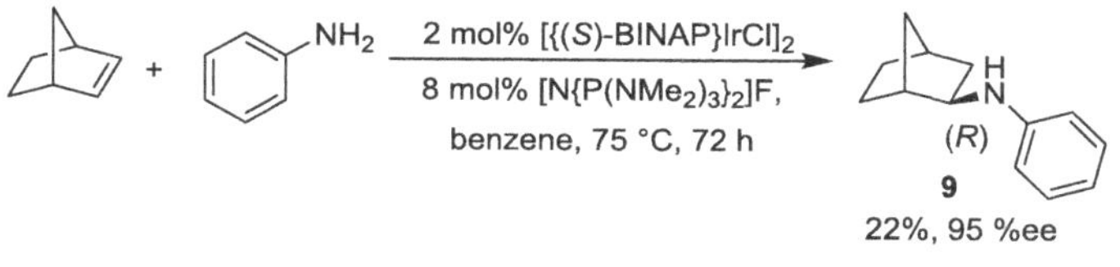

Equation 5. Iridium-catalyzed asymmetric hydroamination of norbornene [39]

Earlier mechanistic studies by Milstein on an achiral iridium-catalyst system indicate that the iridium-catalyzed norbornene hydroamination involves amine

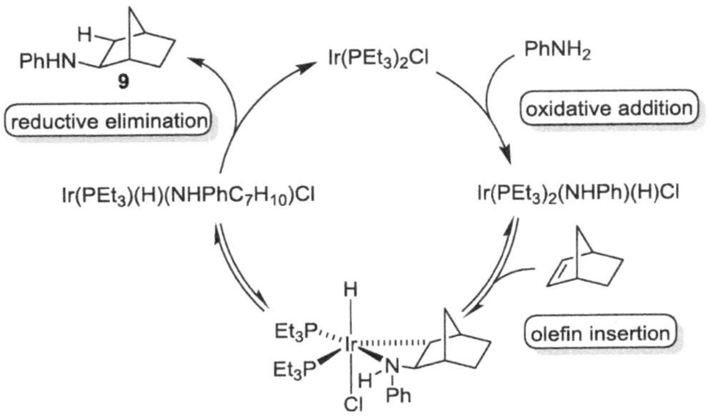

Scheme 6 Proposed mechanism for iridium-catalyzed hydroamination of norbornene via amine activation

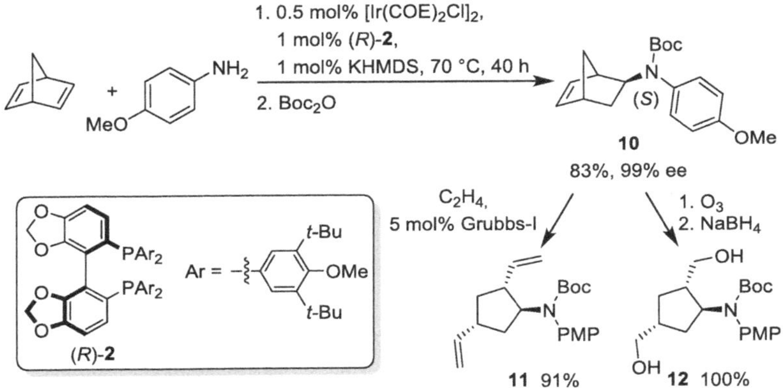

Scheme 7 Stereoselective synthesis of cyclopentylamines via asymmetric hydroamination of norbornadiene [42]

activation as a key step in the catalytic cycle [40], rather than *alkene* activation which is observed for most other late transition metal-catalyzed hydroamination reactions [41]. Thus, the iridium-catalyzed hydroamination of norbornene with aniline is initiated by an oxidative addition of aniline to the metal center, followed by insertion of the strained olefin into the iridium-amido bond (Scheme 6). Subsequent reductive elimination completes the catalytic cycle and gives the hydroamination product **9**. Unfortunately, this catalyst system seems to be limited to highly strained olefins.

This chemistry has been extended to a number of bicyclic alkenes and dienes utilizing various chelating axially chiral bisphosphine iridium-catalysts (Scheme 7) [22, 42]. Further synthetic transformations of the chiral hydroamination product **10** provide access to functionally substituted chiral cyclopentylamines with multiple stereocenters, such as **11** and **12**. It should be noted that alkylamines, such as

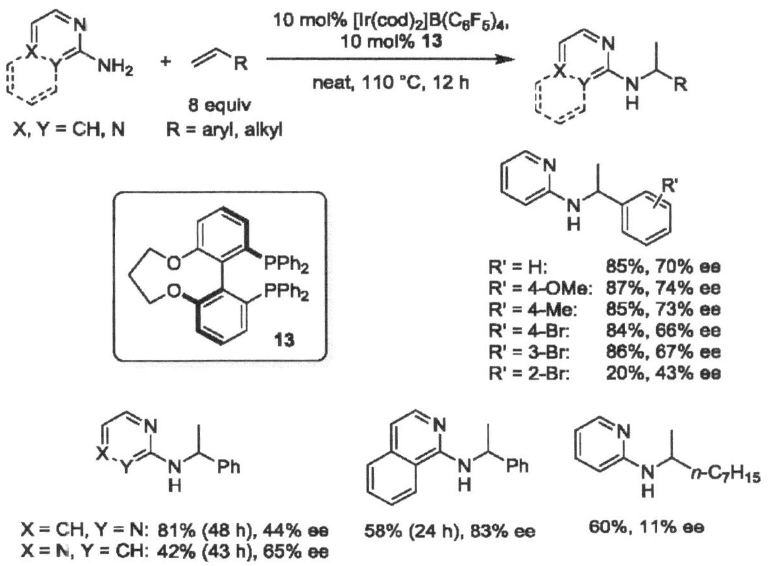

Scheme 8 Iridium-catalyzed asymmetric hydroamination of α-olefins with heteroaromatic amines [43]

octylamine or *N*-methyl aniline, and sterically encumbered aniline derivatives, such as *o*-toluidine or *o*-anisidine, do not undergo hydroamination reactions under these conditions. However, amides and sulfonamides react at 120°C in high yield and greater than 90% enantioselectivity (Eq. 6) [22].

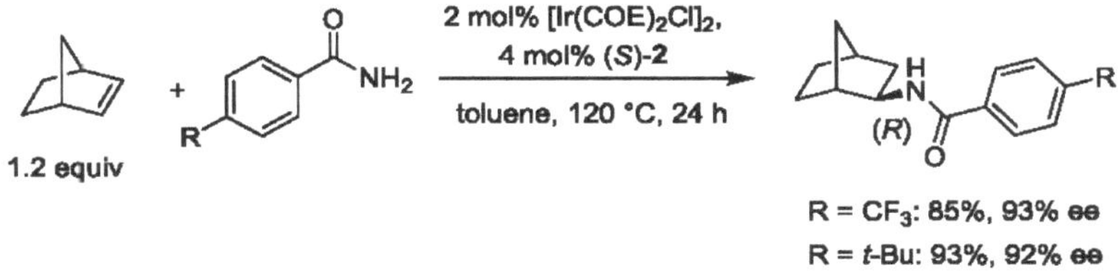

Equation 6. Stereoselective synthesis of bicycloaminoalkanes via asymmetric hydroamination of norbornene [22]

Cationic iridium catalysts also facilitate the Markovnikov hydroamination of vinyl arenes with 2-aminopyridines, 2-aminopyrimidines, 2-aminopyrazins, and related heteroaromatic amines with moderate to good enantioselectivities of up to 83% *ee* using C_3-TUNEPHOS **13** as chiral ligand (Scheme 8) [43]. A variety of electron-donating and electron-withdrawing substituents are tolerated in the vinyl arene; however, *ortho*-substitution does diminish the yield and enantioselectivity. Simple α-olefins, such as 1-nonene or *tert*-butylethylene, also react in moderate yield, albeit low enantioselectivities.

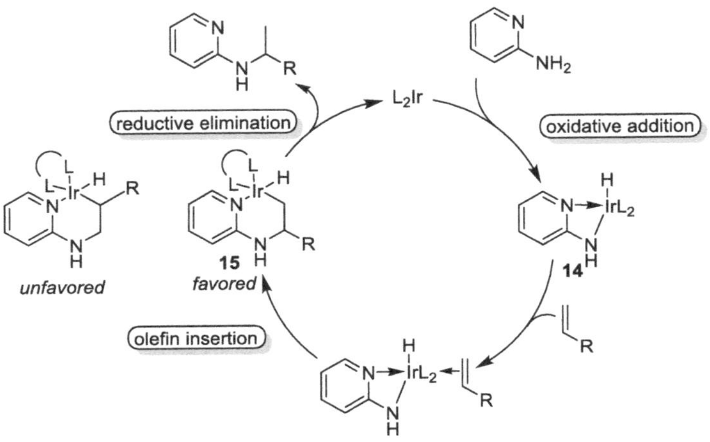

Scheme 9 Proposed mechanism for the iridium-catalyzed regioselective intermolecular hydroamination of alkenes with 2-aminopyridine [43]

The tentative mechanism for this iridium-catalyzed hydroamination of alkenes with heteroaromatic amines has been proposed to proceed analogous to the Milstein mechanism (as shown in Scheme 6) via oxidative addition of the amine to afford the intermediate **14** (Scheme 9). Alkene insertion into the iridium–amide bond yields the more favorable intermediate **15**, which releases the hydroamination product via reductive elimination. The chelation of the pyridine nitrogen to iridium appears to be crucial for the control of regioselectivity. Steric hindrance in the *ortho*-position to the pyridine nitrogen results in formation of a mixture of Markovnikov and anti-Markovnikov product (Eq. 7).

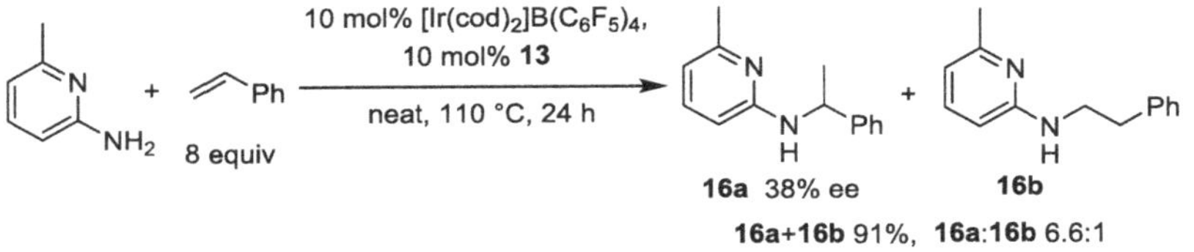

Equation 7. Diminished regiocontrol due to steric hindrance in the iridium-catalyzed hydroamination of alkenes with 2-aminopyridine [43]

2.2 *Enzymatic Intermolecular Hydroamination of Alkenes*

The utilization of biocatalysis for the development of stereoselective synthetic methodology has increased significantly over the last decade [44]. Phenylalanine ammonia lyase (PAL) is an enzyme known to mediate the deamination of phenylalanine into *E*-cinnamic acid and ammonia (Eq. 8) [45].

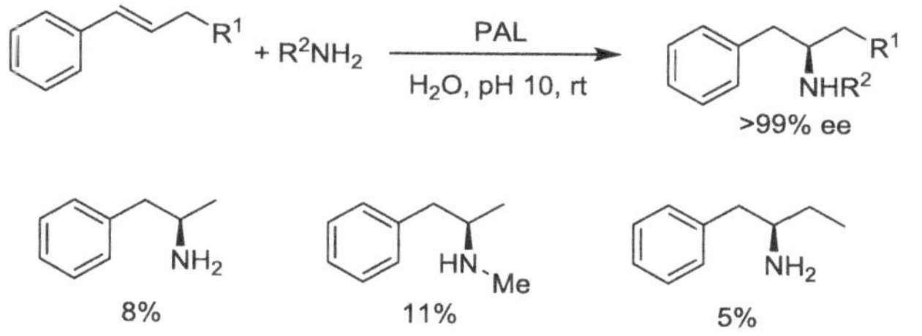

Scheme 10 Enzymatic hydroamination of vinyl arenes [46]

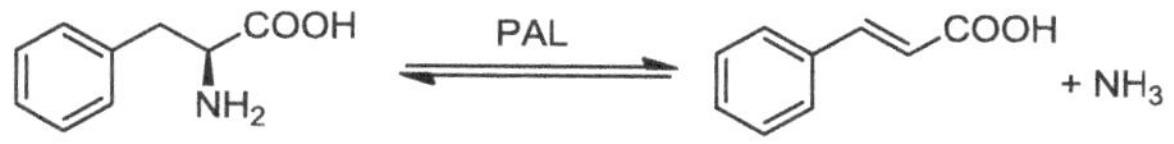

Equation 8. Enzyme-catalyzed deamination of phenylalanine

Conversely, the hydroamination of vinyl arenes catalyzed by PAL from the parsley plant *Petroselinum crispum* (Scheme 10) has been reported recently [46]. Although the method is currently limited to *E*-β-methylstyrene derivatives and very few amines are reactive, the approach can certainly become a solid foundation for future advances in the area.

2.3 Cope-Type Hydroamination

The so-called Cope-type hydroamination, which is also referred to as reverse Cope elimination, provides a conceptually different approach to the addition of a nitrogen-center to a carbon–carbon multiple bond [47]. The hydroxylamine starting material and reaction product in the Cope-type hydroamination both contain nitrogen in a different oxidation level than the amines found in a conventional hydroamination process.

While there are no direct asymmetric intermolecular Cope-type hydroaminations, this reaction can be achieved via a catalyst that reversibly tethers the alkene and the hydroxylamine to facilitate the reaction in a "temporary" intramolecular fashion. Aldehydes can form a temporary tether between an allylic amine and a hydroxylamine. The resulting transient aminal species can undergo intramolecular Cope-type hydroamination in a highly stereocontrolled fashion when a chiral aldehyde, e.g., **17** or **18**, is used (Scheme 11) [48, 49]. The reaction proceeds with high stereoselectivity for electron-rich and electron-poor *N*-benzylhydroxylamines, while aliphatic hydroxylamines are less active and less selective. The enantioselectivities obtained with aldehyde **17** vary significantly with the steric demand and nucleophilicity of the allylic amine, due to a slow epimerization of **17** under catalytic conditions that diminishes selectivities for slower reactions [49]. The bicyclic aldehyde **18** is configurationally more stable and provides high selectivities more reliably.

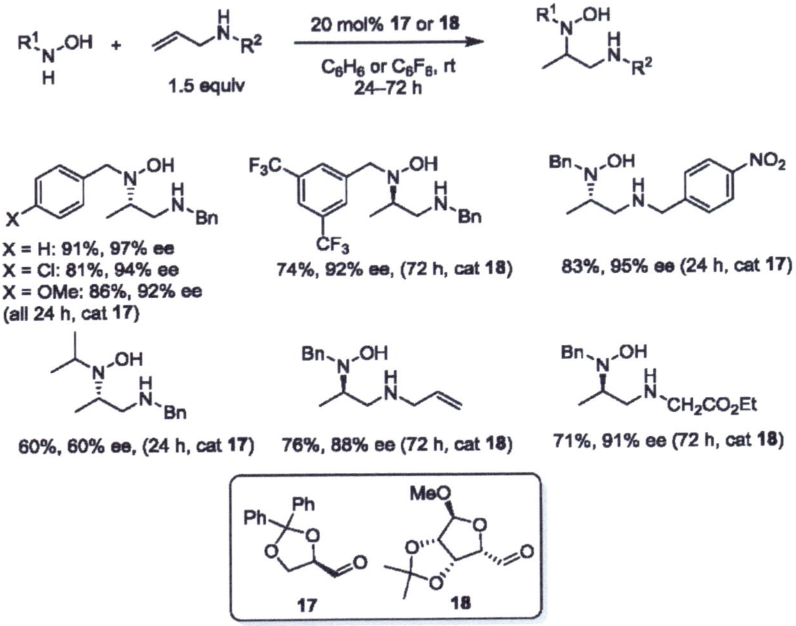

Scheme 11 Asymmetric Cope-type hydroamination with chiral aldehydes [48, 49]

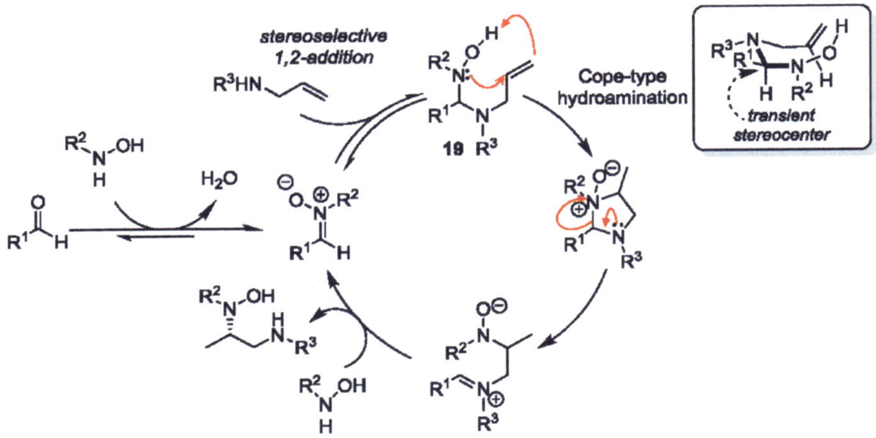

Scheme 12 Proposed mechanism for the asymmetric Cope-type hydroamination via temporary intramolecularity [48–50]

The reaction mechanism (Scheme 12) [50] begins with the condensation of the hydroxylamine and the aldehyde precatalyst to form a nitrone. Addition of the allylamine affords the transient chiral mixed aminal **19**. The subsequent rate-limiting, Cope-type hydroamination step proceeds with high chirality transfer of the stereocenter in **19** via a highly organized bicyclic transition state. It is currently unclear whether the stereoinduction results from stereoselective formation of aminal **19** or from a synergistic effect between the aminal formation and Copy-type hydroamination step.

2.4 *Intramolecular Hydroamination of Aminoalkenes*

2.4.1 Rare Earth Metal-Based Catalysts

Among the hydroamination catalyst systems developed so far, organo rare earth metal complexes have been found to be the most versatile and most active catalysts for the hydroamination of non-activated alkenes, allenes, 1,3-dienes, and alkynes [5, 51, 52]. As noted in Sect. 2.1, intermolecular hydroamination reactions with these catalysts remain challenging. Therefore, most rare earth metal-catalyzed reactions have been performed in an intramolecular fashion producing predominantly pyrrolidine and piperidine derivatives.

The mechanism of the hydroamination/cyclization [53, 54] proceeds through a metal–amido species, which is formed upon protonolysis of a metal–amido, metal–alkyl, or metal–aryl bond in the precatalyst (Scheme 13), followed by insertion of the olefin into the metal–amido bond with a seven-membered chair-like transition state (for $n = 1$). The roughly thermoneutral insertion step [54] is usually rate-determining, giving rise to a zero order rate dependence on substrate concentration and first order rate dependence on catalyst concentration. The resulting metal–alkyl species undergoes fast protonolysis with a second amine molecule, regenerating the metal–amido species and releasing the heterocyclic product. Although the mechanism was established for rare earth metal complexes, alkali and alkaline earth metal-based systems are apparently operating in agreement with this mechanism.

Although the protonolysis step is considered to be fast, a strong primary isotope effect as well as an effect of the isotope substitution on stereoselectivity [53, 55] have been observed, which is indicative of significant N-H bond disruption in the transition state of the rate-determining alkene insertion step. A plausible explanation involves partial proton transfer from a coordinated amine to the α-carbon in the four-membered insertion transition state (Scheme 14). However, an alternative scenario involving reversible olefin insertion followed by rate-determining protonolysis of the metal–alkyl bond has been suggested for some catalyst systems based on magnesium [56] and yttrium [57].

Chiral rare earth metal-based catalysts for the asymmetric hydroamination/cyclization of aminoalkenes were first introduced in 1992 by Marks and co-workers [58, 59]. The C_1-symmetric chiral *ansa*-lanthanocene complexes **20–22** with (+)-neomenthyl, (–)-menthyl, or (–)-phenylmenthyl substituents attached to one of the cyclopentadienyl ligands (Fig. 1) exist in two diastereomeric forms, depending on which diastereotopic face of the cyclopentadienyl ligand coordinates to the metal center. Generally, one of the two diastereomers

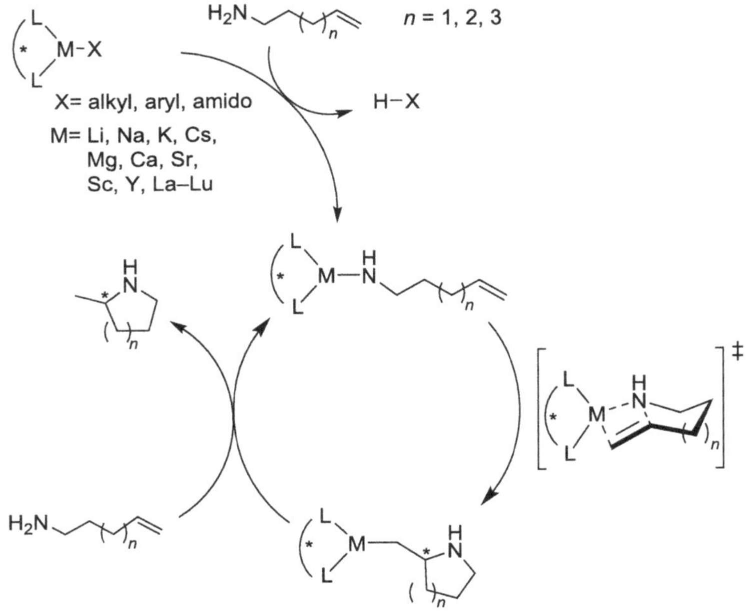

Scheme 13 Proposed mechanism for the hydroamination/cyclization of aminoalkenes using alkali, alkaline earth, and rare earth metal-based catalysts

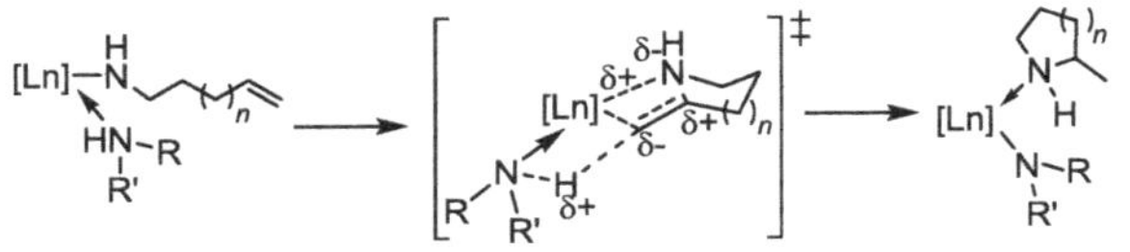

Scheme 14 Proposed concerted protonolysis/insertion via free amine-assisted alkene insertion in hydroamination/cyclization (RR′NH = substrate or hydroamination product) [53]

predominates in solution and most of the complexes can be obtained diastereomerically pure by fractional crystallization.

The lanthanocene catalysts display high catalytic activity, which is proportional to the ionic radius of the rare earth metal (Table 1). The rate of cyclization depends on the ring size of the azacyclic product (5 > 6 >> 7) and the presence of rate enhancing *gem*-dialkyl substituents [60, 61]. The proposed stereomodel for aminopentene substrates (Fig. 2) [59] predicts preferred formation of the (*S*)-pyrrolidine enantiomer due to unfavorable steric interactions between an axial substituent of the substrate with the substituents on the cyclopentadienyl ligands of the (*S*)-lanthanocene diastereomer in the transition state of the cyclization. Hence, dominance of one of the two catalyst epimers is essential to achieve high enantioselectivities.

Unfortunately, the complexes undergo facile epimerization under the conditions of catalytic hydroamination via reversible protolytic cleavage of the metal cyclopentadienyl bond (Scheme 15) [59, 62–64]. Thus, the product enantioselectivity is

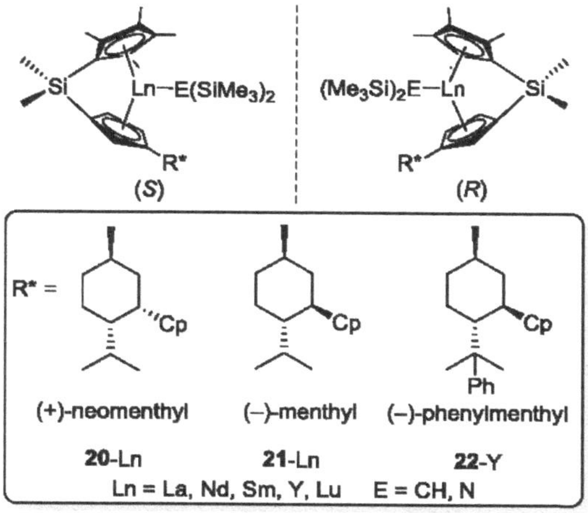

Fig. 1 Chiral lanthanocene precatalysts for asymmetric hydroamination [58, 59]

Table 1 Lanthanocene-catalyzed aminoalkene hydroamination/cyclization [58, 59]

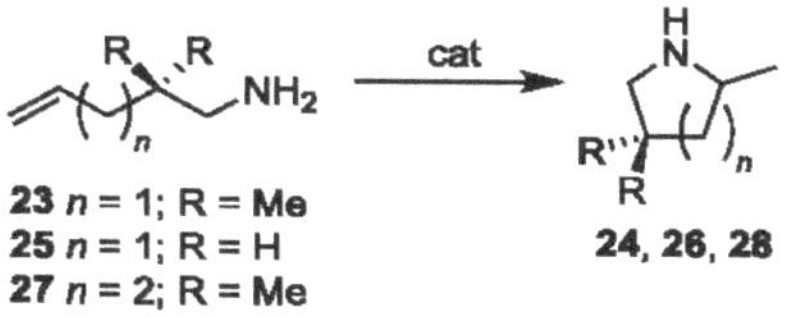

Cat	Substrate	*T*, °C	N_t, h^{-1}	% *ee* (config)
(*R*)-**20**-Sm	**23**	−30		64 (*R*)
(*R*,*S*)-**20**-Y	**23**	25	38	36 (*R*)
(*R*)-**20**-Y	**23**	25	21	40 (*R*)
(*S*)-**21**-Sm	**23**	25	84	53 (*S*)
(*S*)-**21**-Sm	**23**	−30		74 (*S*)
(*R*)-**22**-Y	**23**	25	8	56 (*S*)
(60/40) (*R*,*S*)-**22**-Y	**23**	25		54 (*S*)
(*S*)-**20**-Sm	**25**	25	33	55 (*R*)
(*R*)-**20**-Sm	**25**	25	62	52 (*R*)
(*S*)-**21**-Sm	**25**	0		72 (*S*)
(*R*)-**22**-Y	**25**	25		64 (*S*)
(*R*)-**20**-Sm	**27**	25		17 (*R*)
(*R*)-**21**-Sm	**27**	25	2	15 (*R*)

limited by the catalyst's epimeric ratio in solution and the absolute configuration of the hydroamination product is independent of the diastereomeric purity of the precatalyst. Complexes with a (+)-neomenthyl substituent on the cyclopentadienyl ligand generally produce the (*R*)-(–)-pyrrolidines, whereas (–)-menthyl and (–)-phenylmenthyl substituted complexes yield the (*S*)-(+)-pyrrolidines, which is in agreement with the proposed stereomodel and solution studies on the equilibrium epimer ratios in the presence of simple aliphatic amines.

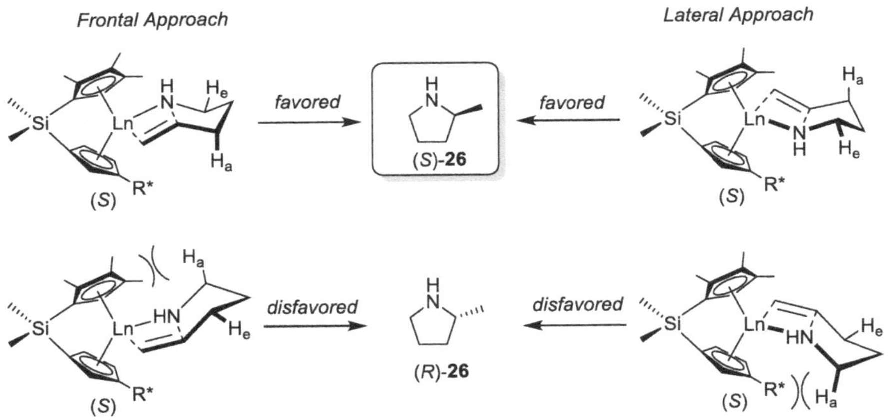

Fig. 2 Proposed stereomodels for the lanthanocene-catalyzed hydroamination/cyclization of pent-4-enylamine (**25**) [59]

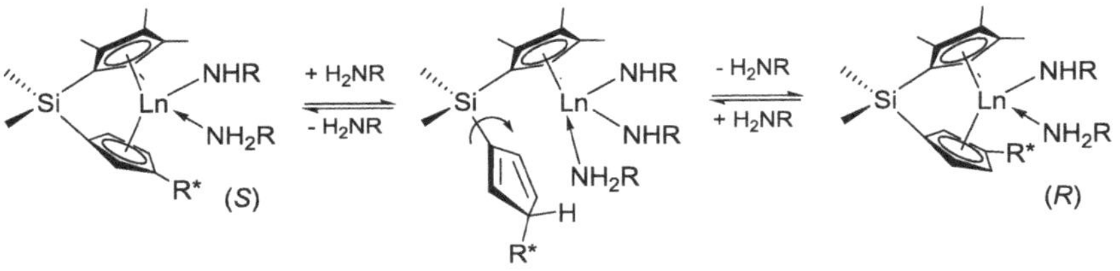

Scheme 15 Amine-induced epimerization of chiral lanthanocene complexes

The highest enantioselectivities of up to 74% *ee* are achieved with the (–)-menthyl-substituted samarium complex (*S*)-**21**-Sm in the formation of pyrrolidine products, whereas the formation of piperidines suffers from significantly diminished selectivities. The extended "wingspan" of the octahydrofluorenyl ligand in complex (*S*)-**29** is supposed to provide an increased enantiofacial discrimination for prochiral substrates, but only isolated examples of improved selectivities are observed (Eq. 9) [62].

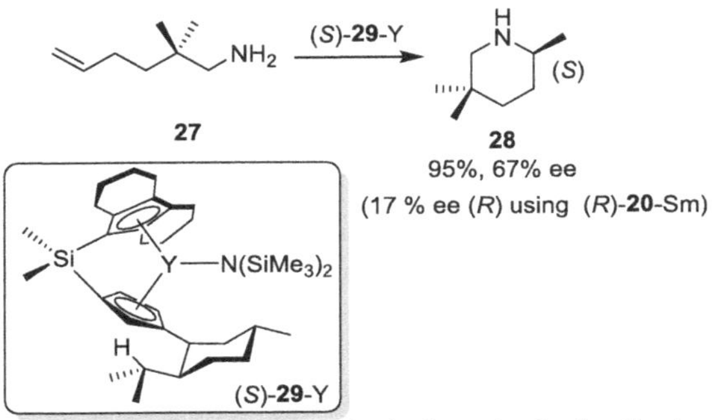

Equation 9. Improved enantioselectivities in the hydroamination/cyclization of aminohexenes using a chiral octahydrofluorenyl yttrocene catalyst [62]

Internal 1,1- or 1,2-disubstituted olefins **30** and **32** are much less reactive for hydroamination and require significantly harsher reaction conditions [63, 65–70], except for aminoalkenes bearing a phenyl substituent attached to the unsaturated carbon–carbon linkage. The formation of pyrrolidines and piperidines often proceeds at comparable rates (Eq. 10), contrasting with the general trend of significant faster five-membered ring formation observed with terminal aminoalkenes [63]. Despite these harsh reaction conditions, moderate enantioselectivities of up to 58% *ee* at 100°C (up to 68% at 60°C) are observed.

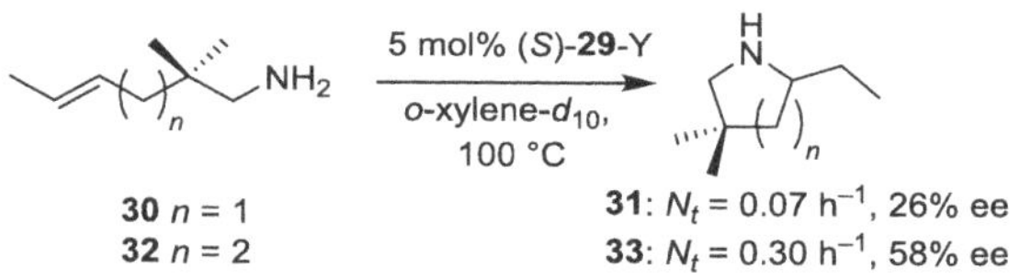

Equation 10. Enantioselective hydroamination/cyclization of internal aminoalkenes [63]

Significant progress in this area has been made since 2003 utilizing non-cyclopentadienyl-based ligand sets [8, 9, 12–14, 16, 24, 55, 69–103], avoiding configurational instability issues of the chiral lanthanocene complexes. However, while enantioselectivities have improved rather significantly, only a few catalyst systems [55, 79] display the same high catalytic activity as the lanthanocene catalysts.

A variety of bisoxazolinato rare earth metal complexes have been studied with regard to their hydroamination/cyclization catalytic activity (Scheme 16) [75]. The precatalysts are conveniently generated in situ from $[Ln\{N(SiMe_3)_2\}_3]$ (Ln = La, Nd, Sm, Y, Lu) and the corresponding bisoxazoline, resulting in minimal reactivity loss, but with similar enantioselectivity, in comparison to the isolated precatalysts. The ligand accelerated catalyst system shows the highest rates for a 1:1 metal to ligand ratio.

Based on a molecular modeling study, the preferred formation of the (*R*) pyrrolidine product results from the approach of the alkene to an empty equatorial coordination site of the bisoxazolinate complex with the amide being bound in the apical position (Fig. 3). The approach of the alkene to an apical coordination site with an equatorial La–N bond is expected to favor slightly formation of the (*S*)-enantiomer. Interestingly, catalysts with aliphatic substituents (*i*-Pr, *t*-Bu) in the 4-position of the bisoxazolinate ligand produce products with opposite configuration, potentially due to a change in the mode of approach of the alkene moiety.

The ate-complexes $[Li(THF)_4][Ln\{(R)\text{-}1,1'\text{-}\{C_{10}H_6N(R)\}_2\}_2]$ ((*R*)-**36**; Ln = Sm, Yb, Y; R = alkyl; Fig. 4) [13, 80–84, 87] are unusual hydroamination catalysts as they lack an obvious leaving amido or alkyl group that is replaced during the initiation step by the substrate. It is very likely that at least one of the amido groups is protonated during the catalytic cycle, analogous to the mechanism proposed for Michael additions and aldol reactions catalyzed by heterobimetallic rare earth metal–alkali metal–BINOL complexes [104, 105].

Scheme 16 Bisoxazolinato rare earth metal catalysts in the hydroamination/ cyclization of aminoalkenes [75]

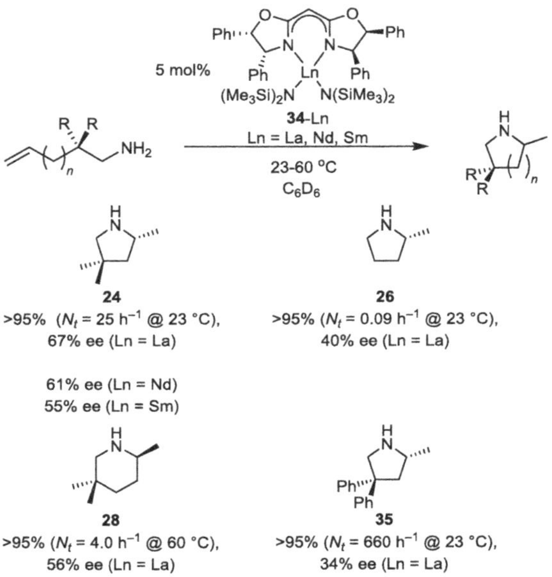

Fig. 3 Stereomodel for enantioselective hydroamination/cyclization using bisoxazolinate lanthanum catalyst **34**-La [75]. The structure of the bisoxazolinate ligand is simplified for clarity

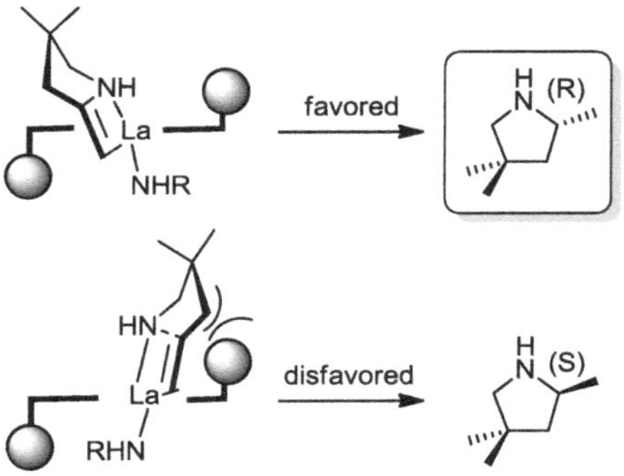

The best catalytic results are obtained using a small rare earth metal (Yb) and a large cyclopentyl-substituent on the diamidobinaphthyl amido ligand (Table 2), though the moderate catalytic activity requires the presence of *gem*-dialkyl substituents in the aminoalkene substrates.

A variety of neutral and anionic alkyl and amido complexes containing only one diamidobinaphthyl ligand, e.g., (*R*)-**37** and (*R*)-**38**, exhibit better catalytic activity while generally retaining enantioselectivities comparable to their corresponding bis (diamidobinaphthyl) complex **36** congener [13, 16, 69, 70, 83, 85, 86, 88–91, 93, 94]. The presence of lithium chloride in (*R*)-**37b** seems to have a positive effect on catalytic activity and enantioselectivity in comparison to the neutral complex (*R*)-**37a**, suggesting a cooperative effect between yttrium and lithium in this species [91]. Inclusion of alkali metal salts is a prominent problem in rare earth

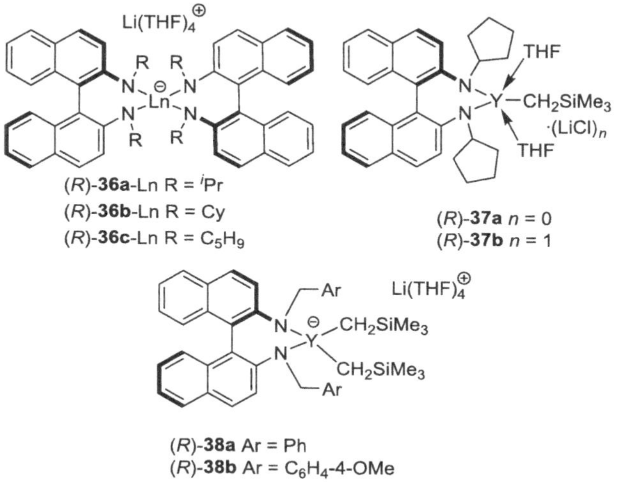

Fig. 4 Diamidobinaphthyl rare earth metal complexes for enantioselective hydroamination/cyclization [82, 83, 86, 87, 89, 91]

Table 2 Catalytic hydroamination/cyclization of aminoalkenes using diamidobinaphthyl complexes

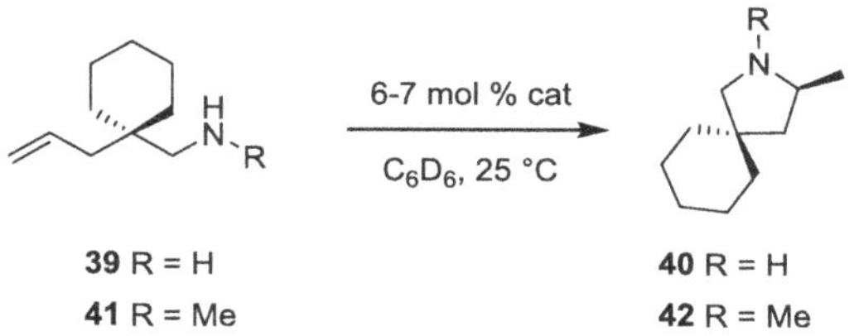

Cat	R	*t*, h	Conv., %	% *ee*	References
(*R*)-**36a**-Y	H	20	100	67	Riegert et al. [83]
(*R*)-**36b**-Yb	H	18	94	65	Riegert et al. [82]
(*R*)-**36c**-Yb	H	20	90	87	Aillaud et al. [87]
(*R*)-**37a**	H	3.3	89	75	Aillaud et al. [86]
(*R*)-**37b**	H	2	90	80	Chapurina et al. [91]
(*R*)-**37a**	Me	0.17	95	83	Queffelec et al. [89]

metal complex synthesis and it is therefore crucial to take salt effects [106] into account when studying catalytic reactions.

In contrast to most other catalyst systems, complex (*R*)-**37a** facilitates the cyclization of aminoalkenes with a secondary amino group faster and with higher enantioselectivity compared to the corresponding primary aminoalkene (Table 2) [89].

The high activity of the bisalkyl ate-complexes (*R*)-**38a**,**b** allow facile cyclization of 1,2-disubstituted and trisubstituted aminoalkenes (Eqs. 11 and 12) with enantioselectivities of up to 77% *ee* at temperatures of 40–110°C.

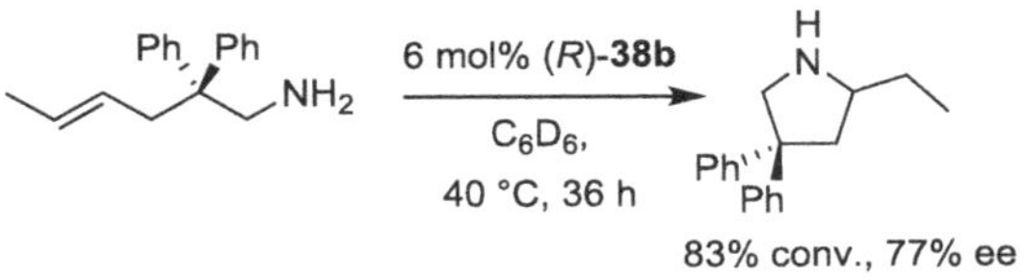

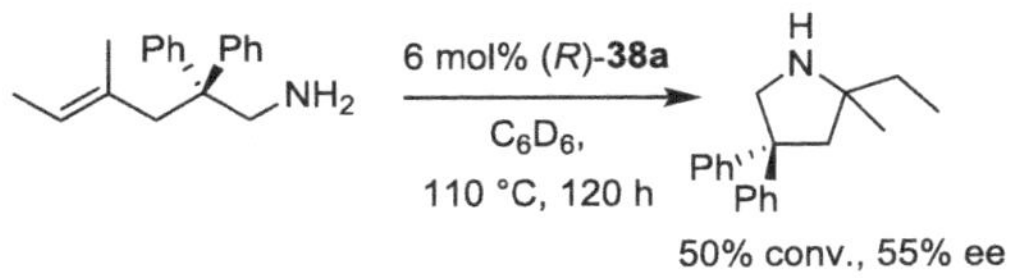

Equations 11 and 12. Enantioselective hydroamination/cyclization of internal aminoalkenes with diamidobinaphthyl yttrium ate complexes [69, 70]

Good to high enantioselectivities for a wide range of aminoalkene substrates, including internal olefins or secondary amines, are achieved using the aminothiophenolate catalyst system (*R*)-**43** (Scheme 17), which is generated in situ [92]. Variation of the steric demand of the silyl substituent attached to the thiophenolate moiety allows facile fine-tuning of the enantiomeric excess, resulting in increasing selectivity with increasing steric hindrance.

While the larger bite angle of the amino(thio)phenolate ligand is believed to improve enantiofacial differentiation through a better "reach around" of the chiral ligand around the metal center [73], the multidentate nature of the ligand also electronically saturates the metal center, resulting in diminished catalytic performance. Enantiomeric excess of up to 89% has been achieved at 30°C, though reactions at this temperature require a long period of time to reach completion.

The tetraazane precatalysts **46** (Scheme 18) may be activated with alkyl lithium reagents, such as *n*-BuLi or $LiCH_2SiMe_3$, to generate a moderately active hydroamination catalyst [103]. Cyclization of Thorpe–Ingold-activated substrates proceeds with up to 90% *ee* and even the *o*-allylbenzylamine can be cyclized at elevated temperature to give the tetrahydroisoquinoline **47** in 75% *ee*. However, aminopentene **25** lacking *gem*-dialkyl activation reacts only sluggishly. It appears that the tetradentate tetraazane ligand framework saturates the metal center in a similar manner to the tetradentate aminothiophenolate ligand set in (*R*)-**43**, reducing its catalytic activity. Similar trends in catalytic activity have been observed for the moderately enantioselective chiral

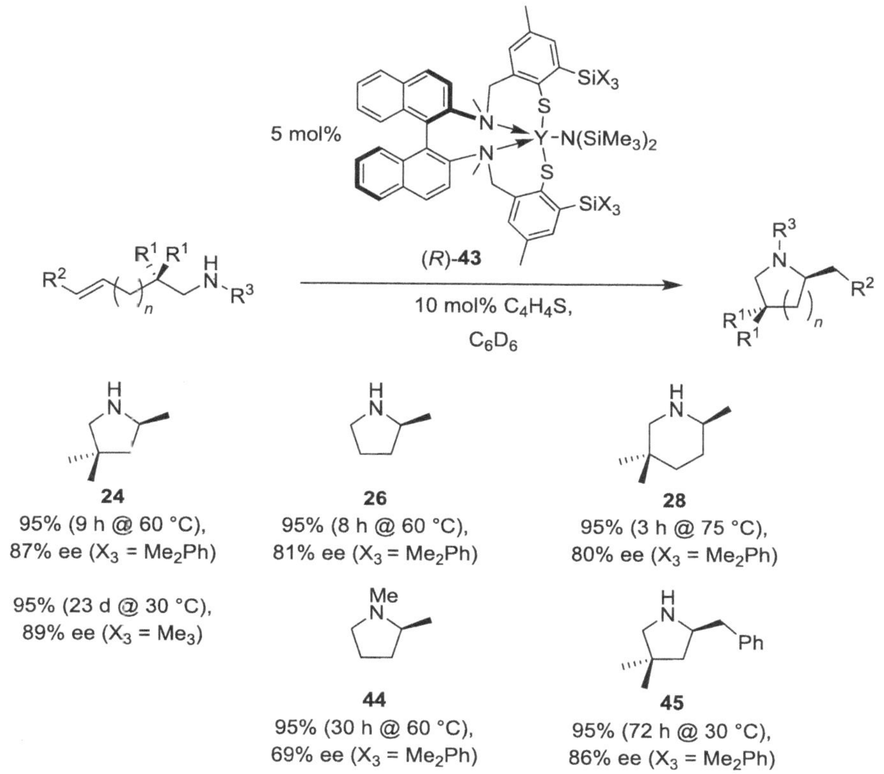

Scheme 17 Catalytic hydroamination/cyclization of aminoalkenes using chiral aminothiophenolate yttrium complexes [92]

benzamidinate complex **48** [102] and the linked-bis(diketiminate) complex **49** [101] (Scheme 19).

Significantly higher catalytic activities are achieved when more electron deficient ligand sets are employed. Binaphtholate aryl complexes (*R*)-**4**-Ln and (*R*)-**50**-Ln, (Ln = Sc, Y, Lu) with sterically demanding tris(aryl)silyl substituents in the 3 and 3′ position show excellent catalytic activity at room temperature, comparable in magnitude to lanthanocene catalysts, and achieve up to 95% *ee* in hydroamination/cyclization reactions of aminoalkenes (Scheme 20) [77, 79]. The sterically demanding tris(aryl)silyl substituents in the diolate complexes play a pivotal role not only to achieve high enantioselectivities but also to prevent undesired complex aggregation [76, 78] and reduce detrimental amine binding of the substrate and product to the catalytic active metal centers [79].

Increasing the denticity of the ligand, e.g., by incorporation of an additional amine donor, diminishes the catalytic activity. The NOBIN-derived aminodiolate complexes (*S*)-**53** are somewhat less active than the binaphtholate complexes (*R*)-**4**

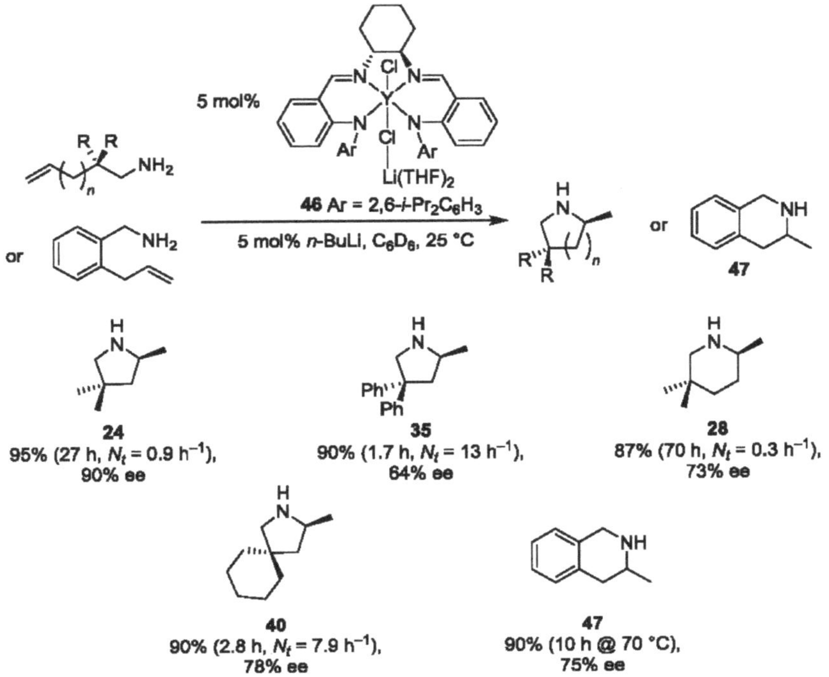

Scheme 18 Asymmetric hydroamination/cyclization of aminoalkenes catalyzed by a tetraazane rare earth metal complex [103]

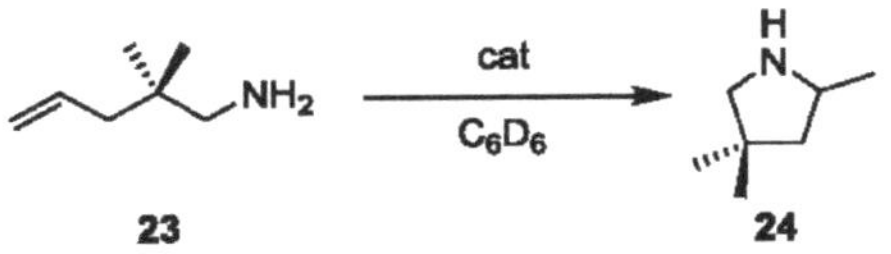

cat (mol%)	T, °C	t, h	Conv., %	% ee (config.)
(S)-**48** (5)	60	35	92	75 (S)
(R,R)-**49** (9.6)	25	50	>95	76 (R)

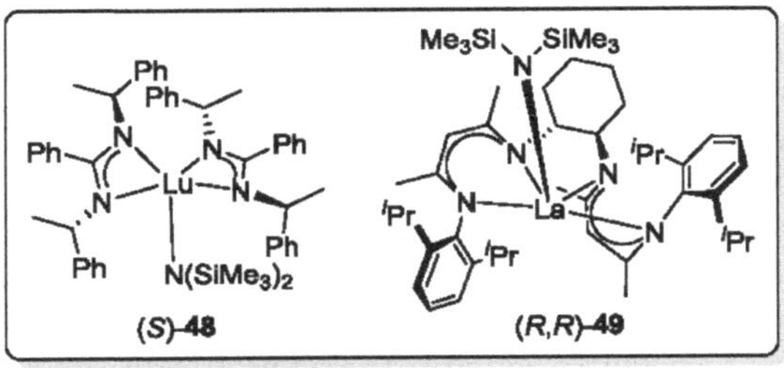

Scheme 19 Catalytic hydroamination/cyclization of 2,2-dimethylpent-4-en-1-amine (**23**) using the chiral benzamidinate complex (S)-**48** [102] and the chiral linked-bis(diketiminate) complex (R,R)-**49** [101]

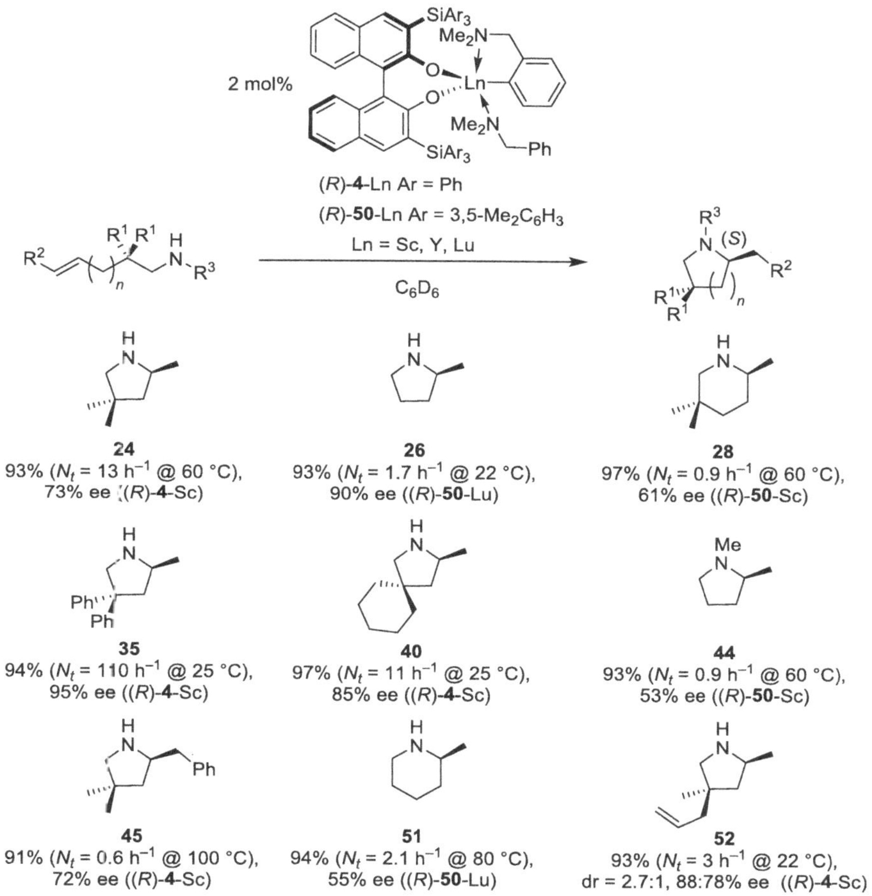

Scheme 20 Catalytic asymmetric hydroamination/cyclization of aminoalkenes using 3,3′-bis (trisarylsilyl)-substituted binaphtholate rare earth metal complexes [77, 79]

and (*R*)-**50** (Scheme 21) [24]. While enantioselectivities of up to 92% *ee* can be achieved with complex (*S*)-**53**, modifications to the aminodiolate ligand set reveals that the selectivity has a nonlinear dependency on the steric demand of the silyl-substituents.

The zwitterionic oxazolinylborato yttrium complex **55** displays good activity and excellent enantioselectivity in the cyclization of a variety of *gem*-disubstituted aminopentenes (Scheme 22) [55]. The observed kinetic isotope effect and the isotopic perturbation of enantioselectivity was initially interpreted as an indication of a concerted insertion/protonolysis mechanism (Scheme 14); however, a computational study suggests that a stepwise mechanism involving a reversible olefin insertion and subsequent irreversible and rate determining protonolysis prevails energetically [57].

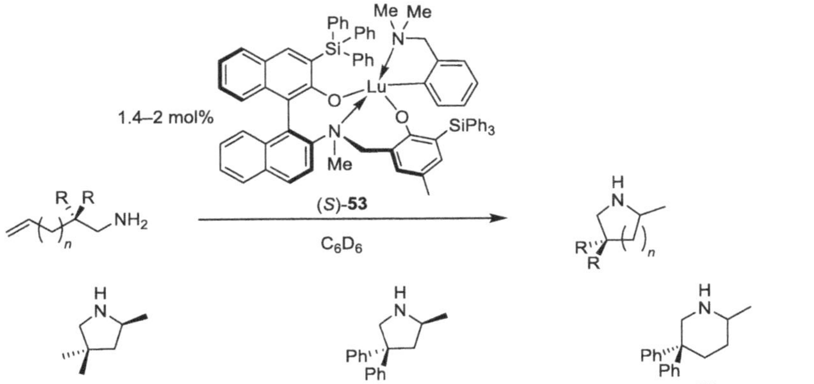

24
>95% (7 h, N_t = 12 h^{-1} @ 40 °C), 62% ee (1.4 mol% (*S*)-**53**)

35
>95% (0.25 h, N_t = 200 h^{-1} @ 25 °C), 92% ee (1.4 mol% (*S*)-**53**)

54
>95% (2.7 h, N_t = 20 h^{-1} @ 25 °C), 73% ee (2 mol% (*S*)-**53**)

Scheme 21 Catalytic asymmetric hydroamination/cyclization of aminoalkenes using NOBIN-based aminodiolate rare earth metal complexes [24]

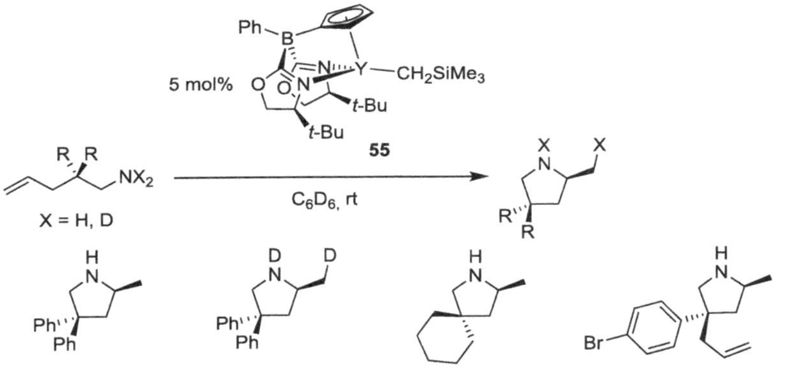

35
93% (10 min), 94% ee

35-d_2
94% ee

k_H/k_D = 2.6(4)

40
95% (10 min), 93% ee

56
94% (15 min), dr = 1.9:1, 95:92% ee

Scheme 22 Hydroamination/cyclization of *gem*-disubstituted aminopentenes using a chiral zwitterionic yttrium cyclopentadienyl-bis(oxazolidinyl)borate complex [55]

2.4.2 Alkali Metal-Based Catalysts

While hydroamination catalysts based on transition metals have been studied intensively over the last 2 decades, only a limited number of reports on alkali metal-based hydroamination catalysts have emerged, although the first reports date back more than 60 years [107]. In particular, the application of chiral alkali metal complexes in asymmetric hydroamination of non-activated aminoalkenes has drawn little attention to date [108–110].

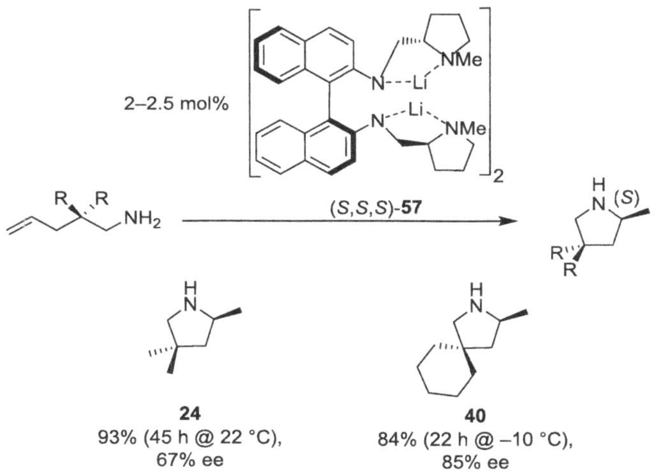

Scheme 23 Lithium-catalyzed asymmetric hydroamination/cyclization of aminoalkenes [108, 111]

The proline derived diamidobinaphthyl dilithium salt (*S*,*S*,*S*)-**57** is dimeric in the solid state and can be prepared via deprotonation of the corresponding tetraamine with *n*-BuLi. It represents the first example of a chiral main group metal-based catalyst for asymmetric intramolecular hydroamination reactions of aminoalkenes [108]. The unique reactivity of (*S*,*S*,*S*)-**57**, which allows reactions at or below ambient temperatures with product enantioselectivities of up to 85% *ee* (Scheme 23) [111], is believed to derive from the close proximity of the two lithium centers chelated by the proline substituents. More simple lithium amides require significantly higher reaction temperatures and give inferior selectivities.

More recently, a lithium-based catalyst system utilizing the cyclopentyl-substituted diaminobinaphthyl ligand **58** has been shown to catalyze the asymmetric hydroamination/cyclization (Eq. 13) [110]. While enantioselectivities remain low to moderate, the selectivities seem to be unaffected by the twofold excess (with respect to ligand **58**) of the alkyl lithium reagent.

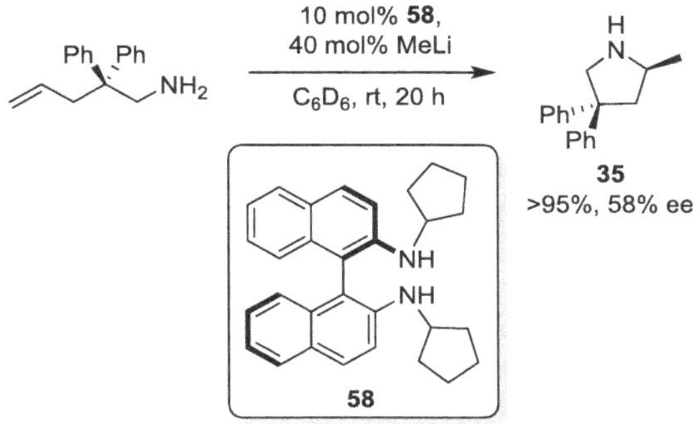

Equation 13. Enantioselective hydroamination/cyclization of an aminoalkene with a diamidobinaphthyl lithium catalyst [110]

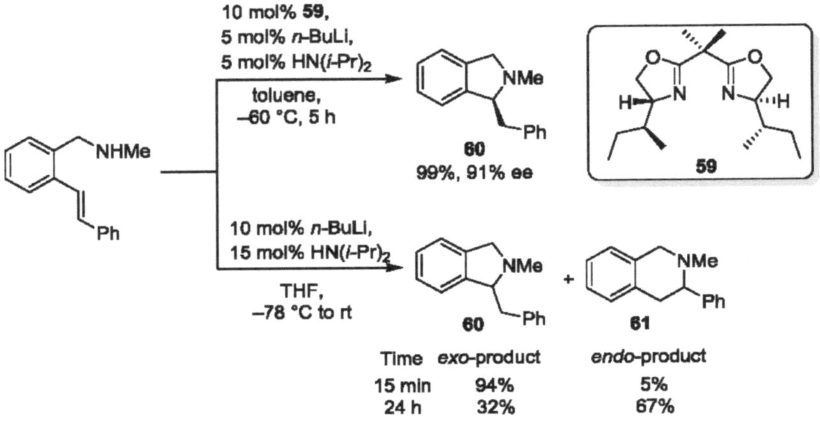

Scheme 24 Kinetic vs thermodynamic control in the lithium-catalyzed cyclization of aminostilbenes [109]

The asymmetric hydroamination/cyclization of amino-substituted stilbenes has been studied utilizing chiral bisoxazoline lithium catalysts [109]. Enantioselectivities reaching as high as 91% *ee* are achieved when the reactions are performed in toluene at –60°C under kinetic control to give the *exo*-cyclization product **60** (Scheme 24). However, the hydroamination/cyclization reaction in THF solution is reversible, producing predominantly the thermodynamically favored *endo*-cyclization product **61** after equilibration for 24 h at room temperature.

2.4.3 Alkaline Earth Metal-Based Catalysts

The chemistry of alkaline earth metal catalysts is significantly limited by the existence of facile Schlenk equilibria [112–114], which may lead to the formation of achiral catalytically active species. A number of chiral alkaline earth metal hydroamination catalysts have been developed (Fig. 5) [115–121]. However, most catalysts fail to reach even moderate enantioselectivities in aminoalkene hydroamination/cyclization (Table 3) owing to the presence of facile ligand redistribution processes occurring under catalytic conditions.

The bis(amido) magnesium complex (*S*,*S*,*S*)-**62** [115] as well as the chiral diketiminato calcium compound **63** [116] show very low selectivity for the intramolecular hydroamination, apparently as a result of facile ligand redistribution reactions (Table 3, entries 1 and 2). Chiral calcium ethylenediamine [120] and bisimidazoline [121] complexes also suffer from facile ligand redistribution, but the amidobisoxazoline complex (*R*)-**64** is more resilient, giving enantioselectivities of up to 50% *ee* (Table 3, entry 3) [119].

Although the tris(oxazolinyl)borate **65** is stable to ligand redistribution, only low selectivities of up to 36% *ee* are obtained (Table 3, entry 5) [117]. In marked

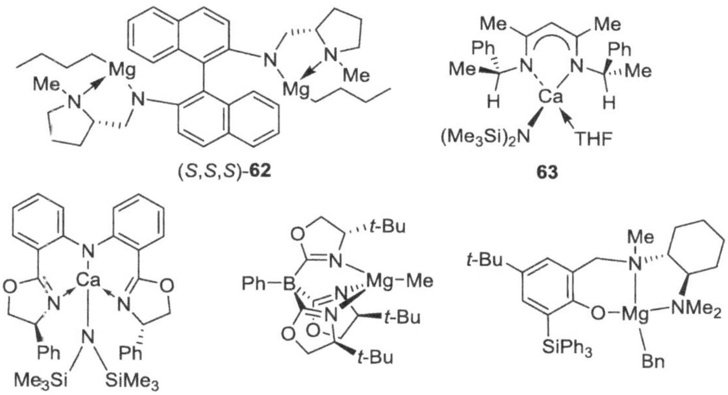

Fig. 5 Selected examples of chiral alkaline earth metal catalysts [115–119]

Table 3 Alkaline earth metal-catalyzed asymmetric hydroamination/cyclization of aminopentenes

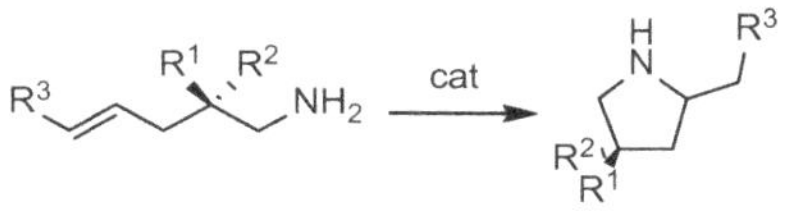

Entry	R^1, R^2	R^3	Cat (mol%)	*T*, °C	*t*, h	Yield, %	% *ee* (config)	References
1	Me, Me	H	**62** (5)	100	22	99	4 (*S*)	Horrillo-Martínez et al. [115]
2	Ph, Ph	H	**63** (10)	20	1	98	10 (*R*)	Buch et al. [116]
3	Ph, Ph	H	**64** (20)	50	24	>99	50	Nixon et al. [119]
4	Me, Me	H	**65** (10)	80	120	80	27 (*R*)	Neal et al. [117]
5	$-(CH_2)_5-$	H	**65** (10)	60	26	93	36 (*R*)	Neal et al. [117]
6	Me, Me	H	**66** (5)	22	10	97	79 (*S*)	Zhang et al. [118]
7	$-(CH_2)_5-$	H	**66** (2)	22	4.5	99	85 (*S*)	Zhang et al. [118]
8	Ph, Ph	Ph	**66** (2)	−20	12	98	93 (*S*)	Zhang et al. [118]

contrast, the chiral magnesium (amino)phenolate **66** [118] displays selectivities of up to 93% *ee* as well as superior reactivity (Table 3, entries 6–8). This promising example illustrates that alkaline earth metal catalysts can reach reactivity and selectivity levels comparable to rare earth metal systems.

2.4.4 Group 4 Metal-Based Catalysts

The chemistry of organometallic Group 4 metal compounds is well developed thanks to their importance in polyolefin synthesis. Hence, their application in

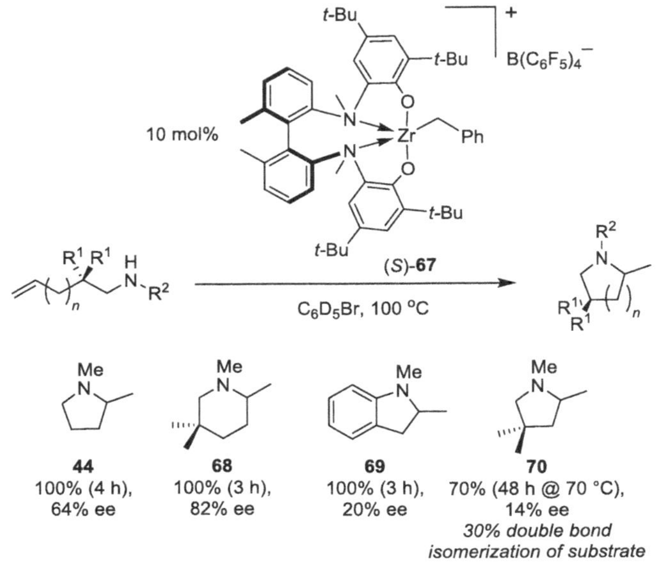

Scheme 25 Hydroamination/cyclization of secondary aminoalkenes using a cationic chiral zirconium catalyst system [133]

catalytic asymmetric hydroamination reactions is highly desirable. Group 4 metal complexes are commonly less sensitive and easier to prepare than rare earth metal complexes. Most important of all, many potential precatalysts or catalyst precursors are commercially available.

Group 4 metal-based catalysts have been studied intensively in hydroamination reactions involving alkynes and allenes over the last 2 decades [122–126], whereas (achiral) hydroamination reactions involving aminoalkenes have been observed much later [127–129]. With only a few exceptions, the reactivity of these catalysts is significantly lower than that of rare earth, alkali, and alkaline earth metal-based catalysts. In most instances *gem*-dialkyl activation [60, 61] of the aminoalkene substrate is required for catalytic turnover. Due to the harsh reaction conditions, the hydroaminoalkylation of aminoalkenes can compete with the hydroamination reaction under certain circumstances [130–132].

The first chiral Group 4 metal catalyst system for asymmetric hydroamination/cyclization of aminoalkenes was based on the cationic aminophenolate complex (*S*)-**67** [133]. Secondary aminoalkenes react readily to yield hydroamination products with enantioselectivities of up to 82% *ee* (Scheme 25). For catalyst solubility reasons, reactions are commonly performed at 100°C in bromobenzene using 10 mol% catalyst loading. The mechanism of this cationic system is thought to proceed similar to the σ-bond metathesis mechanism of rare earth metal-based catalyst systems (Scheme 13). Aminoalkenes with a primary amino group do not react, which is thought to be caused by facile α-deprotonation of the catalytic active

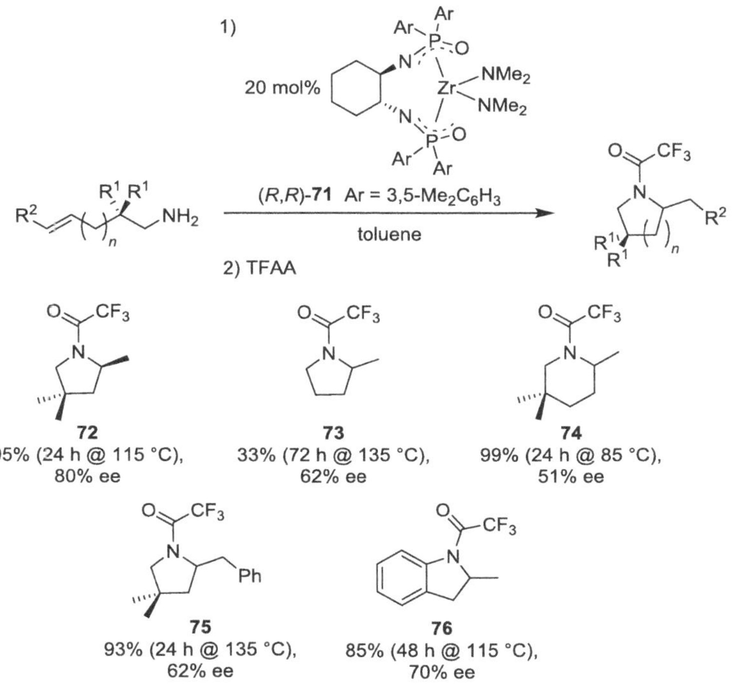

Scheme 26 Hydroamination/cyclization of primary aminoalkenes using a neutral bis(phosphinic amido) zirconium catalyst system [136]

cation metal–amido species leading to an unreactive metal-imido species [127, 133, 134]. The cationic catalyst systems are also prone to double bond isomerization via C–H activation that can significantly reduce product enantioselectivity and yield [127, 133, 135].

In contrast to the cationic Group 4 metal hydroamination catalysts, their neutral counterparts will generally react only with primary aminoalkenes. For example, the chiral bis(phosphinic amido) zirconium complex (*R*,*R*)-**71** (Scheme 26) is found to be superior in reactivity and enantioselectivity for the cyclization of primary aminoalkenes in comparison to a wide range of diamido, diolate, and aminoalcoholate titanium, zirconium, and hafnium complexes [136]. Cyclization of aminopentenes proceed with enantioselectivities as high as 80% *ee*, but formation of six-membered rings, e.g., **74**, are somewhat less selective. Preliminary mechanistic studies indicate that this catalyst system undergoes slow ligand redistribution reactions, leading to chiral catalytically inactive as well as achiral catalytically active species.

A variety of related axially chiral sulfonamide and bis(thiophosphinic amido) titanium and zirconium complexes achieve moderate enantioselectivities of up to 38% *ee* [137, 138].

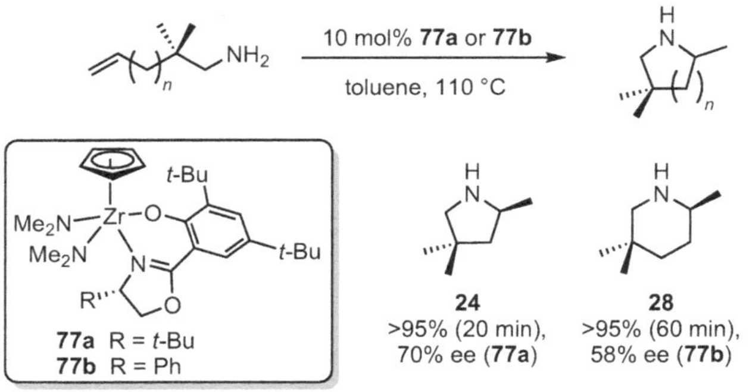

Scheme 27 Hydroamination/cyclization of primary aminoalkenes using a neutral bis(phosphinic amido) zirconium catalyst system [136]

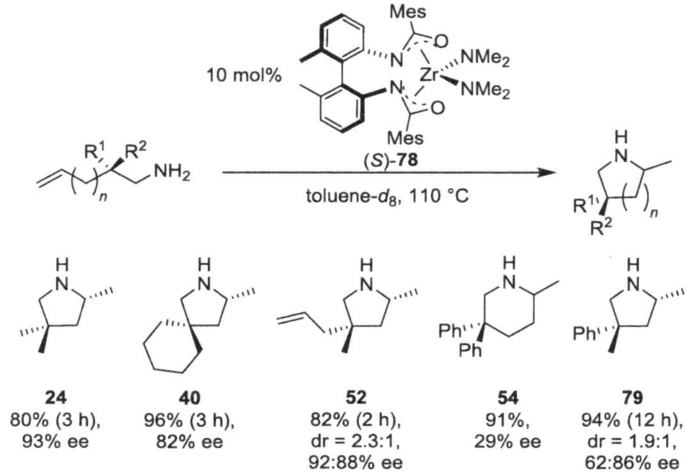

Scheme 28 Hydroamination/cyclization of primary aminoalkenes using a neutral bis(amidate) zirconium catalyst system [140–142]

The half-sandwich zirconium salicyloxazoline complexes **77a** and **77b** have been reported to cyclize *gem*-dimethyl-substituted primary aminoalkenes with good activity at 110°C, but moderate enantioselectivities of up to 70% *ee* (Scheme 27) [139].

Much higher enantioselectivities of up to 93% *ee* may be achieved for aminopentene substrates using the chiral bis(amidate) zirconium complex (*S*)-**78** (Scheme 28) [140–144]. However, the cyclization of aminohexene substrates proceeds with significantly reduced selectivities.

Although the binaphthalenedicarboxamide zirconium complex (*S*)-**80** [145] and related biphenyl complexes [146] closely resemble the structure of complex (*S*)-**78**, the altered connectivity of the two amidate moieties to the axially chiral ligand backbone favors a more open κ[1] binding mode of the *N*-mesityl amidate ligands. This κ[1] binding mode renders the metal more electron-deficient and the catalyst system more active. The increased reactivity may be utilized to reduce catalyst loadings down to 0.5 mol% and reduce reaction temperatures to 70°C in the formation of pyrrolidines [145]. However, the more remote arrangement of the *N*-aryl substituents results in significant lower selectivities in comparison to (*S*)-**78**. Nevertheless, the increased reactivity allows cyclization of aminoheptene **81** to give the azepane **82** in 60% *ee* (Eq. 14). This result is particularly noteworthy as aminohexenes and aminoheptenes are frequently observed to undergo hydroaminoalkylation (via α-C-H activation) instead of hydroamination (via N-H activation) [130–132]. Related dicarboxamide complexes based on a biphenyl backbone cyclize aminopentenes with up to 74% *ee* [146]. Depending on the steric demand of the substituent on the amide nitrogen atom, the latter system rearranges to a homochiral dimer with significantly reduced catalyst activity while retaining enantioselectivity [146].

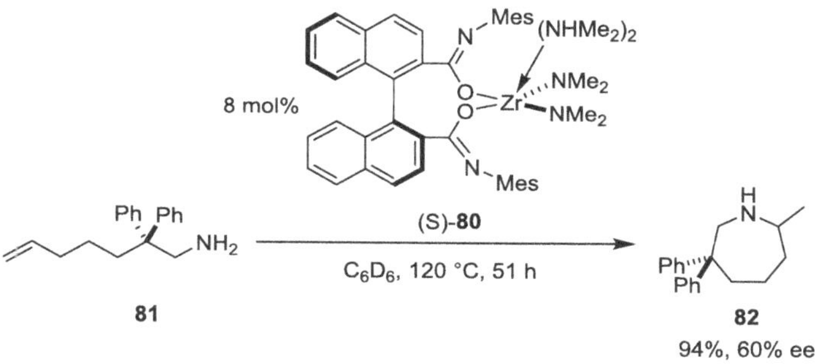

Equation 14. Enantioselective azepane synthesis via zirconium-catalyzed asymmetric hydroamination/cyclization of an aminoheptene [145]

The mechanism of alkene hydroamination is much less well understood than the mechanism for alkyne and allene hydroamination [147–154]. Based on the observation that most neutral Group 4 metal alkene hydroamination catalysts are unreactive towards secondary aminoalkene substrates, a mechanism analogous to that for alkyne and allene hydroamination involving metal-imido species as catalytically active species has been proposed (Scheme 29) [128, 129, 136, 140]. The metal-imido species is generated via reversible α-elimination of an amine from a bis(amido) precursor. The formation of the imido species can be the rate-determining step, as indicated by kinetic studies with a half-sandwich zirconium salicyloxazoline catalyst system [155]. Subsequently, the imido species undergoes a most likely reversible [149, 151] [2 + 2]-cycloaddition with the alkene moiety. The resulting azametalacyclobutane is protolytically cleaved to regenerate the metal-imido species and release the hydroamination product. However, a few (achiral) neutral Group 4 metal catalyst systems have been reported to catalyze cyclization of secondary aminoalkenes and it has been suggested that a lanthanide-

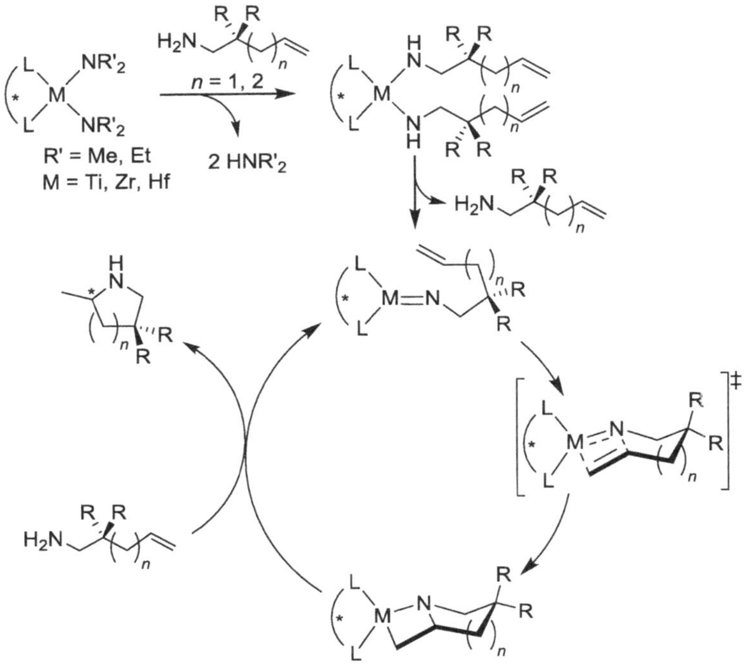

Scheme 29 Proposed mechanism for the hydroamination/cyclization of primary aminoalkenes involving Group 4 metal-imido species

like σ-bond metathesis mechanism (Scheme 13) is operational in these cases [156–160].

The chiral zwitterionic zirconium cyclopentadienyl-bis(oxazolinyl)borate complex **83** [55, 161, 162] possesses significantly improved reactivity compared to other Group 4 metal hydroamination catalysts. Thorpe–Ingold-activated substrates can be cyclized at temperatures as low as –30°C with enantioselectivities of up to 99% *ee* for *gem*-disubstituted aminopentenes (Scheme 30) [161, 162]. The catalyst tolerates a number of aprotic functional groups, such as aryl halides (**56**) and acetals (**84**). Substrates incorporating 1,2-disubstituted alkenes are cyclized with excellent enantioselectivity, e.g., **86**. Increasing the ring size results in significantly diminished selectivities for the piperidine **54**, whereas the azepane **82** is formed again in outstanding 91% *ee*, thus indicating that there is a nonlinear relationship between ring size and selectivity.

A pronounced primary kinetic isotope effect and isotopic perturbation of enantioselectivity resulting in higher selectivities for the *N*-deuterated substrates have been attributed to a concerted alkene insertion/protonolysis step.

A DFT study for a tethered bis(ureate) zirconium catalyst system indicates the general validity of a concerted proton-assisted cyclization pathway [163]. However, similar studies on tris(oxazolinyl)phenylborate magnesium [56] and cyclopentadienyl-

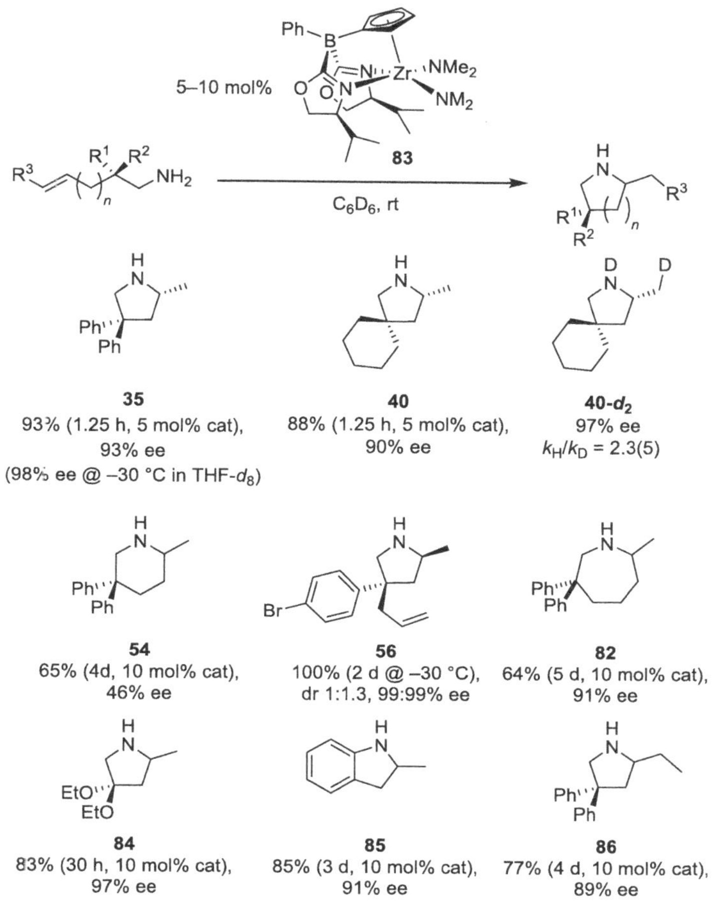

Scheme 30 Room temperature hydroamination/cyclization of *gem*-disubstituted aminoalkenes using a chiral zwitterionic zirconium cyclopentadienyl-bis(oxazolidinyl)borate complex [161, 162]

bis(oxazolinyl)borate yttrium [57] catalyst systems favor an alternative mechanistic scenario involving reversible olefin insertion and rate-determining protonolysis.

2.4.5 Group 5 Metal-Based Catalysts

Group 5 metal complexes exhibit catalytic activity comparable to neutral Group 4 metal complexes and scope and limitations are quite similar. Generally only Thorpe–Ingold-activated primary aminoalkenes react, suggesting that these metals operate via a metal-imido [2+2] cycloaddition mechanism similar to Group 4 metals (Scheme 29). However, the information on a possible reaction mechanism is scarce and a study on the tantalum-catalyzed hydroamination of alkynes was unable to confirm this mechanistic scenario [164–166]. Some of the Group 5 metal

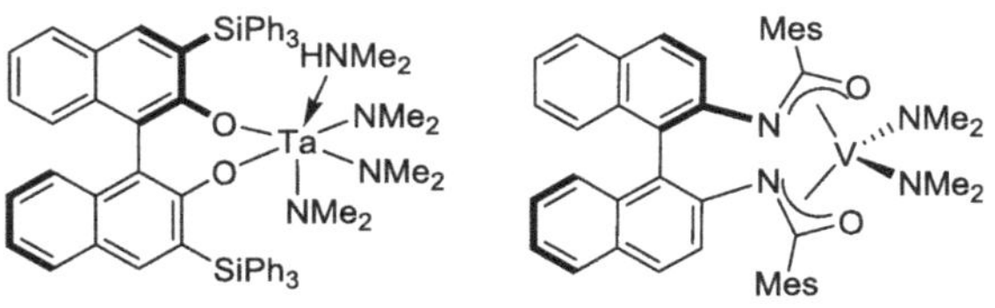

(*R*)-**87** (*R*)-**88**

Fig. 6 Selected Group 5 metal catalysts for asymmetric hydroamination of aminoalkenes (Mes = 2,4,6-$Me_3C_6H_2$) [169, 170]

complexes which are active in asymmetric hydroamination of aminoalkenes are also catalytically active in the (intermolecular) hydroaminoalkylation of alkenes with secondary amines with high chemoselectivity and high enantioselectivities of up to 98% *ee* [130, 167–171].

The 3,3′-silyl-disubstituted binaphtholate tantalum complex (*R*)-**87** (Fig. 6) catalyzes the cyclization of Thorpe–Ingold-activated aminoalkenes [169]. Enantioselectivities of up to 81% *ee* (Eq. 15) are achieved with reaction rates at a similar order of magnitude to most Group 4 metal complexes. The chiral bis(amidate) vanadium(IV) complex (*R*)-**88** also shows appreciable catalytic activity and enantioselectivity (Eq. 16) [170]. It is unclear whether the catalyst retains the +4 oxidation state and related vanadium(III) complexes also show comparable activity. However, the corresponding bis(amidate) niobium(V) and tantalum(V) complexes are catalytically inactive.

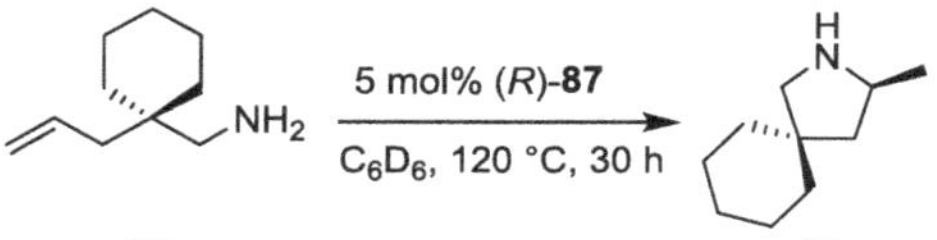

39 **40**
>95% conv.,
81% ee (*S*)

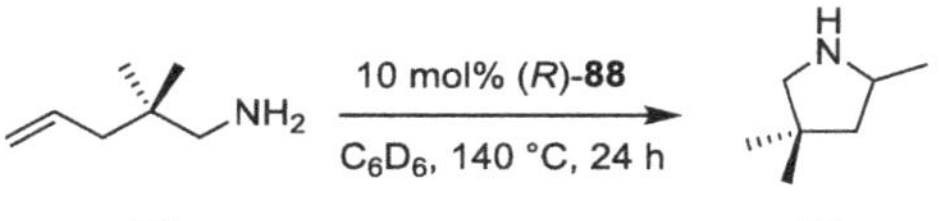

23 **24**
89% conv.,
76% ee

Equations 15 and 16. Asymmetric hydroamination/cyclization of aminopentenes with chiral Group 5 metal complexes [169, 170]

2.4.6 Late Transition Metal-Based Catalysts

The limitations of late transition metal catalysts in the intramolecular hydroamination of aminoalkenes pose a significant challenge with respect to enantioselective transformations and examples are scarce. The asymmetric rhodium-catalyzed hydroamination of *N*-benzyl aminopentenes has been achieved with the chiral biaryl phosphines **89a** and **89b** with enantioselectivities of up to 91% *ee* (Scheme 31) [172]. Catalyst loadings are moderate (typically 5 mol%) and reaction temperatures range from 50 to 100°C. The nature of the protecting group can have a pronounced influence on product enantioselectivities, e.g., higher enantioselectivities are obtained in some cases when using the sterically more bulky *N*-(2-methyl)benzyl group. Unprotected primary aminoalkenes generally show poorer reactivity. The method is significantly less efficient and less enantioselective for the synthesis of piperidines from *N*-benzyl aminohexenes.

Based on a mechanistic study of related achiral rhodium complexes [173], the mechanism involves a catalytic active cationic rhodium species where the biaryldialkylphosphine ligand binds in a κ^1,η^6 chelating fashion (Scheme 32). Olefin coordination to this 16 valence electron species is followed by reversible nucleophilic attack of the amine to the coordinated alkene. The olefin complex is the resting state and protonation of the β-aminoalkyl-rhodium species is turnover limiting.

N-Alkenyl ureas can be cyclized at room temperature with moderate enantioselectivities utilizing the diasteromerically pure *tropos* BIPHEP-gold(I) complex (*S*)-**90** with an Au–Au interaction (Eq. 17) [174]. Catalysts lacking this Au–Au interaction exhibit lower enantioselectivities. The chiral counteranion also has a synergistic effect on catalyst selectivity.

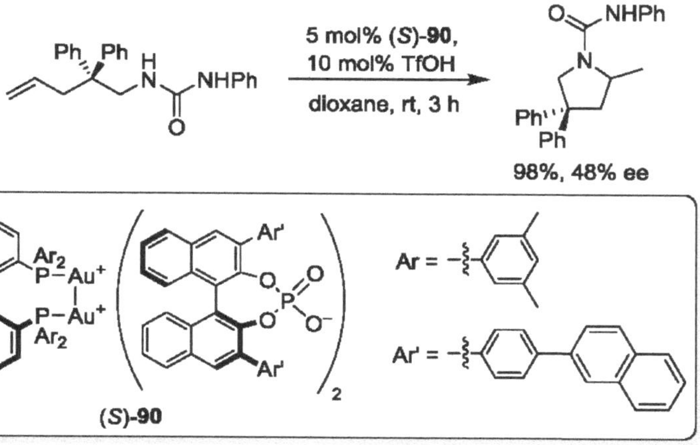

Equation 17. Asymmetric hydroamination/cyclization of an alkenyl urea with a diasteromerically pure BIPHEP-gold complex [174]

The copper-catalyzed hydroamination/cyclization of *N*-sulfonyl-2-allylanilines has been achieved using a bisoxazoline copper(II) triflate catalyst and stoichiometric

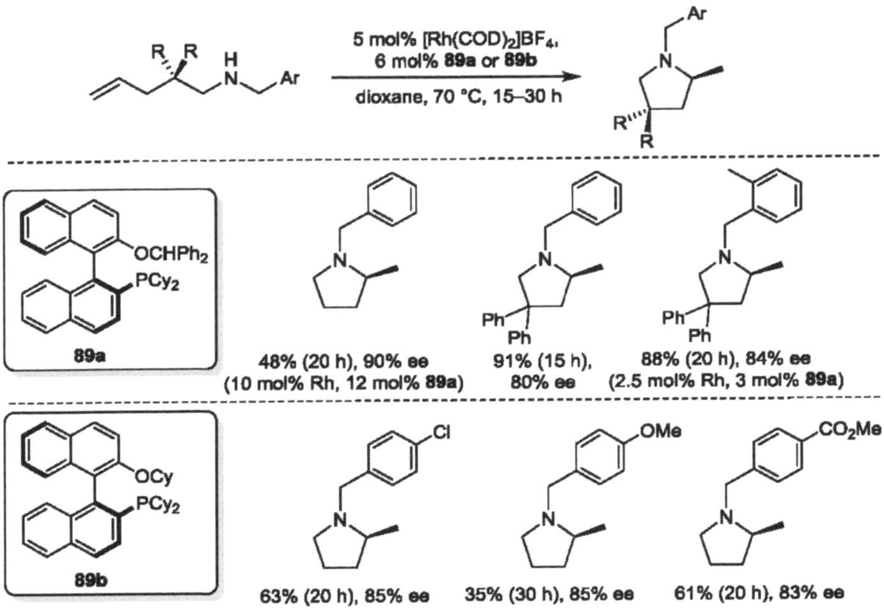

Scheme 31 Rhodium-catalyzed enantioselective hydroamination/cyclization of aminopentenes [172]

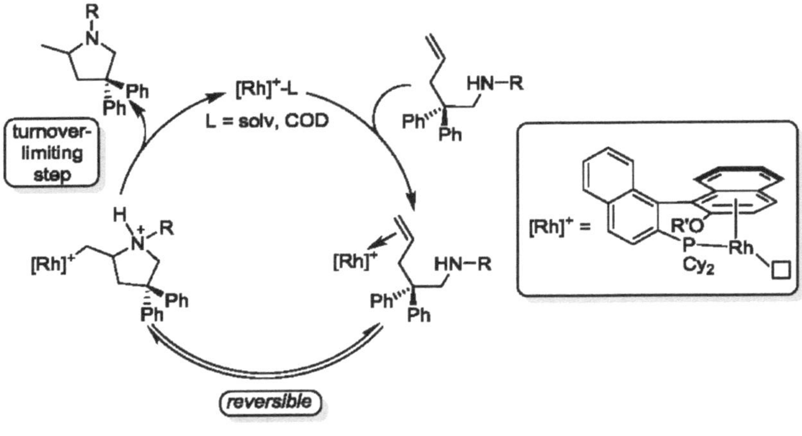

Scheme 32 Proposed mechanism for the rhodium-catalyzed hydroamination/cyclization of aminopentenes, based on a mechanistic study on a related achiral catalyst system [173]

amounts of cyclohexadiene as hydrogen donor, MnO_2 as oxidant, and a proton trap (Scheme 33) [175]. The resulting 2-methylindolines **91** are obtained in up to 90% *ee*. However, formation of the carbamination by-product for arylsulfonamide substrates, e.g., **92**, and the necessity to use reagents in stoichiometric amounts, make this

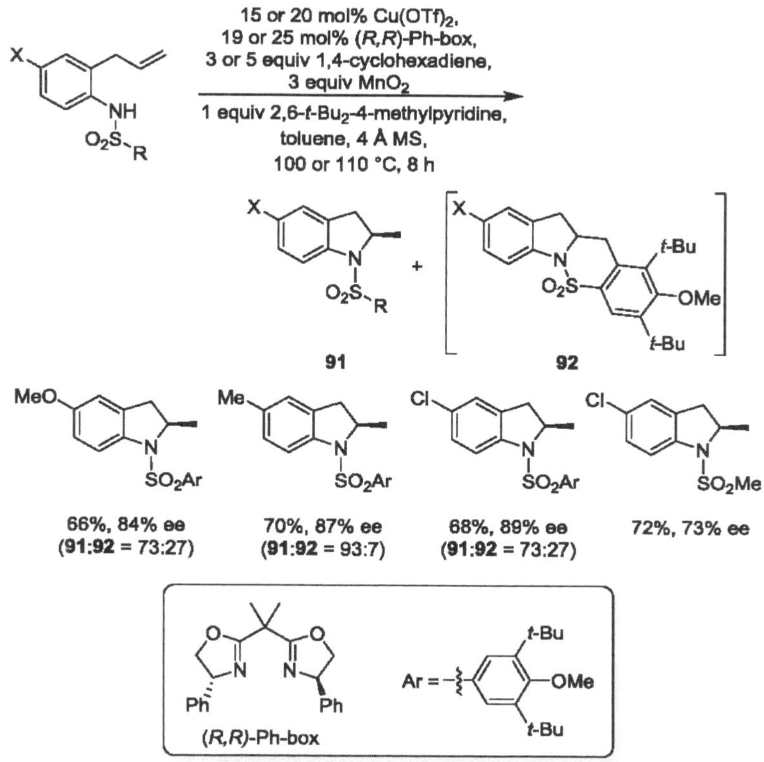

Scheme 33 Copper-catalyzed enantioselective hydroamination/cyclization of *N*-sulfonyl-2-allylanilines [175]

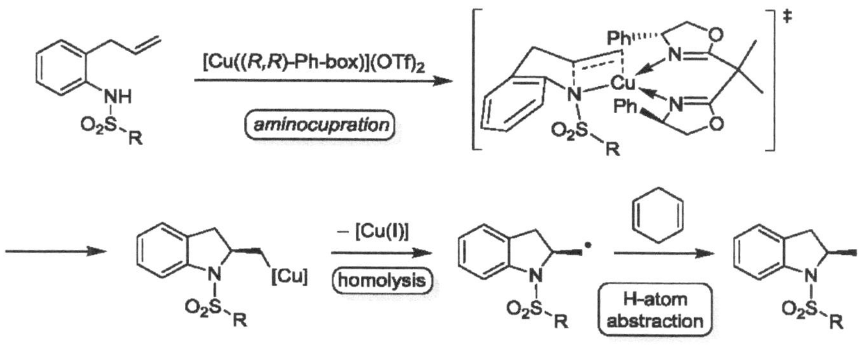

Scheme 34 Proposed mechanism of copper-catalyzed enantioselective hydroamination/cyclization of *N*-sulfonyl-2-allylanilines [175–177]

transformation overall less desirable. The reaction is believed to proceed via addition of the sulfonamide nitrogen to the alkene in a *cis* aminocupration transition state (Scheme 34) [176, 177]. The resulting unstable copper(II)-alkyl undergoes homolysis.

Subsequently the primary organic radical is trapped by hydrogen atom abstraction from 1,4-cyclohexadiene and the copper(I) species is reoxidized by MnO_2.

2.4.7 Organocatalytic Asymmetric Hydroamination of Aminoalkenes

In the last decade, the field of asymmetric organocatalysis has seen significant progress. In particular, the application of Brønsted acids in metal-free enantioselective catalysis is rapidly increasing [178–180]. Several research groups have recently demonstrated that strong Brønsted acids can be used in both *intra-* [181–185] and *inter*molecular [186–188] hydroamination reactions, although protection of the amino group is often necessary.

The chiral phosphoric acid diester (*R*)-**93** catalyzes the asymmetric hydroamination of an *N*-benzyl-protected aminopentene (Eq. 18) [184]. Although both selectivity and catalytic activity for the hydroamination are rather low, the method itself remains very promising. However, the organocatalytic hydroamination of aminodienes and aminoallenes has been achieved with high stereoselectivity (see Sects.3.2 and 4.2) [189].

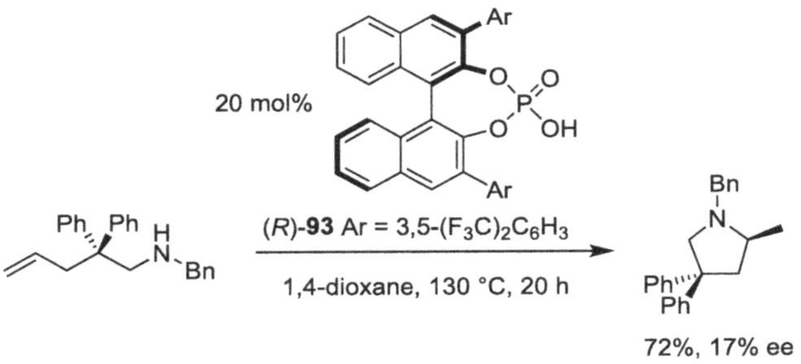

Equation 18. Organocatalytic asymmetric hydroamination/cyclization of a secondary aminoalkene [184]

2.4.8 Cope-Type Intramolecular Hydroamination

The polar transition state of Cope-type hydroamination can be stabilized through H-bonding interaction. H-bond donors, such as the thiourea **94**, can therefore accelerate Cope-hydroamination and allow facile control of the stereochemistry (Scheme 35) [190]. Various electron-rich and electron-poor γ-styryl hydroxyl-amines undergo facile cyclization at 3°C with high degree of stereoinduction. However, electron-donating groups on the vinyl arene ring decrease the rate of the reaction. The cyclization of substrates lacking *gem*-dialkyl substituents requires a higher reaction temperature (30°C) and proceeds with diminished enantios-electivity. The reaction also proceeds with slightly lower selectivities in the absence of an aromatic substituent at the alkene terminus.

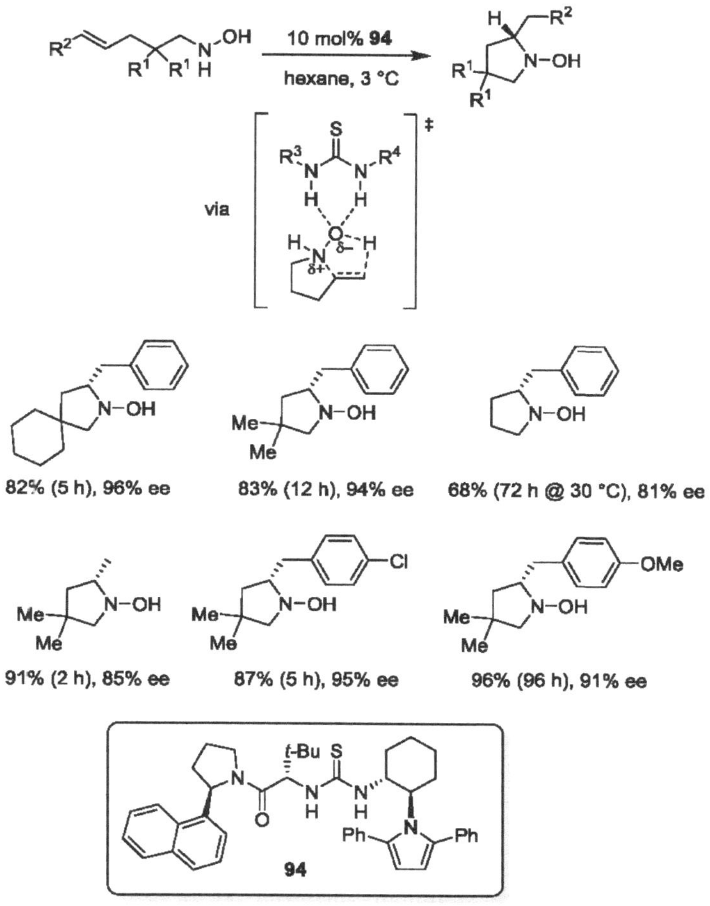

Scheme 35 Asymmetric thiourea-catalyzed intramolecular Cope-type hydroamination [190]

3 Hydroamination of Dienes

3.1 Intermolecular Hydroamination of Dienes

The palladium-catalyzed asymmetric hydroamination of cyclohexadiene with arylamines utilizing a variant of Trost's ligand (*R*,*R*)-**95** proceeds with high enantioselectivities under mild conditions (Scheme 36) [34]. This catalyst system can be applied to a broader range of anilines with consistent high enantioselectivities under neutral conditions at room temperature, but rather moderate activity. Other chiral bidentate phosphine ligands, such as (*R*)-BINAP, (*S*,*S*)-DIOP, or (*S*,*S*)-BDPP give only inferior enantioselectivities. The mechanism is believed to follow a similar pathway as proposed for palladium-catalyzed hydroamination of vinyl arenes (Scheme 4) [30].

Scheme 36 Palladium-catalyzed asymmetric hydroamination of cyclohexadiene [34]

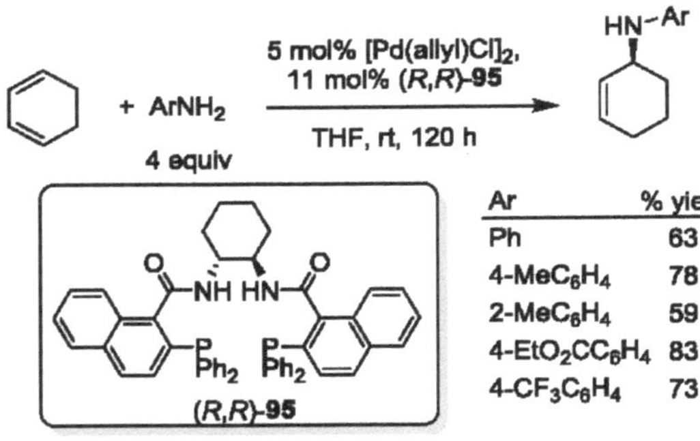

Ar	% yield	% ee
Ph	63	92
4-MeC_6H_4	78	86
2-MeC_6H_4	59	90
4-$EtO_2CC_6H_4$	83	95
4-$CF_3C_6H_4$	73	95

3.2 *Intramolecular Hydroamination of Aminodienes*

The cyclization of aminodienes can be achieved readily with rare earth metal and alkali metal catalysts due to the transient formation of an η^3-allyl intermediate. Protonation leads to (*E*)/(*Z*)-alkenylpyrrolidines and alkenylpiperidines, and, under certain conditions, also allyl isomers (Scheme 37). Cyclizations with chiral lanthanocenes generally produce the (*E*)-olefins with high (*E*)-selectivity ($E/Z \geq 93{:}7$) [191, 192]. The reaction rates are higher for aminodienes compared to the corresponding aminoalkenes, despite increased steric encumbrance of the cyclization transition state (Scheme 37). However, in most cases, the increased reactivity is at the expense of enantioselectivity. The aminooctadiene **96** is an exception with 63% *ee* observed in a benzene solution at 25°C (71% *ee* in methylcyclohexane at 0°C) using (*S*)-**29**-Sm, which provides facile access to (+)-coniine **98** after hydrogenolysis of the Cbz-protected vinylpiperidine **97** (Scheme 38) [192].

The hydroamination/cyclization of terminal aminodienes can also be catalyzed by chiral diamidobinaphthyl dilithium salts with up to 72% *ee* (Eq. 19) [110, 193]. Although the (*E*)/(*Z*)-selectivity of the product is moderate in most cases, both diastereoisomers can be obtained with comparable enantiomeric excess.

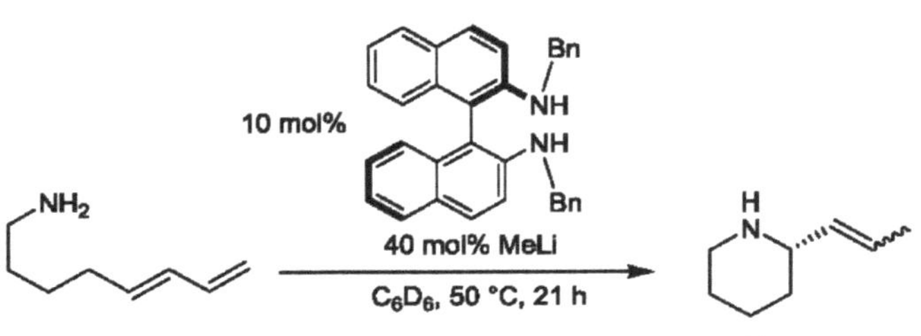

96

99
62%, *E*:*Z* = 89:11
72% ee (*E*)/55% ee (*Z*)

Equation 19. Lithium-catalyzed asymmetric hydroamination/cyclization of aminodienes [110, 193]

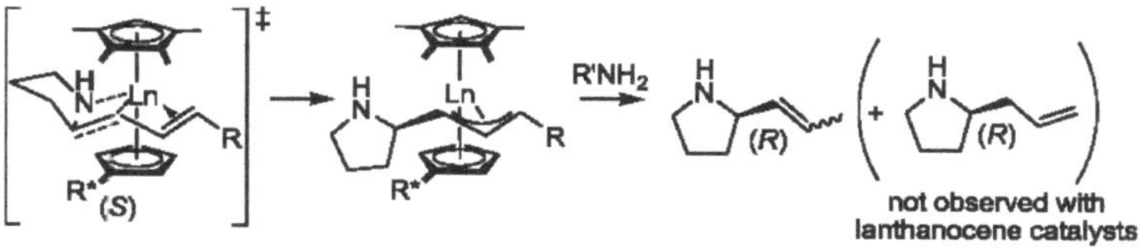

Scheme 37 Stereomodel for the lanthanocene-catalyzed hydroamination/cyclization of aminodienes. The silicon linker bridging the two cyclopentadienyl ligands was omitted for the sake of clarity

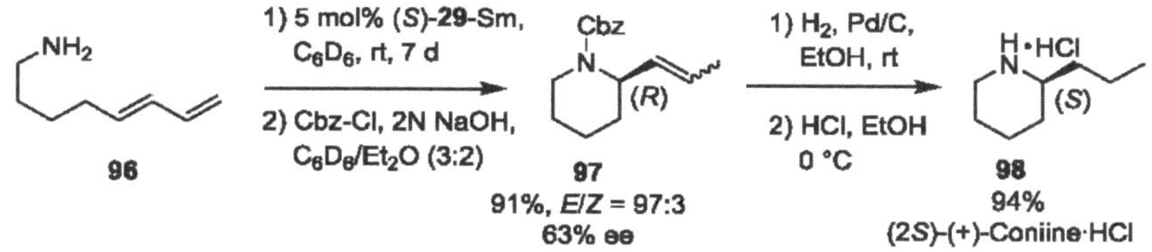

Scheme 38 Synthesis of (+)-coniine·HCl via enantioselective aminodiene hydroamination/cyclization [192]

Late transition metal complexes also seem to be predisposed to the cyclization of *N*-protected aminodienes. However, only a limited number of enantioselective examples have been reported: The cyclization of a dienylsulfonamide is catalyzed by a chiral palladium phosphine complex (Eq. 20) [194]. Unfortunately, the resulting enantioselectivities remain low and a significant excess of the bidentate RENORPHOS ligand is required.

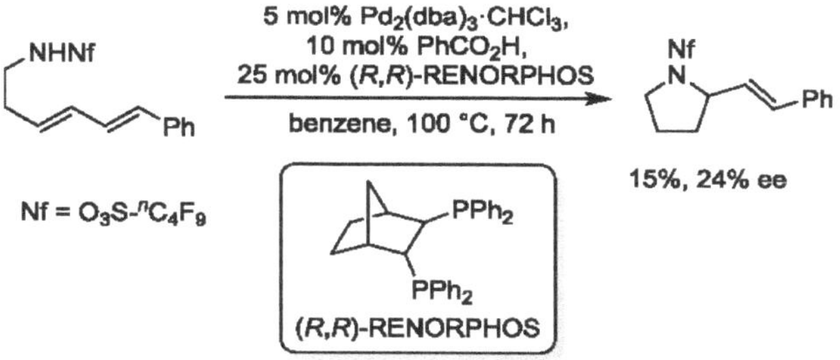

Equation 20. Palladium-catalyzed asymmetric hydroamination/cyclization of a dienylsulfonamide [194]

Gold catalysts are also quite efficient in the cyclization of aminodienes (Table 4) [195]. The reaction rates are significantly increased by addition of stoichiometric amounts of an alcohol. The best results are obtained in the presence of (–)-menthol which also significantly improves the regioselectivity in favor of the (*E*)-alkenylpyrrolidines and (*E*)-alkenylpiperidines.

The two different regioisomers **100a** and **100b** are believed to be formed via two different mechanisms (Scheme 39). Product **100a** is thought to form through a

Table 4 Gold-catalyzed asymmetric hydroamination of protected aminodienes [195]

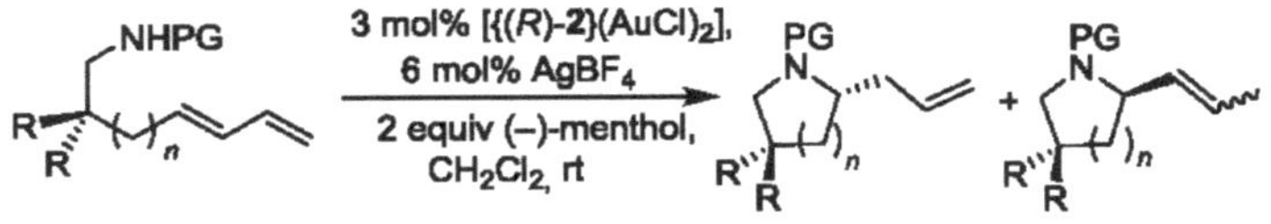

100a **(E)/(Z)-100b**

Substrate	*t*, h	Yield, % (**100a**:(*Z*)-**100b**:(*E*)-**100b**)	% *ee* ((*E*)-**100b**)
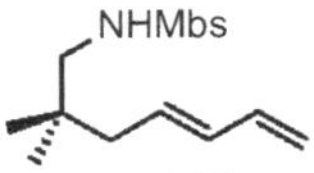	24	95 (1:<0.25:6.3)	95
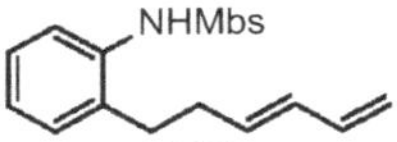	72	77 (1:3.9:8.2)	97
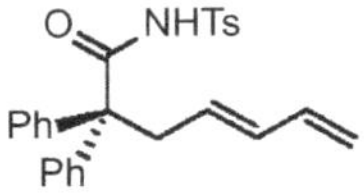	6	99 (1:0.03:1.5)	94
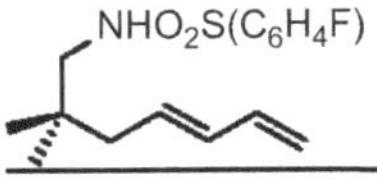	24	67 (1:<0.1:7.5)	91

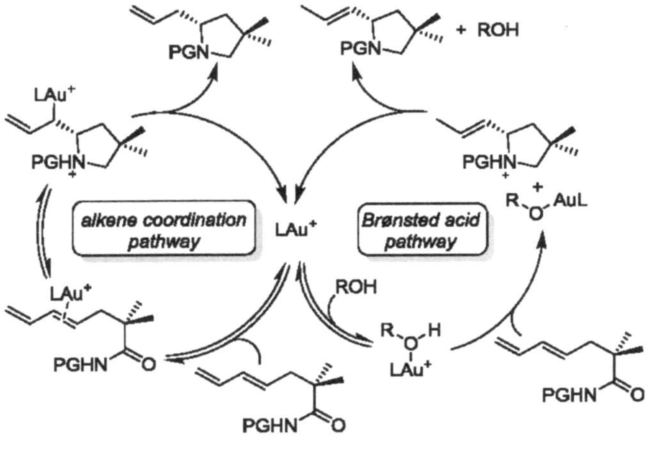

Scheme 39 Two mechanistically distinct pathways leading to different regioisomers in the gold-catalyzed hydroamination/cyclization of aminodienes in the presence or absence of an alcohol [195]

classic mechanism involving olefin coordination to gold, followed by nucleophilic attack of the amine nitrogen and subsequent protonation of the gold–carbon bond. The alkenylazacycle **100b** is formed in a chiral Brønsted acid mechanistic pathway in which gold acts as a Lewis acid by binding to (–)-menthol and increasing its Brønsted acidity. Protonation of the diene is followed by nucleophilic ring-closure.

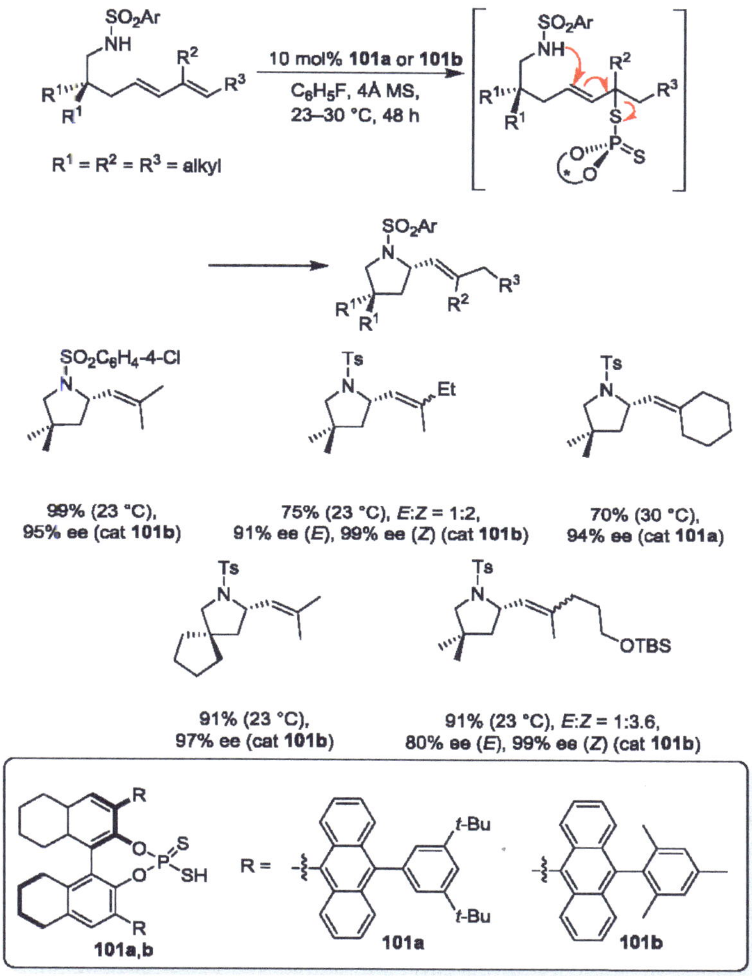

Scheme 40 Organocatalytic asymmetric hydroamination/cyclization of *N*-sulfonyl aminodienes [189]

The efficient organocatalytic hydroamination of *N*-sulfonyl-protected aminodienes has been achieved utilizing chiral dithiophosphoric acids (Scheme 40) [189, 196]. Enantioselectivities as high as 99% *ee* may be obtained in the formation of five-membered rings and even a *tert*-butyldimethylsilyl-protected ether was stable under these conditions.

The dithiophosphoric acid-catalyzed reaction is thought to be mechanistically distinct from simple Brønsted acid-catalyzed processes in that it proceeds via an initial *syn*-addition of the dithiophosphoric acid to the distal alkene. The resulting allylic dithiophosphate ester intermediate undergoes intramolecular *syn*-S_N2'-type substitution by the sulfonylamide nucleophile, displacing and releasing the dithiophosphoric acid catalyst. The high degree of organization of this cyclization transition state is the reason for the high level of enantioselectivity.

4 Hydroamination of Allenes

4.1 *Intermolecular Hydroamination of Allenes*

In contrast to the more developed intramolecular asymmetric hydroamination of aminoallenes (see Sect. 4.2), the first intermolecular asymmetric hydroaminations of allenes have been reported only recently [197, 198].

Gold catalysts are very efficient in the hydroamination of allenes with electron deficient, protected amines. The axial chiral MeOBIPHEP-ligated bis(gold(I))-complex (*S*)-**1** catalyzes the addition of carbamates to 1,3-disubstituted allenes at room temperature after activation with silver tetrafluoroborate (Scheme 41) [197]. The allylic carbamates are obtained in good to excellent enantioselectivities and a variety of electron-withdrawing and electron-donating groups in the allene are tolerated. However, 1,3-disubstituted allenes with substituents larger than methyl at both termini react sluggishly and with low enantioselectivity.

It should be noted that allene racemization by the cationic gold catalyst is fast (<5 min) in comparison to the hydroamination reaction. Furthermore, the enantioselectivity of the allylic carbamate hydroamination product decreases slightly with increasing conversion, presumably due to product poisoning of the catalyst system.

Monosubstituted allenes undergo hydroamination with aniline derivatives in moderate to high yield and enantioselectivities up to 90% *ee* using a Josiphos rhodium(I) catalyst system (Scheme 42) [198]. The reactions make use of ethanol as co-solvent to improve yield and enantioselectivity. A variety of functional groups, such as alkyls, halides, (thio)ethers, esters, ketones, unprotected phenols, and alcohols, are tolerated in the aromatic amine, but electron deficient anilines result in diminished yields and slightly reduced enantioselectivities. The allene can also be varied significantly, including TBS-protected alcohols and phthalimide functionalities.

Mechanistic studies suggest that the reaction proceeds via oxidative addition of the aniline to rhodium(I) (Scheme 43). Deuterium labeling experiments indicate that the resulting amido rhodium(III) hydride **103** can undergo reversible and non-productive hydrometalation of the terminal alkene moiety leading to **104**. However, hydrometalation of the internal alkene moiety generates a π-allyl-rhodium species **105**, which releases the branched allylic amine hydroamination product via reductive elimination.

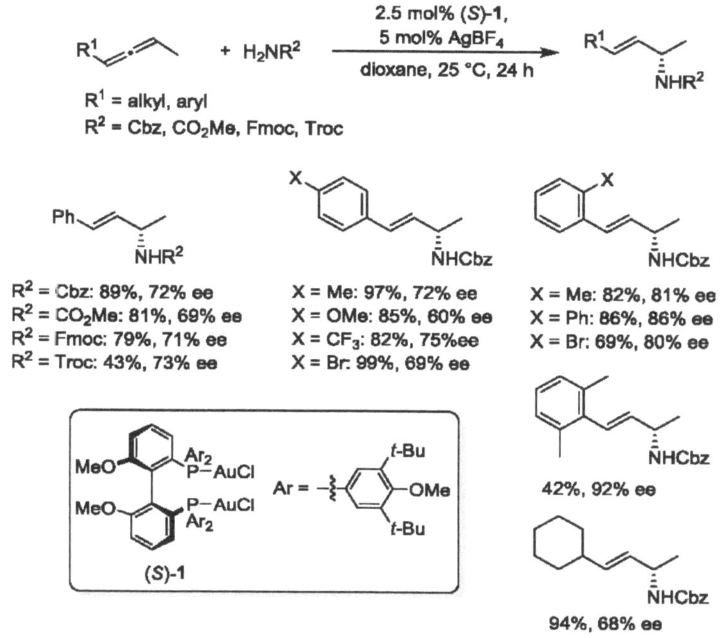

Scheme 41 Enantioselective gold-catalyzed hydroamination of allenes with carbamates [197]

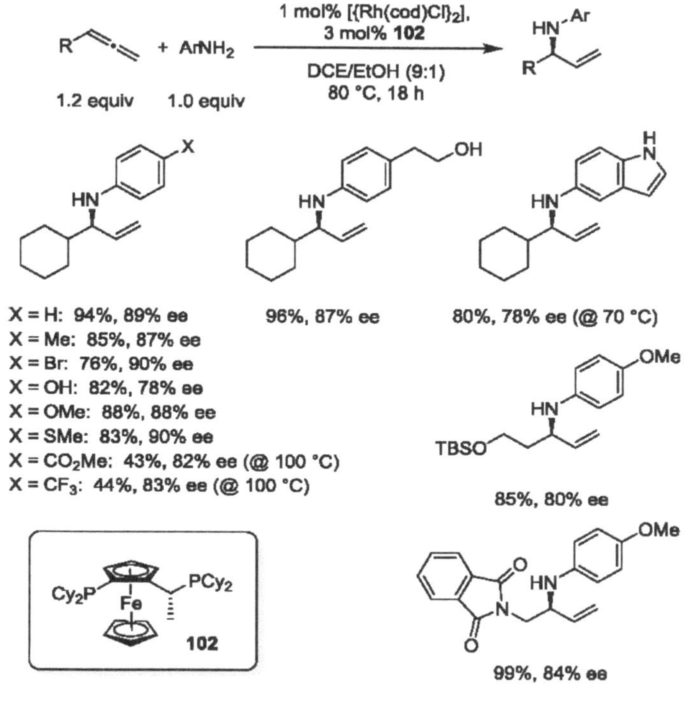

Scheme 42 Enantioselective rhodium-catalyzed hydroamination of allenes with anilines [198]

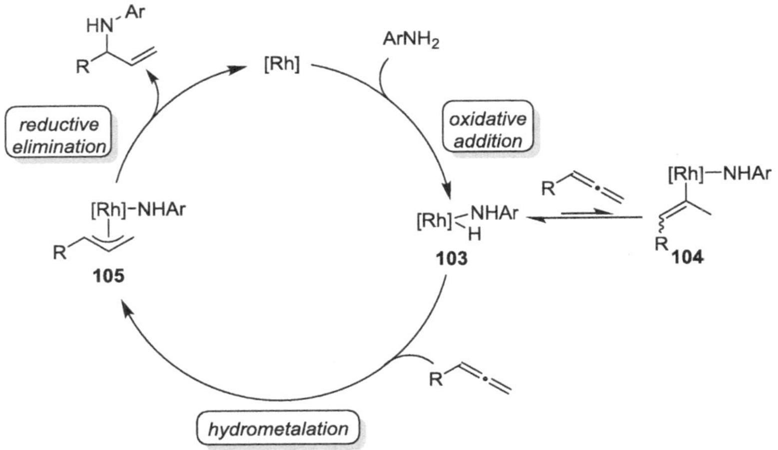

Scheme 43 Proposed mechanism for the enantioselective rhodium-catalyzed hydroamination of allenes with anilines [198]

4.2 Intramolecular Hydroamination of Aminoallenes

The cyclization of aminoallenes can proceed via two pathways, generating two regioisomeric products (Scheme 44). Formation of an imine via the endocyclic pathway **A** usually predominates for monosubstituted aminoallenes, whereas vinylpyrrolidines (pathway **B**) are generated preferentially when 1,3-disubstituted or trisubstituted aminoallenes are employed.

Hence, application of chiral amino-alcoholate titanium and tantalum complexes in the cyclization of aminoallene **107** produces the desired vinylpyrrolidine **108** in low to good enantioselectivities [199–202]. The highest enantioselectivity of 80% *ee* is observed with the apparently monomeric tantalum catalyst **106** [202]. Catalysts derived from L-amino alcohols produce preferentially the (*S*) enantiomer. The *gem*-diphenyl substituents in **106** are crucial to achieve high enantioselectivities. Lack of steric demand on the alcohol carbon atom results in low enantioselectivities and precatalysts are prone to dimerization.

The highly stereoselective cyclization of protected aminoallenes has been demonstrated utilizing a dinuclear gold(I)-phosphine complex with a coordinating *p*-nitrobenzoate (OPNB) counteranion (Scheme 45) [203]. The most enantioselective catalytically active species is believed to be monocationic with one of the counteranions remaining coordinated to the dinuclear gold complex. Similarly, *N*-allenyl carbamates and *N*-allenyl ureas have been cyclized using (*S*)-**1**/$AgClO_4$ with up to 93% *ee* [204, 205]. Gold catalysts based on an axially chiral *N*-heterocyclic carbene ligand have reported significantly lower enantioselectivities of only 44% *ee* in the cyclization of *N*-allenyl carbamates [206].

Similarly, double-protected β-allenyl hydrazines and *N*-Boc-protected β- and γ-allenyl hydroxylamines can be cyclized by binuclear gold(I)-phosphine catalysts

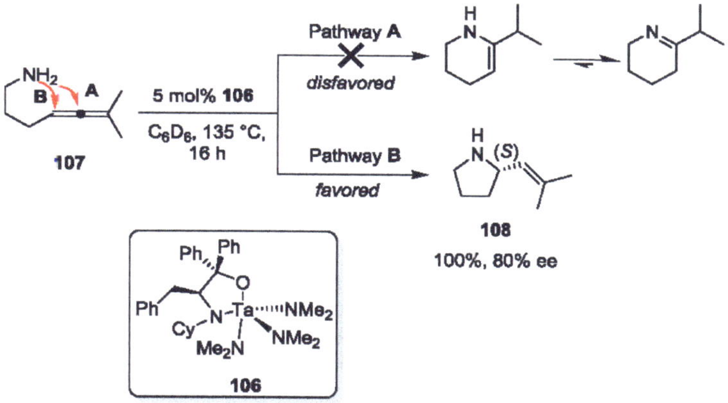

Scheme 44 Two pathway mechanism for the hydroamination/cyclization of aminoallenes catalyzed by a neutral Group 5 metal complex [202]

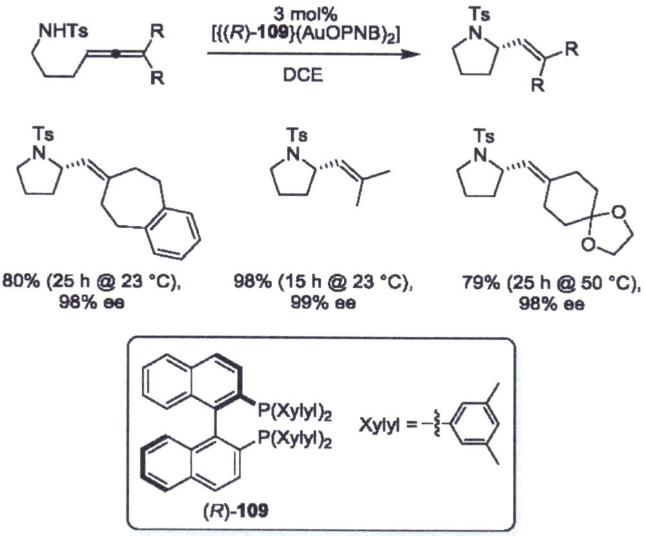

Scheme 45 Gold-catalyzed asymmetric hydroamination/cyclization of sulfonyl-protected aminoallenes [203]

to yield pyrazolidines, isoxazolidines, and tetrahydrooxazines in good to high yield and excellent enantioselectivities (Scheme 46) [207].

The synergistic effect between an achiral catalyst and its chiral counteranion can be exploited to achieve high stereoselectivities. For example, the cyclization of γ-allenylsulfonamides using the achiral gold(I)-precatalyst $PhMe_2PAuCl$ proceeds with excellent enantioselectivities in the presence of the chiral counteranion (*R*)-**110** (Scheme 47) [208].

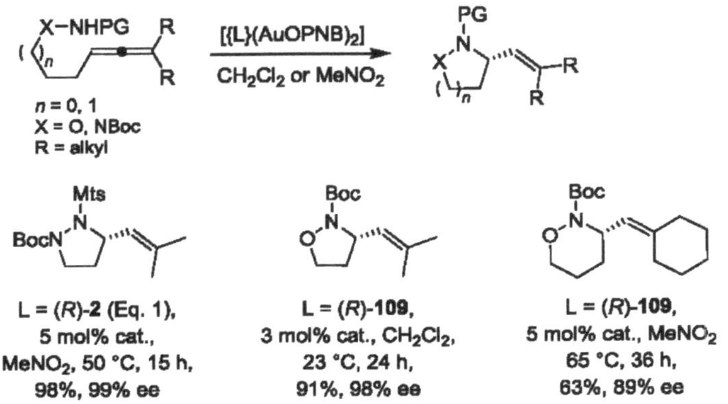

Scheme 46 Gold-catalyzed asymmetric intramolecular hydroamination of hydrazines and hydroxylamines [207]

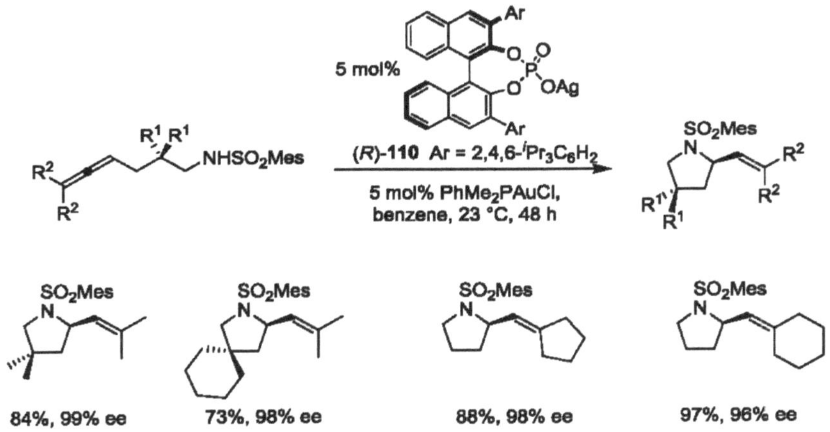

Scheme 47 Chiral counteranion effect in the gold-catalyzed hydroamination/cyclization of aminoallenes [208]

The influence of nuclearity has been investigated in the mono-, bis-, and trimetallic gold catalysts **111–113** (Scheme 48) [209]. The monometallic complex **111** cyclizes γ-allenyl sulfonamides to produce vinylpyrrolidines in racemic form or with low enantiomeric excess. Interestingly, the absolute configuration of vinylpyrrolidines prepared with **111** is opposite to the configuration observed with the other two catalysts. The C_2-symmetric dinuclear complex **112** provides the vinylpyrrolidines with comparable to slightly higher enantiomeric excess (but opposite configuration compared to **111**). The highest activity and selectivity is observed with the C_3-symmetric trinuclear complex **113**. Complex **113** exhibits a short intramolecular Au-Au interaction and it can be speculated that this secondary force is responsible for the enhanced catalytic activity and enantioselectivity.

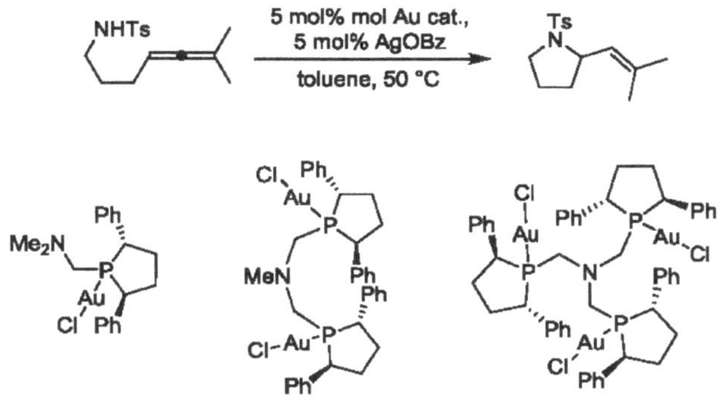

111 –21% ee (120 h) **112** 22% ee (120 h) **113** 94% ee (48 h)

Scheme 48 Chiral mono-, di-, and trinuclear phospholane gold(I) complexes and their reactivity in the hydroamination of aminoallenes [209]

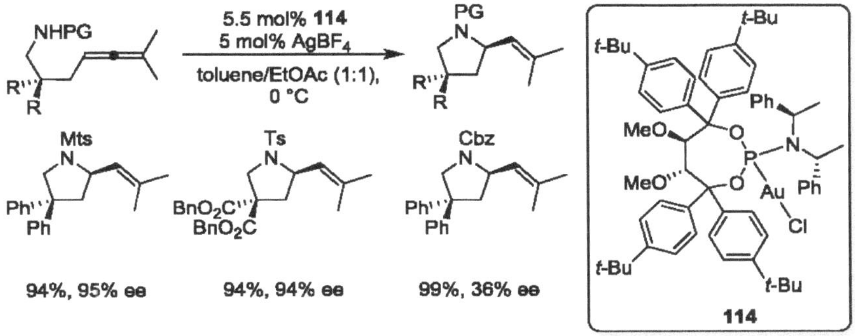

Scheme 49 Asymmetric hydroamination of aminoallenes catalyzed by a mononuclear phosphoramidite gold(I) complex [210]

Mononuclear gold complexes are also able to achieve high activity and enantioselectivity as documented by the monometallic phosphoramidite-gold(I) complex **114** which exhibits excellent activity and enantioselectivity in the cyclization of Thorpe–Ingold activated γ-allenyl sulfonamides (Scheme 49) [210]. However, the cyclization of a corresponding Cbz-protected substrate proceeds with low optical purity. Crystallographic and computational studies show that the monodentate phosphoramidite ligand forms a quasi C_3-symmetric conic binding pocket. Three phenyl rings form the walls of the cavity surrounding the gold atom, which allows an effective chirality transfer from the ligand on the substrate during the catalytic process.

Similar to the cyclization of aminodienes, chiral dithiophosphoric acids are also efficient in the enantioselective cyclization of γ-allenyl sulfonamides and β-allenyl hydroxylamines to yield the corresponding vinyl pyrrolidines and vinyl isoxazolidines in good to high yield and up to 97% *ee* (Scheme 50) [189]. The mechanism is believed to be similar to that proposed for the organocatalytic hydroamination of aminodienes (Scheme 40).

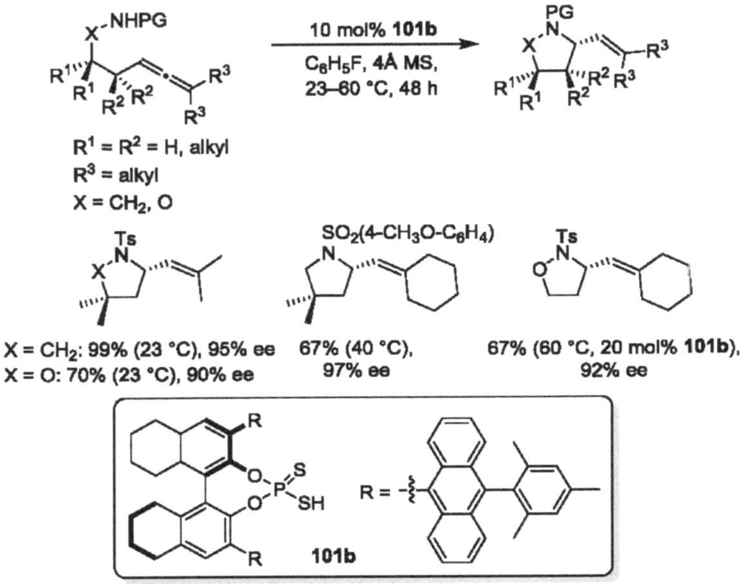

Scheme 50 Organocatalytic enantioselective hydroamination of aminoallenes [189]

5 Hydroamination of Alkynes

Intermolecular hydroamination of alkynes, a process with a relatively low activation barrier, has not been used for the synthesis of chiral amines, since the achiral Schiff base is the major reaction product (Scheme 1b). However, protected aminoalkynes may undergo an interesting intramolecular allylic cyclization using a palladium catalyst with a chiral Norphos-derived ligand (Scheme 51) [194, 211, 212]. Relatively high catalyst loadings are required in order to achieve high enantioselectivities of up to 95% *ee*. An excess of the phosphine ligand is required due to its gradual oxidation under catalytic conditions to the phosphine oxide.

6 Hydroamination with Enantiomerical Pure Amines

6.1 Hydroaminations Using Achiral Catalysts

The diastereoselective cyclization of chiral aminoalkenes, such as α-substituted hex-5-enylamines, can be utilized in the synthesis of piperidine-based alkaloids, such as (–)-pinidinol (Eq. 21) [213]. Depending on the catalyst structure, the cyclization proceeds with excellent *cis* diastereoselectivity, which can be explained

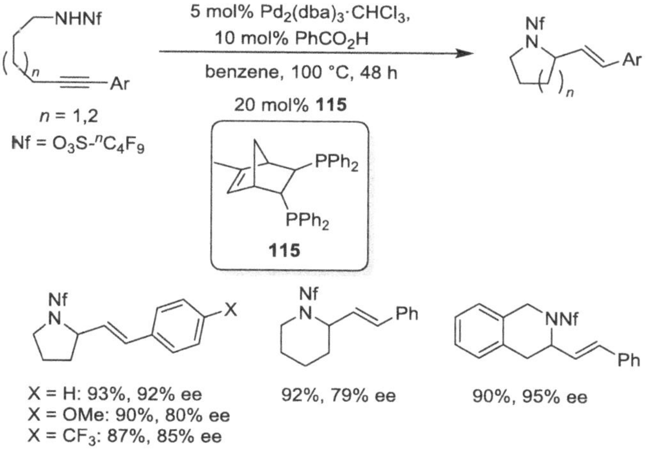

Scheme 51 Palladium-catalyzed asymmetric intramolecular hydroamination of aminoalkynes [211]

with unfavorable 1,3-diaxial interactions in the chair-like cyclization transition state leading to the *trans*-product (Fig. 7).

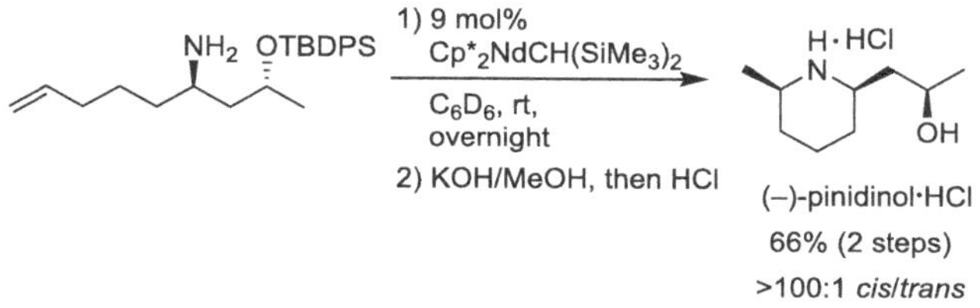

Equation 21. Diastereoselective synthesis of (–)-pinidinol [213]

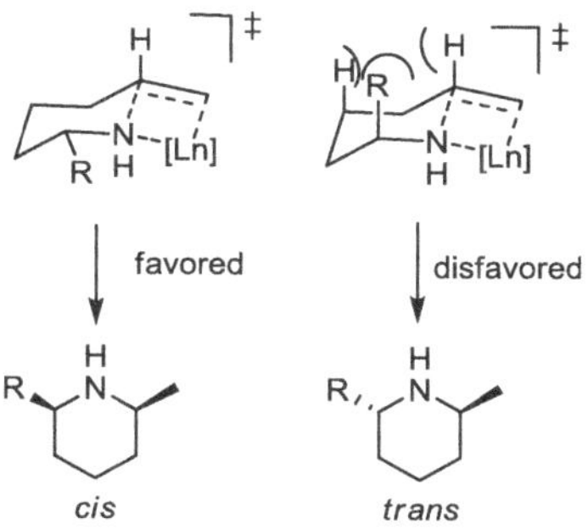

Fig. 7 Proposed cyclization transition states for the preferred formation of *cis*-2,6-disubstituted piperidines

The hydroamination/cyclization of chiral aminoallenes has also been utilized in the synthesis of various alkaloid skeletons [214]. The pyrrolidine alkaloid (+)-197B (Scheme 52), as well as the indolizidine alkaloid (+)-xenovenine (Scheme 53) have been prepared using a hydroamination/cyclization reaction as a key step. Reaction of the aminoallene–alkene **116** containing an allene and a terminal alkene

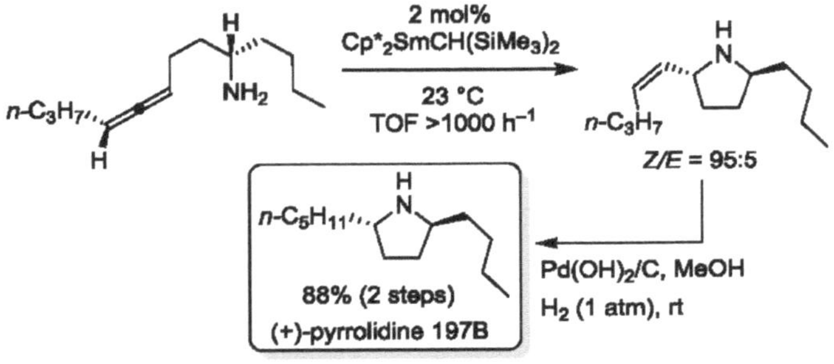

Scheme 52 Diastereoselective synthesis of pyrrolidine (+)-197B via hydroamination/cyclization of an aminoallene [214]

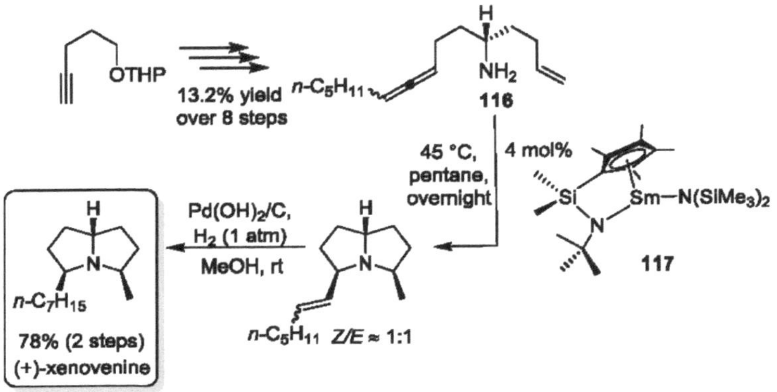

Scheme 53 Diastereoselective synthesis of (+)-xenovenine via hydroamination/bicyclization of the aminoallene-alkene **116** [214]

moiety positioned at equal distance to the amino group requires the sterically open constrained-geometry catalyst **117** to achieve facile and stereospecific bicyclization, while sterically more encumbered lanthanocene catalysts react selectively at the allene moiety and leave the alkene moiety untouched.

Lithium-catalyzed hydroamination reactions have been utilized in the synthesis of opium alkaloids and benzomorphans [215–217]. The synthesis of *O*-methyl metazocine (**119a**) and *O*-methyl pentazocine (**119b**) proceeds via base-catalyzed isomerization/hydroamination, in which the alkene **118** is initially stereoselectively isomerized to the vinyl arene **120** followed by intramolecular hydroamination (Scheme 54) [215]. The analogous synthesis of (–)-codeine via intramolecular hydroamination of the vinyl arene **121** requires irradiation with a 150-W tungsten bulb in the presence of LDA to induce the cyclization (Eq. 22) [216, 217]. The addition is facilitated by single electron transfer [218–220], which is promoted by irradiation. No hydroamination is observed in the absence of irradiation even under refluxing conditions. This lack of reactivity can be attributed to the more electronically rich aromatic ring of **121** in comparison to the benzomorphan precursor **118**.

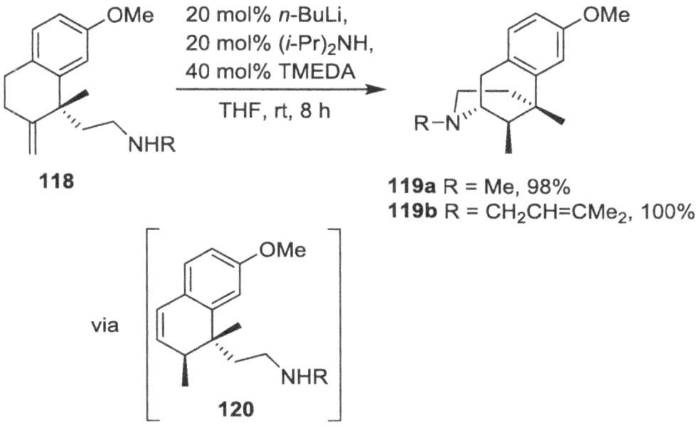

Scheme 54 Synthesis of *O*-methyl metazocine (**119a**) and *O*-methyl pentazocine (**119b**) via alkene isomerization/hydroamination [215]

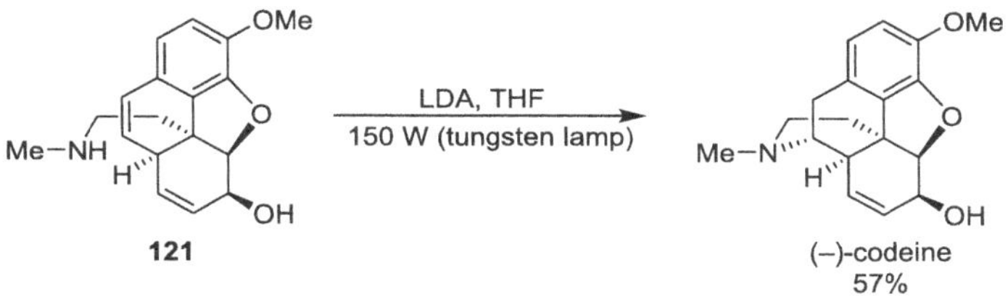

Equation 22. Synthesis of (–)-codeine via base-catalyzed hydroamination/cyclization [216, 217]

The addition of aniline to a chiral allene has been demonstrated to proceed with significant axial chirality transfer in the presence of simple gold salts (Scheme 55) [221]. Note, however, that cationic gold species are known to racemize chiral allenes [222].

(*R*)-(–)-Coniine has been successfully synthesized via silver tetrafluoroborate-catalyzed diastereospecific hydroamination of a secondary chiral aminoallene (Scheme 56) [223]. Utilizing the same methodology, the quinolizidine alkaloid (–)-clavepictine A exhibiting antimicrobial, antifungal, and antitumor activity has been prepared via silver nitrate-catalyzed diastereoselective cyclization of aminoallene **122** (Scheme 57) [224]. Subsequent deacylation provides access to (+)-clavepictine B.

Hydroamination reactions involving alkynes and enantiomerically pure chiral amines can produce novel chiral amine moieties after single-pot reduction of the Schiff base intermediate **123** (Scheme 58) [225]. Unfortunately, partial racemization of the amine stereocenter is observed with many titanium-based hydroamination catalysts, even in the absence of an alkyne substrate. No racemization is observed when the sterically hindered Cp*$_2$TiMe$_2$ or the constrained geometry catalyst Me$_2$Si(C$_5$Me$_4$)(*t*-BuN)Ti(NMe$_2$)$_2$ is used in the catalytic reaction. The addition of pyridine also suppresses the racemization mostly.

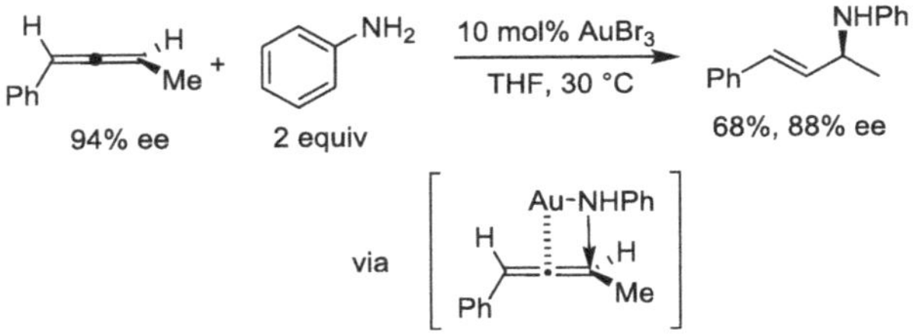

Scheme 55 Stereoselective gold-catalyzed intermolecular hydroamination of a chiral allene [221]

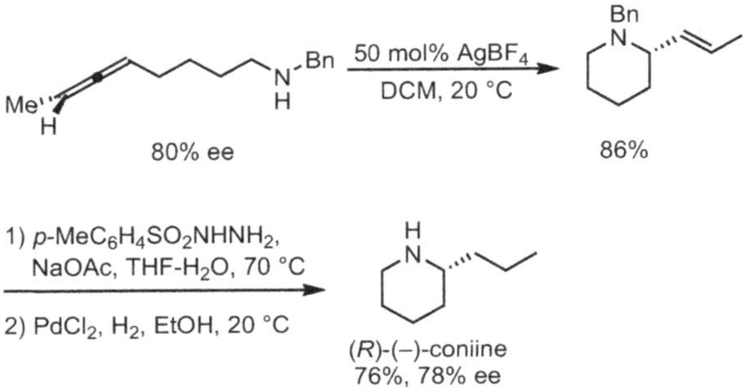

Scheme 56 Synthesis of (*R*)-(–)-coniine via silver-catalyzed diastereoselective cyclization of an aminoallene [223]

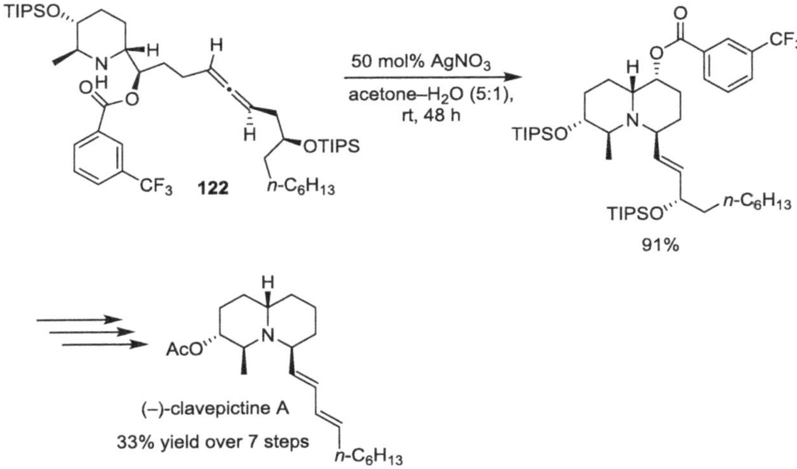

Scheme 57 Synthesis of (–)-clavepictine A via silver-catalyzed diastereoselective cyclization of aminoallene **122** [224]

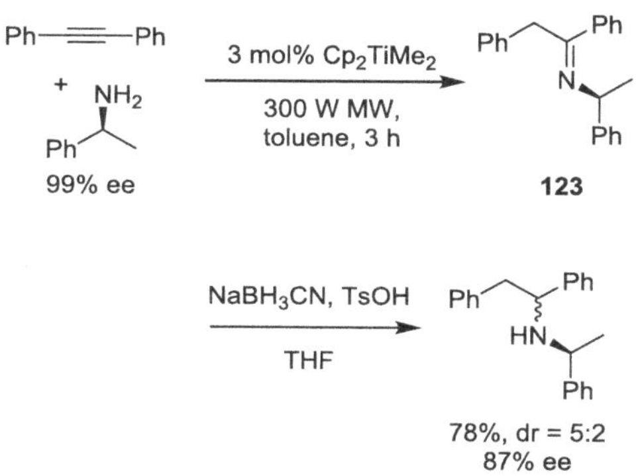

Scheme 58 Partial racemization observed in the titanium-catalyzed hydroamination of tolane with 1-phenylethylamine [225]

6.2 Kinetic Resolution of Chiral Aminoalkenes and Aminoallenes

The efficient kinetic resolution of chiral aminoalkenes has been achieved utilizing the binaphtholate complexes (*R*)-**4**-Ln and (*R*)-**50**-Ln (Table 5) [77, 79, 226]. Various chiral aminopentenes have been kinetically resolved with resolution factors f (defined as $f = K^{\text{dias}} \times k_{\text{fast}}/k_{\text{slow}}$, where K^{dias} is the Curtin–Hammett equilibrium constant between the two diastereomeric substrate/catalyst complexes and $k_{\text{fast}}/k_{\text{slow}}$ the ratio between the faster and the slower reaction rate constant) as high as 19 and enantiomeric excess for recovered starting material reaching $\geq$80% *ee* at conversions close to 50%. The 2,5-disubstituted pyrrolidines are obtained in good to excellent *trans* diastereoselectivity, depending on the steric hindrance of the α-substituent. Kinetic resolution of **124e** using 1 mol% (*R*)-**4**-Lu yields enantiopure (*S*)-**124e** ($\geq$99% *ee*) in 33% re-isolated yield at 64% conversion [79].

The preferred formation of (2*S*,5*S*)-**125a** using (*R*)-binaphtholate complexes can be attributed to impeded cyclization of (*R*)-**124a** due to unfavorable steric interactions of the vinylic methylene protons with a trisarylsilyl substituent in the chair-like transition state (Fig. 8). Fast exchange between matched and mismatched aminoalkene prior to cyclization is imperative for an effective kinetic resolution process. Kinetic analysis of the kinetic resolution process has revealed that the Curtin–Hammett equilibrium favors the matched substrate/catalyst combination in aminopentene substrates **124e**–**124g** containing α-aryl substituents [79, 226], whereas aliphatic substituents in **124a**–**124d** shift this equilibrium in favor of the mismatched substrate/catalyst combination [226].

The formation of the *trans*-2,5-disubstituted pyrrolidine is strongly favored in case of the matched substrate/catalyst pair due to an equatorial placement of the α-substituent in the cyclization transition state (Fig. 8, transition state **C**). The mismatched substrate/catalyst pair often shows diminished diastereoselectivity, because the α-substituent can be placed in an axial position (Fig. 8,

Table 5 Catalytic kinetic resolution of chiral aminopentenes [77, 79, 226]

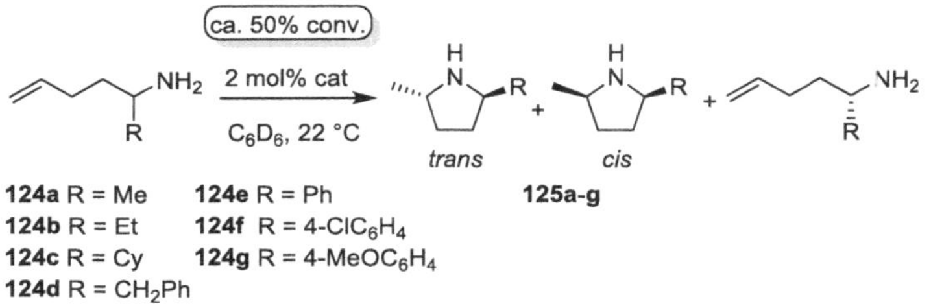

Substrate	Cat	*t*, h	Conv., %	*trans:cis*	% *ee* of recov. start. mat.	*f*
124a	(*R*)-**50**-Y	26	52	13:1	80	16
124b	(*R*)-**50**-Y	6	51	20:1	57	5.9
124c	(*R*)-**4**-Lu	23	47	8:1	51	6.0
124d	(*R*)-**4**-Y	9	50	20:1	42	3.6
124e	(*R*)-**4**-Lu	15[a]	52	≥50:1	83	19
124e	(*R*)-**4**-Lu	27[b]	64	n.r.	99	n.r.
124f	(*R*)-**50**-Y	10[a]	51	≥50:1	80	19
124g	(*R*)-**4**-Y	8[a]	50	≥50:1	78	19

[a]Reaction at 40°C
[b]Using 1.3 mol% catalyst at 45°C
n.r. = not reported

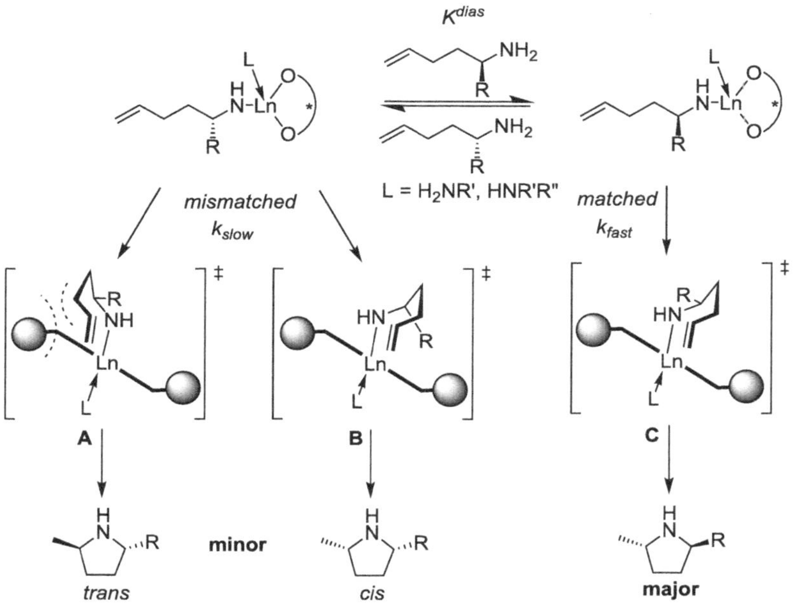

Fig. 8 Proposed stereomodel for kinetic resolution of chiral aminopentenes with an equatorial approach of the olefin. The structure of the 3,3′-tris(aryl)silyl-substituted binaphtholate ligand is simplified for clarity

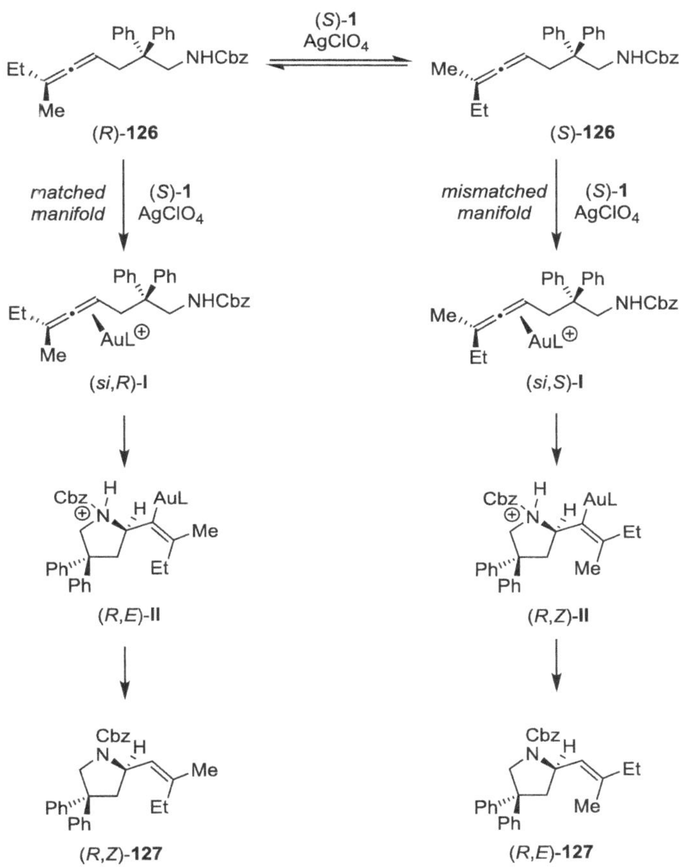

Scheme 59 Proposed mechanism for the gold-catalyzed dynamic kinetic resolution of aminoallenes

transition state **B**) in order to minimize steric interactions of the substrate with the sterically demanding trisarylsilyl substituents of the binaphtholate ligand (Fig. 8, transition state **A**).

Cationic gold(I) complexes are well known to racemize allenes [222], which can be exploited in the facile dynamic kinetic resolution of axially chiral *N*-(γ-allenyl) carbamates with trisubstituted allenyl groups [227]. A mixture of the dinuclear gold(I) phosphine complex (*S*)-**1** and $AgClO_4$ catalyzes the cyclization of the Cbz-protected aminoallene **126** to yield predominantly the (*Z*)-vinylpyrrolidine (*R*,*Z*)-**127** with excellent enantioselectivity (Eq. 23).

Initial mechanistic studies suggest that the (*E*)-vinylpyrrolidine (*R*,*E*)-**127** is formed in the mismatched reaction manifold (Scheme 59). In the matched manifold, the cationic gold(I)-species is believed to coordinate preferentially to the *si* face of the allene substrate ((*si*,*R*)-**I**). Subsequent attack of the tethered nucleophilic carbamate nitrogen leads to the σ-alkenyl-gold complex (*R*,*E*)-**II**.

Finally, protonolysis of the Au–C bond proceeds with retention of configuration, releasing the (*R*,*Z*)-vinylpyrrolidine product.

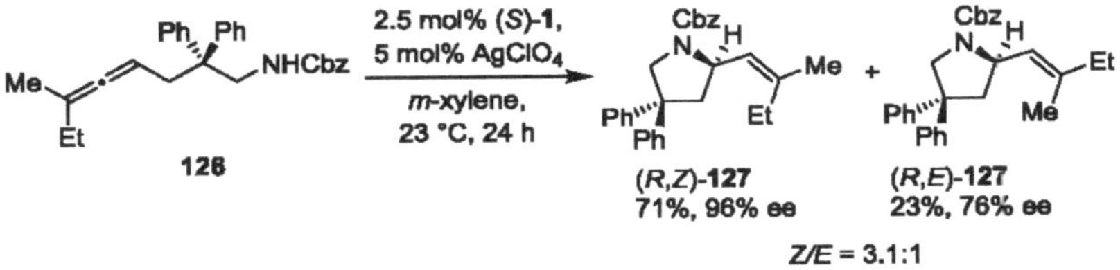

Equation 23. Gold-catalyzed dynamic kinetic resolution of aminoallenes [227]

7 Synthesis of Chiral Amines via Reaction Sequences Involving Hydroamination

One-pot multistep reaction sequences [228–232] are an efficient method to introduce additional complexity and versatility to the hydroamination reaction. With the exception of the examples mentioned in Sect. 5, the hydroamination of alkynes cannot be applied directly to the synthesis of chiral amines, as unsaturated Schiff bases or enamines are major products of an alkyne hydroamination. However, since an asymmetric reduction and hydrosilylation of imines are well developed areas, sequential hydroamination/asymmetric reduction and hydroamination/asymmetric hydrosilylation of an alkyne can serve as alternative routes to the desired saturated amines. Notably, since alkynes are in general more reactive than alkenes, these methods can be more effective than asymmetric hydroamination of alkenes.

An intramolecular hydroamination/asymmetric hydrosilylation sequence has been accomplished using the chiral catalyst (*S*,*S*)-(ebthi)$TiMe_2$ (ebthi = ethylenebis(tetrahydroindenyl)). The aminoalkynes **128a**,**b** undergo efficient hydroamination (>95% conv.), followed by enantioselective hydrosilylation in up to 66% *ee* and moderate overall yields (Eq. 24) [233].

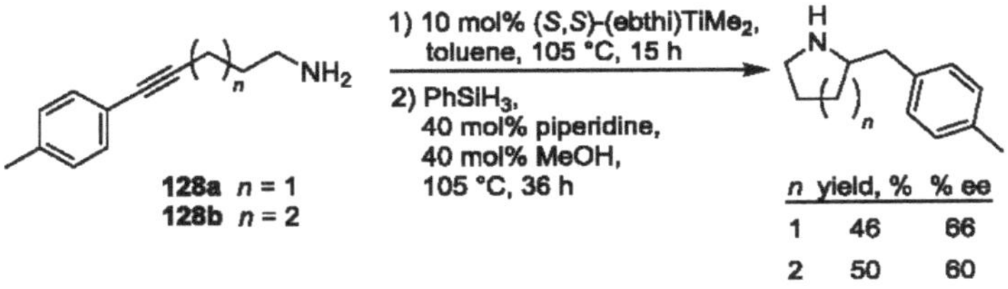

Equation 24. Titanium-catalyzed tandem hydroamination/hydrosilylation of aminoalkynes [233]

Similarly, the hydroamination of an aminoalkyne can be put into sequence with an asymmetric hydrogenation, as demonstrated in the synthesis of the opium alkaloids (*S*)-(+)-laudanosine and (*S*)-(–)-xylopinine (Scheme 60) [234]. Sonogashira coupling

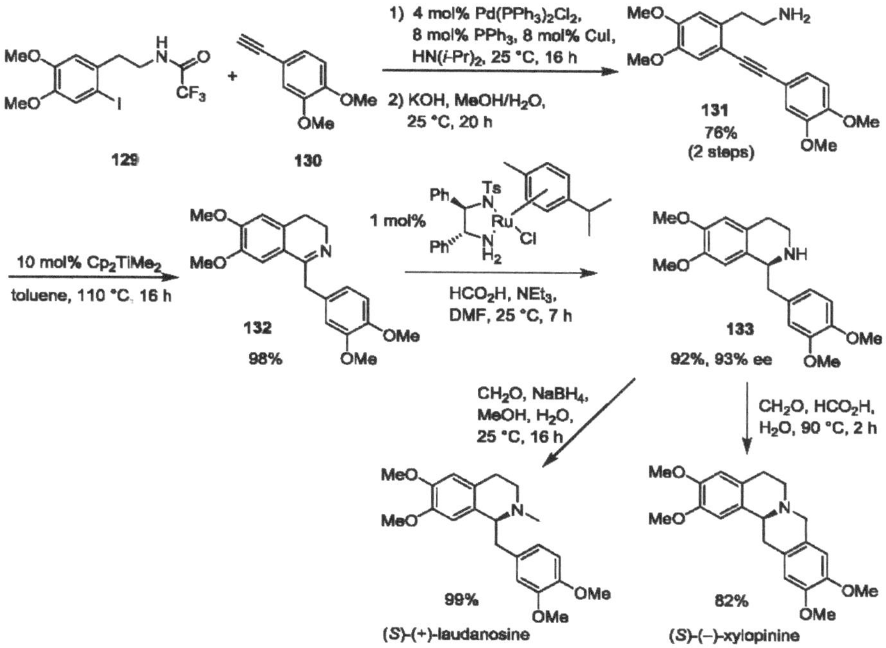

Scheme 60 Synthesis of opium alkaloids involving titanium-catalyzed hydroamination/cyclization [234]

of the aryl iodide **129** and the phenyl acetylene **130** furnishes the aminoalkyne **131** after deprotection of the amide. Highly regioselective titanium-catalyzed hydroamination/cyclization of **131** leads to the 3,4-dihydroisoquinoline **132**, which is reduced by asymmetric transfer hydrogenation using Noyori's chiral ruthenium catalyst with high enantioselectivity. Reductive amination of the 1,2,3,4-tetrahydroisoquinoline **133** leads then directly to (*S*)-(+)-laudanosine, while Pictet–Spengler cyclization of **133** produces (*S*)-(–)-xylopinine. Note, however, the hydroamination product **132** is isolated and purified prior to the hydrogenation step.

The sequential hydroamination/asymmetric hydrogenation has been extended to the stereoselective synthesis of 3-substituted morpholines in good yield and excellent enantioselectivities (Scheme 61) [235]. This sequence can be carried out as a one-pot procedure without the need to isolate the hydroamination product intermediate. A variety of alkyl and aryl substituents with different functional groups are tolerated. While related piperidines and piperazines can be prepared in a similar manner, the resulting products are obtained with low to modest enantiomeric excess.

In a more general approach to the same reaction sequence, enantiomerically enriched secondary amines can be prepared with excellent enantiomeric excess through gold-catalyzed intermolecular hydroamination of alkynes with anilines followed by an organocatalytic asymmetric transfer hydrogenation (Scheme 62) [236]. The transfer hydrogenation is performed with a chiral Brønsted acid and the

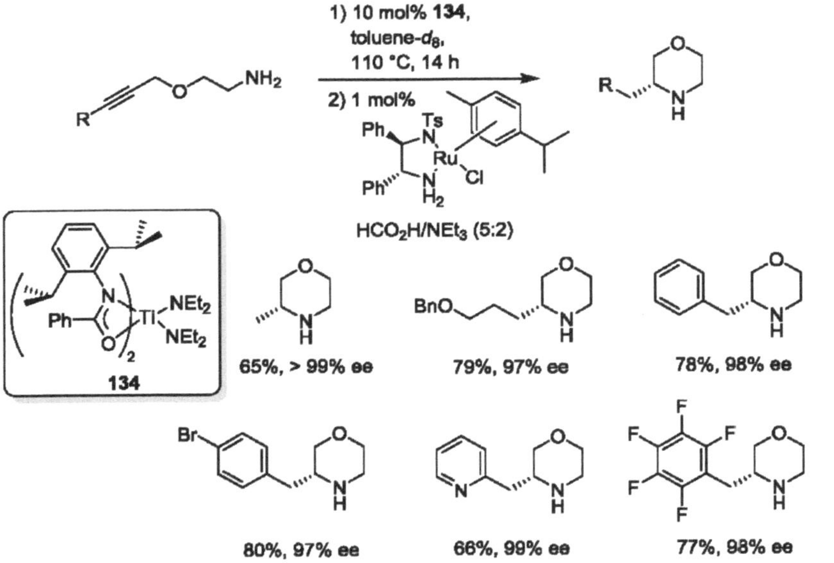

Scheme 61 Synthesis of 3-substituted morpholines via sequential alkyne hydroamination/asymmetric transfer hydrogenation [235]

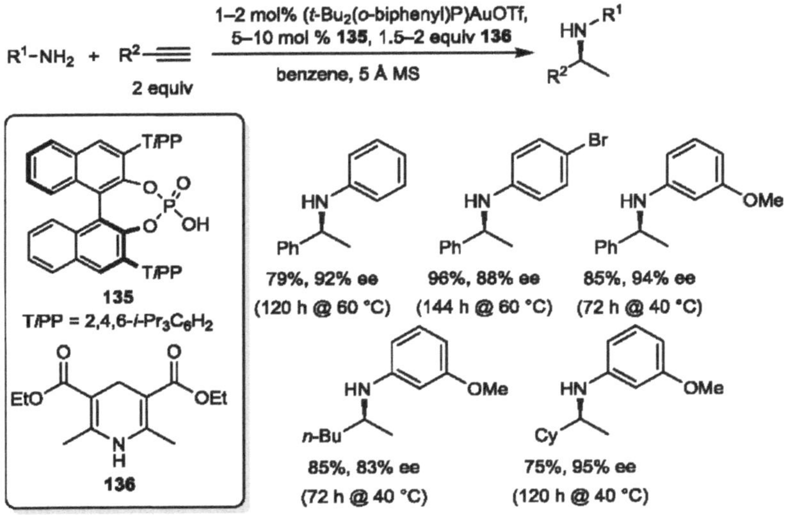

Scheme 62 Tandem intermolecular hydroamination/asymmetric transfer hydrogenation [236]

readily available Hantzsch ester **136** as hydrogen source. The catalysis works for a wide variety of aryl, alkenyl, and aliphatic alkynes as well as anilines with different electronic properties.

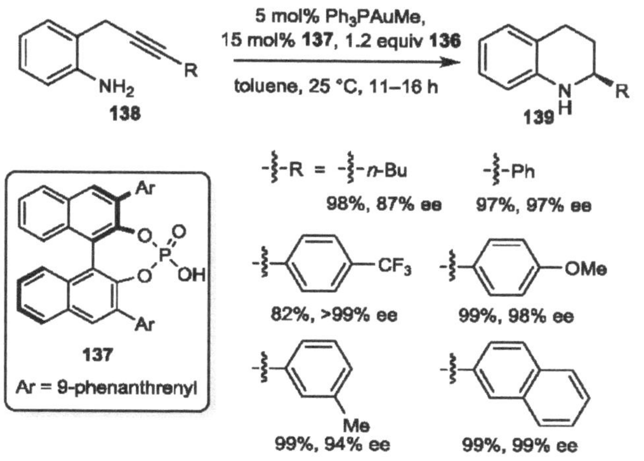

Scheme 63 Tandem intramolecular hydroamination/asymmetric transfer hydrogenation [237]

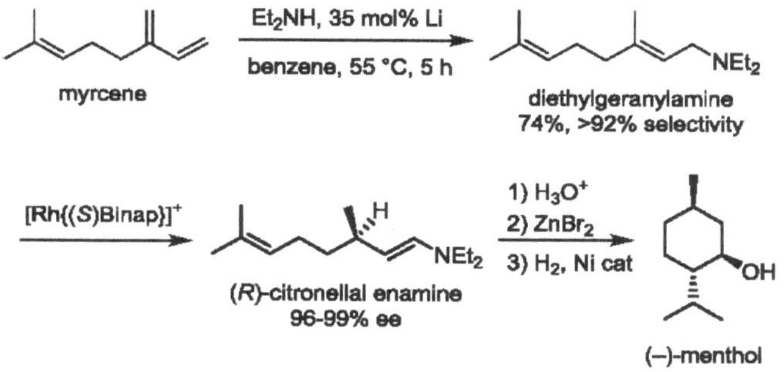

Scheme 64 Utilization of a hydroamination/asymmetric isomerization sequence in the Takasago menthol synthesis [238]

An analogous binary catalyst system can be applied successfully to the intramolecular cyclization of 2-(2-propynylanilines) **138** to yield tetrahydroquinolines **139** with excellent enantioselectivities (Scheme 63) [237]. The reaction is also applicable to a variety of 2-(2-propynylanilines) with substituents in the 4- and 5-position of the aniline ring.

The base-catalyzed regioselective hydroamination of myrcene (Scheme 64) is performed industrially on a multi-ton scale as part of the Takasago menthol synthesis [238]. The hydroamination is followed by an asymmetric isomerization of diethylgeranylamine to (*R*)-citronellal enamine utilizing a BINAP-rhodium catalyst. Hydrolysis of the citronellal enamine, followed by a zinc bromide-catalyzed intramolecular ene reaction and subsequent nickel-catalyzed hydrogenation of the resulting isopulegol provides (–)-menthol with high enantiopurity.

A new strategy for the synthesis of cyclic *N*,*O*-acetals involves sequential hydroamination and subsequent ring-closing metathesis. The palladium-catalyzed

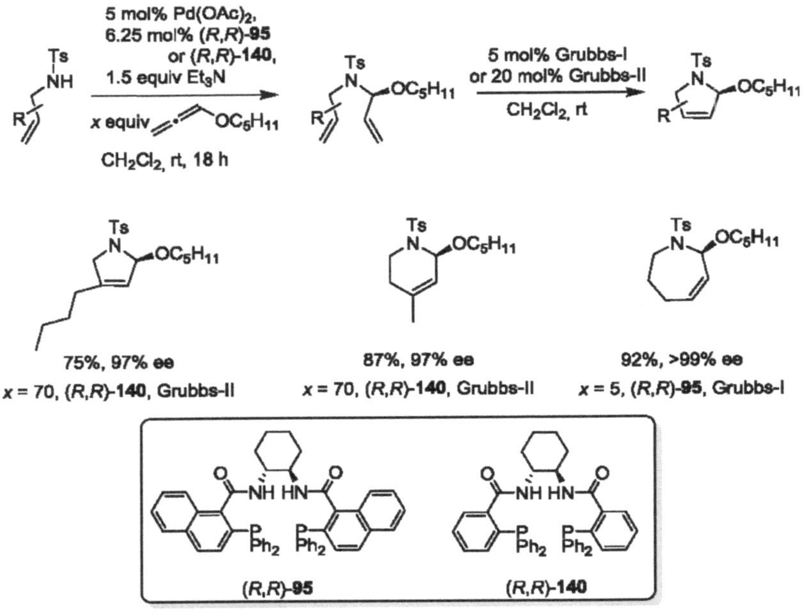

Scheme 65 Synthesis of cyclic *N*,*O*-acetals via sequential asymmetric hydroamination/ring-closing metathesis [239]

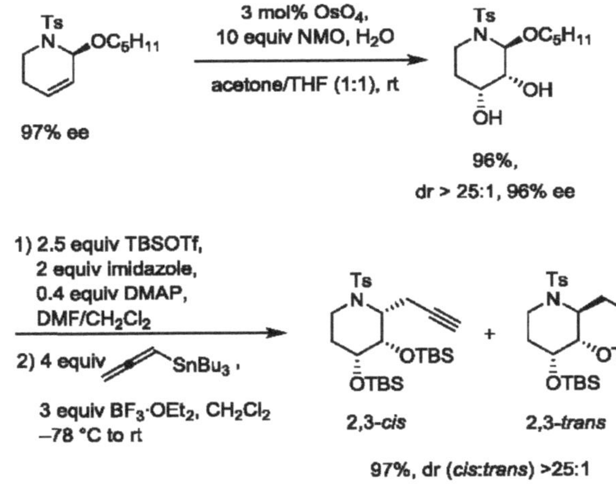

Scheme 66 Stereoselective dihydroxylation and alkylation of cyclic *N*,*O*-acetals [239]

asymmetric addition of *N*-tosyl aminoalkenes to an alkoxyallene generates an acyclic *N*,*O*-acetal, which is then subjected to ruthenium-catalyzed ring-closing metathesis (Scheme 65) [239]. The double bond in the cyclic *N*,*O*-acetal is readily modified further via stereoselective dihydroxylation (Scheme 66), epoxidation, or hydrogenation. Modification of the *N,O*-acetal via its iminium ion allows installation of various alkyl substituents at the 2 position of the azacycle.

8 Conclusions

The development of chiral catalysts for asymmetric hydroamination reactions has made tremendous progress over the last two decades and many transformations can now be performed with enantioselectivities exceeding 90% *ee*. However, significant challenges remain. While first reports on catalytic systems for the stereoselective addition of simple, non-activated amines to unactivated alkenes have surfaced, the reactivities and selectivities of these systems leave a lot to be desired. None of the reported catalyst systems is generally applicable to a wide variety of substrates with consistent high stereochemical induction and tolerance of a multitude of functional groups as well as air and moisture. Certainly, late transition metal-based catalysts, and to some extent enzymatic and organocatalytic approaches, show promising leads that could fill this void. Alkaline earth metal-based catalysts show promising activity, but stereocontrol remains a significant challenge for this class of complex due to facile ligand redistribution processes. Rare earth metal complexes as well as the newest generation of Group 4 metal catalysts remain among the most active and most versatile catalyst systems, but it seems certain that new and powerful catalyst systems will evolve in the near future.

Acknowledgment Generous financial support by the National Science Foundation through a NSF CAREER Award (CHE 0956021) is gratefully acknowledged.

References

1. Müller TE, Beller M (1998) Chem Rev 98:675
2. Müller TE, Hultzsch KC, Yus M, Foubelo F, Tada M (2008) Chem Rev 108:3795
3. Brunet JJ, Neibecker D (2001) In: Togni A, Grützmacher H (eds) Catalytic heterofunctionalization from hydroamination to hydrozirconation. Wiley-VCH, Weinheim, p 91
4. Doye S (2009) In: Schaumann E, Enders D (eds) Science of synthesis, vol 40a. Georg Thieme Verlag, Stuttgart, p 241
5. Reznichenko AL, Hultzsch KC (2013) Top Organomet Chem 43:51
6. Nishina N, Yamamoto Y (2013) Top Organomet Chem 43:115
7. Roesky PW, Müller TE (2003) Angew Chem Int Ed 42:2708
8. Hultzsch KC (2005) Adv Synth Catal 347:367
9. Hultzsch KC (2005) Org Biomol Chem 3:1819
10. Reznichenko AL, Hultzsch KC (2010) In: Nugent T (ed) Chiral amine synthesis: methods, developments and applications. Wiley-VCH, Weinheim, p 341
11. Reznichenko AL, Hultzsch KC (2010) In: De Vries JG, Molander GA, Evans PA (eds) Science of synthesis, stereoselective synthesis, vol 1. Georg Thieme Verlag, Stuttgart, p 689
12. Aillaud I, Collin J, Hannedouche J, Schulz E (2007) Dalton Trans 5105
13. Hannedouche J, Collin J, Trifonov A, Schulz E (2011) J Organomet Chem 696:255

14. Zi G (2009) Dalton Trans 9101
15. Zi G (2011) J Organomet Chem 696:68
16. Hannedouche J, Schulz E (2013) Chem Eur J 19:4972
17. Steinborn D, Taube R (1986) Z Chem 26:349
18. Taube R (2002) In: Cornils B, Herrmann WA (eds) Applied homogeneous catalysis with organometallic compounds, vol 1. Wiley-VCH, Weinheim, p 513
19. Li-Wen Xu C-GX (2005) Eur J Org Chem 633
20. Hii KK (2006) Pure Appl Chem 78:341
21. Zhang Z, Lee SD, Widenhoefer RA (2009) J Am Chem Soc 131:5372
22. Sevov CS, Zhou J, Hartwig JF (2012) J Am Chem Soc 134:11960
23. Reznichenko AL, Nguyen HN, Hultzsch KC (2010) Angew Chem Int Ed 49:8984
24. Reznichenko AL, Hultzsch KC (2013) Organometallics 32:1394
25. Kawatsura M, Hartwig JF (2000) J Am Chem Soc 122:9546
26. Li K, Horton PN, Hursthouse MB, Hii KK (2003) J Organomet Chem 665:250
27. Hu A, Ogasawara M, Sakamoto T, Okada A, Nakajima K, Takahashi T, Lin W (2006) Adv Synth Catal 348:2051
28. Nettekoven U, Hartwig JF (2002) J Am Chem Soc 124:1166
29. Utsunomiya M, Hartwig JF (2003) J Am Chem Soc 125:14286
30. Johns AM, Utsunomiya M, Incarvito CD, Hartwig JF (2006) J Am Chem Soc 128:1828
31. Besson L, Goré J, Cazes B (1995) Tetrahedron Lett 36:3857
32. Al-Masum M, Meguro M, Yamamoto Y (1997) Tetrahedron Lett 38:6071
33. Meguro M, Yamamoto Y (1998) Tetrahedron Lett 39:5421
34. Löber O, Kawatsura M, Hartwig JF (2001) J Am Chem Soc 123:4366
35. Minami T, Okamoto H, Ikeda S, Tanaka R, Ozawa F, Yoshifuji M (2001) Angew Chem Int Ed 40:4501
36. Pawlas J, Nakao Y, Kawatsura M, Hartwig JF (2002) J Am Chem Soc 124:3669
37. Sakai N, Ridder A, Hartwig JF (2006) J Am Chem Soc 128:8134
38. Landis CR, Halpern J (1987) J Am Chem Soc 109:1746
39. Dorta R, Egli P, Zürcher F, Togni A (1997) J Am Chem Soc 119:10857
40. Casalnuovo AL, Calabrese JC, Milstein D (1988) J Am Chem Soc 110:6738
41. Beller M, Breindl C, Eichberger M, Hartung CG, Seayad J, Thiel OR, Tillack A, Trauthwein H (2002) Synlett 1579
42. Zhou J, Hartwig JF (2008) J Am Chem Soc 130:12220
43. Pan S, Endo K, Shibata T (2012) Org Lett 14:780
44. Gröger H (2010) In: Ojima I (ed) Catalytic asymmetric synthesis. Wiley, Hoboken, p 269
45. Hyun MW, Yun YH, Kim JY, Kim SH (2011) Mycobiology 39:257
46. Hauer B, Schneider N, Drew D, Ditrich K, Turner N, Nestl BM (2011) (BASF Aktiengesellschaft, Germany). WO2011/012,632
47. Cooper NJ, Knight DW (2004) Tetrahedron 60:243
48. MacDonald MJ, Schipper DJ, Ng PJ, Moran J, Beauchemin AM (2011) J Am Chem Soc 133:20100
49. MacDonald MJ, Hesp CR, Schipper DJ, Pesant M, Beauchemin AM (2013) Chem Eur J 19:2597
50. Guimond N, MacDonald MJ, Lemieux V, Beauchemin AM (2012) J Am Chem Soc 134:16571
51. Hong S, Marks TJ (2004) Acc Chem Res 37:673
52. Reznichenko AL, Hultzsch KC (2010) Struct Bond 137:1
53. Gagné MR, Stern CL, Marks TJ (1992) J Am Chem Soc 114:275
54. Motta A, Lanza G, Fragalà IL, Marks TJ (2004) Organometallics 23:4097
55. Manna K, Kruse ML, Sadow AD (2011) ACS Catal 1:1637
56. Tobisch S (2011) Chem Eur J 17:14974
57. Tobisch S (2012) Dalton Trans 41:9182

58. Gagné MR, Brard L, Conticello VP, Giardello MA, Stern CL, Marks TJ (1992) Organometallics 11:2003
59. Giardello MA, Conticello VP, Brard L, Gagné MR, Marks TJ (1994) J Am Chem Soc 116:10241
60. Kirby JA (1981) Adv Phys Org Chem 17:183
61. Jung ME, Piizzi G (2005) Chem Rev 105:1735
62. Douglass MR, Ogasawara M, Hong S, Metz MV, Marks TJ (2002) Organometallics 21:283
63. Ryu J-S, Marks TJ, McDonald FE (2004) J Org Chem 69:1038
64. Vitanova DV, Hampel F, Hultzsch KC (2007) J Organomet Chem 692:4690
65. Molander GA, Dowdy ED (1998) J Org Chem 63:8983
66. Molander GA, Dowdy ED (1999) J Org Chem 64:6515
67. Ryu J-S, Marks TJ, McDonald FE (2001) Org Lett 3:3091
68. Kim YK, Livinghouse T (2002) Angew Chem Int Ed 41:3645
69. Chapurina Y, Hannedouche J, Collin J, Guillot R, Schulz E, Trifonov A (2010) Chem Commun 46:6918
70. Chapurina Y, Ibrahim H, Guillot R, Kolodziej E, Collin J, Trifonov A, Schulz E, Hannedouche J (2011) J Org Chem 76:10163
71. Hultzsch KC, Gribkov DV, Hampel F (2005) J Organomet Chem 690:4441
72. O'Shaughnessy PN, Scott P (2003) Tetrahedron Asymmetry 14:1979
73. O'Shaughnessy PN, Knight PD, Morton C, Gillespie KM, Scott P (2003) Chem Commun 1770
74. O'Shaughnessy PN, Gillespie KM, Knight PD, Munslow I, Scott P (2004) Dalton Trans 2251
75. Hong S, Tian S, Metz MV, Marks TJ (2003) J Am Chem Soc 125:14768
76. Gribkov DV, Hultzsch KC, Hampel F (2003) Chem Eur J 9:4796
77. Gribkov DV, Hultzsch KC (2004) Chem Commun 730
78. Gribkov DV, Hampel F, Hultzsch KC (2004) Eur J Inorg Chem 4091
79. Gribkov DV, Hultzsch KC, Hampel F (2006) J Am Chem Soc 128:3748
80. Collin J, Daran J-D, Schulz E, Trifonov A (2003) Chem Commun 3048
81. Collin J, Daran J-D, Jacquet O, Schulz E, Trifonov A (2005) Chem Eur J 11:3455
82. Riegert D, Collin J, Meddour A, Schulz E, Trifonov A (2006) J Org Chem 71:2514
83. Riegert D, Collin J, Daran J-D, Fillebeen T, Schulz E, Lyubov D, Fukin G, Trifonov A (2007) Eur J Inorg Chem 1159
84. Aillaud I, Wright K, Collin J, Schulz E, Mazaleyrat J-P (2008) Tetrahedron Asymmetry 19:82
85. Hannedouche J, Aillaud I, Collin J, Schulz E, Trifonov A (2008) Chem Commun 3552
86. Aillaud I, Lyubov D, Collin J, Guillot R, Hannedouche J, Schulz E, Trifonov A (2008) Organometallics 27:5929
87. Aillaud I, Collin J, Duhayon C, Guillot R, Lyubov D, Schulz E, Trifonov A (2008) Chem Eur J 14:2189
88. Aillaud I, Collin J, Hannedouche J, Schulz E, Trifonov A (2010) Tetrahedron Lett 51:4742
89. Queffelec C, Boeda F, Pouilhés A, Meddour A, Kouklovsky C, Hannedouche J, Collin J, Schulz E (2011) ChemCatChem 3:122
90. Aillaud I, Olier C, Chapurina Y, Collin J, Schulz E, Guillot R, Hannedouche J, Trifonov A (2011) Organometallics 30:3378
91. Chapurina Y, Guillot R, Lyubov D, Trifonov A, Hannedouche J, Schulz E (2013) Dalton Trans 42:507
92. Kim JY, Livinghouse T (2005) Org Lett 7:1737
93. Kim H, Kim YK, Shim JH, Kim M, Han M, Livinghouse T, Lee PH (2006) Adv Synth Catal 348:2609
94. Lovick HM, An DK, Livinghouse TS (2011) Dalton Trans 40:7697
95. Meyer N, Zulys A, Roesky PW (2006) Organometallics 25:4179
96. Yu X, Marks TJ (2007) Organometallics 26:365
97. Heck R, Schulz E, Collin J, Carpentier J-F (2007) J Mol Cat A 268:163
98. Xiang L, Wang Q, Song H, Zi G (2007) Organometallics 26:5323

99. Zi G, Xiang L, Song H (2008) Organometallics 27:1242
100. Wang Q, Xiang L, Song H, Zi G (2008) Inorg Chem 47:4319
101. Vitanova DV, Hampel F, Hultzsch KC (2011) J Organomet Chem 696:321
102. Benndorf P, Jenter J, Zielke L, Roesky PW (2011) Chem Commun 47:2574
103. Zhang Y, Yao W, Li H, Mu Y (2012) Organometallics 31:4670
104. Shibasaki M, Yoshikawa N (2002) Chem Rev 102:2187
105. Yamagiwa N, Matsunaga S, Shibasaki M (2004) Angew Chem Int Ed 43:4493
106. Hevia E, Mulvey RE (2011) Angew Chem Int Ed 50:6448
107. Seayad J, Tillack A, Hartung CG, Beller M (2002) Adv Synth Catal 344:795
108. Horrillo Martínez P, Hultzsch KC, Hampel F (2006) Chem Commun 2221
109. Ogata T, Ujihara A, Tsuchida S, Shimizu T, Kaneshige A, Tomioka K (2007) Tetrahedron Lett 48:6648
110. Deschamp J, Collin J, Hannedouche J, Schulz E (2011) Eur J Org Chem 3329
111. Horrillo-Martínez P (2008) Ph.D thesis, Universität Erlangen-Nürnberg
112. Harder S (2010) Chem Rev 110:3852
113. Barrett AGM, Crimmin MR, Hill MS, Procopiou PA (2010) Proc R Soc A 466:927
114. Crimmin MR, Hill MS (2013) Top Organomet Chem 45:191
115. Horrillo-Martínez P, Hultzsch KC (2009) Tetrahedron Lett 50:2054
116. Buch F, Harder S (2008) Z Naturforsch 63b:169
117. Neal SR, Ellern A, Sadow AD (2011) J Organomet Chem 696:228
118. Zhang X, Emge TJ, Hultzsch KC (2012) Angew Chem Int Ed 51:394
119. Nixon TD, Ward BD (2012) Chem Commun 48:11790
120. Wixey JS, Ward BD (2011) Chem Commun 47:5449
121. Wixey JS, Ward BD (2011) Dalton Trans 40:7693
122. Pohlki F, Doye S (2003) Chem Soc Rev 32:104
123. Bytschkov I, Doye S (2003) Eur J Org Chem 935
124. Doye S (2004) Synlett 1653
125. Odom AL (2005) Dalton Trans 225
126. Severin R, Doye S (2007) Chem Soc Rev 36:1407
127. Gribkov DV, Hultzsch KC (2004) Angew Chem Int Ed 44:5542
128. Bexrud JA, Beard JD, Leitch DC, Schafer LL (2005) Org Lett 7:1959
129. Kim H, Lee PH, Livinghouse T (2005) Chem Commun 5205
130. Roesky PW (2009) Angew Chem Int Ed 48:4892
131. Müller C, Saak W, Doye S (2008) Eur J Org Chem 2731
132. Bexrud JA, Eisenberger P, Leitch DC, Payne PR, Schafer LL (2009) J Am Chem Soc 131:2116
133. Knight PD, Munslow I, O'Shaughnessy PN, Scott P (2004) Chem Commun 894
134. Kissounko DA, Epshteyn A, Fettinger JC, Sita LR (2006) Organometallics 25:1076
135. Gribkov DV (2005) Ph.D thesis, Universität Erlangen-Nürnberg
136. Watson DA, Chiu M, Bergman RG (2006) Organometallics 25:4731
137. Zi G, Zhang F, Liu X, Ai L, Song H (2010) J Organomet Chem 695:730
138. Xiang L, Zhang F, Zhang J, Song H, Zi G (2010) Inorg Chem Commun 13:666
139. Gott AL, Clarke AJ, Clarkson GJ, Scott P (2008) Chem Commun 1422
140. Wood MC, Leitch DC, Yeung CS, Kozak JA, Schafer LL (2007) Angew Chem Int Ed 46:354
141. Wood MC, Leitch DC, Yeung CS, Kozak JA, Schafer LL (2009) Angew Chem Int Ed 48:6938
142. Wood MC, Leitch DC, Yeung CS, Kozak JA, Schafer LL (2010) Angew Chem Int Ed 49:6475
143. Gott AL, Clarke AJ, Clarkson GJ, Scott P (2007) Organometallics 26:1729
144. Zi G, Zhang F, Xiang L, Chen Y, Fang W, Song H (2010) Dalton Trans 39:4048
145. Reznichenko AL, Hultzsch KC (2010) Organometallics 29:24
146. Ayinla RO, Gibson T, Schafer LL (2011) J Organomet Chem 696:50
147. Walsh PJ, Baranger AM, Bergman RG (1992) J Am Chem Soc 114:1708

148. Baranger AM, Walsh PJ, Bergman RG (1993) J Am Chem Soc 115:2753
149. Walsh PJ, Hollander FJ, Bergman RG (1993) Organometallics 12:3705
150. Lee SY, Bergman RG (1995) Tetrahedron 51:4255
151. Polse JL, Andersen RA, Bergman RG (1998) J Am Chem Soc 120:13405
152. Pohlki F, Doye S (2001) Angew Chem Int Ed 40:2305
153. Straub BF, Bergman RG (2001) Angew Chem Int Ed 40:4632
154. Tobisch S (2007) Chem Eur J 13:4884
155. Allan LEN, Clarkson GJ, Fox DJ, Gott AL, Scott P (2010) J Am Chem Soc 132:15308
156. Stubbert BD, Marks TJ (2007) J Am Chem Soc 129:6149
157. Majumder S, Odom AL (2008) Organometallics 27:1174
158. Leitch DC, Payne PR, Dunbar CR, Schafer LL (2009) J Am Chem Soc 131:18246
159. Leitch D, Turner C, Schafer L (2010) Angew Chem Int Ed 49:6382
160. Leitch DC, Platel RH, Schafer LL (2011) J Am Chem Soc 133:15453
161. Manna K, Xu S, Sadow AD (2011) Angew Chem Int Ed 50:1865
162. Manna K, Everett WC, Schoendorff G, Ellern A, Windus TL, Sadow AD (2013) J Am Chem Soc 135:7235
163. Tobisch S (2012) Inorg Chem 51:3786
164. Anderson LL, Arnold J, Bergman RG (2004) Org Lett 6:2519
165. Anderson LL, Arnold J, Bergman RG (2006) Org Lett 8:2445
166. Anderson LL, Schmidt JAR, Arnold J, Bergman RG (2006) Organometallics 25:3394
167. Eisenberger P, Ayinla RO, Lauzon JMP, Schafer LL (2009) Angew Chem Int Ed 48:8361
168. Zi G, Zhang F, Song H (2010) Chem Commun 46:6296
169. Reznichenko AL, Emge TJ, Audörsch S, Klauber EG, Hultzsch KC, Schmidt B (2011) Organometallics 30:921
170. Zhang F, Song H, Zi G (2011) Dalton Trans 40:1547
171. Reznichenko AL, Hultzsch KC (2012) J Am Chem Soc 134:3300
172. Shen X, Buchwald SL (2010) Angew Chem Int Ed 49:564
173. Liu Z, Yamamichi H, Madrahimov ST, Hartwig JF (2011) J Am Chem Soc 133:2772
174. Kojima M, Mikami K (2012) Synlett 23:57
175. Turnpenny BW, Hyman KL, Chemler SR (2012) Organometallics 31:7819
176. Sherman ES, Fuller PH, Kasi D, Chemler SR (2007) J Org Chem 72:3896
177. Paderes MC, Belding L, Fanovic B, Dudding T, Keister JB, Chemler SR (2012) Chem Eur J 18:1711
178. Dalko PI, Moisan L (2004) Angew Chem Int Ed 43:5138
179. Seayad J, List B (2005) Org Biomol Chem 3:719
180. Akiyama T (2007) Chem Rev 107:5744
181. Schlummer B, Hartwig JF (2002) Org Lett 4:1471
182. Haskins CM, Knight DW (2002) Chem Commun 2724
183. Ackermann L, Kaspar LT, Althammer A (2007) Org Biomol Chem 5:1975
184. Ackermann L, Althammer A (2008) Synlett 995
185. McKinney Brooner RE, Widenhoefer RA (2011) Chem Eur J 17:6170
186. Anderson LL, Arnold J, Bergman RG (2005) J Am Chem Soc 127:14542
187. Marcseková K, Doye S (2007) Synthesis 145
188. Yang L, Xu L-W, Xia C-G (2008) Tetrahedron Lett 49:2882
189. Shapiro ND, Rauniyar V, Hamilton GL, Wu J, Toste FD (2011) Nature 470:245
190. Brown AR, Uyeda C, Brotherton CA, Jacobsen EN (2013) J Am Chem Soc 135:6747
191. Hong S, Marks TJ (2002) J Am Chem Soc 124:7886
192. Hong S, Kawaoka AM, Marks TJ (2003) J Am Chem Soc 125:15878
193. Deschamp J, Olier C, Schulz E, Guillot R, Hannedouche J, Collin J (2010) Adv Synth Catal 352:2171
194. Lutete LM, Kadota I, Yamamoto Y (2004) J Am Chem Soc 126:1622
195. Kanno O, Kuriyama W, Wang ZJ, Toste FD (2011) Angew Chem Int Ed 50:9919
196. Dion I, Beauchemin AM (2011) Angew Chem Int Ed 50:8233

197. Butler KL, Tragni M, Widenhoefer RA (2012) Angew Chem Int Ed 51:5175
198. Cooke ML, Xu K, Breit B (2012) Angew Chem Int Ed 51:10876
199. Hoover JM, Petersen JR, Pikul JH, Johnson AR (2004) Organometallics 23:4614
200. Hickman AJ, Hughs LD, Jones CM, Li H, Redford JE, Sobelman SJ, Kouzelos JA, Johnson AR (2009) Tetrahedron Asymmetry 20:1279
201. Near KE, Chapin BM, McAnnally-Linz DC, Johnson AR (2011) J Organomet Chem 696:81
202. Hansen MC, Heusser CA, Narayan TC, Fong KE, Hara N, Kohn AW, Venning AR, Rheingold AL, Johnson AR (2011) Organometallics 30:4616
203. LaLonde RL, Sherry BD, Kang EJ, Toste FD (2007) J Am Chem Soc 129:2452
204. Zhang Z, Bender CF, Widenhoefer RA (2007) Org Lett 9:2887
205. Li H, Lee DL, Widenhoefer RA (2011) J Organomet Chem 696:316
206. Liu L-J, Wang F, Wang W, Zhao M-X, Shi M (2011) Beilstein J. Org Chem 7:555
207. LaLonde RL, Wang ZJ, Mba M, Lackner AD, Toste FD (2010) Angew Chem Int Ed 49:598
208. Hamilton GL, Kang EJ, Mba M, Toste FD (2007) Science 317:496
209. Rodríguez L-I, Roth T, Fillol JL, Wadepohl H, Gade LH (2012) Chem Eur J 18:3721
210. Teller H, Corbet M, Mantilli L, Gopakumar G, Goddard R, Thiel W, Fürstner A (2012) J Am Chem Soc 134:15331
211. Patil NT, Lutete LM, Wu H, Pahadi NK, Gridnev ID, Yamamoto Y (2006) J Org Chem 71:4270
212. Narsireddy M, Yamamoto Y (2008) J Org Chem 73:9698
213. Molander GA, Dowdy ED, Pack SK (2001) J Org Chem 66:4344
214. Arredondo VM, Tian S, McDonald FE, Marks TJ (1999) J Am Chem Soc 121:3633
215. Trost BM, Tang W (2003) J Am Chem Soc 125:8744
216. Trost BM, Tang W (2002) J Am Chem Soc 124:14542
217. Trost BM, Tang W, Toste FD (2005) J Am Chem Soc 127:14785
218. Ashby EC, Goel AB, DePriest RN (1981) J Org Chem 46:2429
219. Ashby EC, Goel AB, DePriest RN (1981) Tetrahedron Lett 22:4355
220. Newcomb M, Burchill MT, Deeb TM (1988) J Am Chem Soc 110:6528
221. Nishina N, Yamamoto Y (2006) Angew Chem Int Ed 45:3314
222. Sherry BD, Toste FD (2004) J Am Chem Soc 126:15978
223. Lathbury D, Gallagher T (1986) J Chem Soc Chem Commun 114
224. Ha JD, Cha JK (1999) J Am Chem Soc 121:10012
225. Pohlki F, Bytschkov I, Siebeneicher H, Heutling A, König WA, Doye S (2004) Eur J Org Chem 1967
226. Reznichenko AL, Hampel F, Hultzsch KC (2009) Chem Eur J 15:12819
227. Zhang Z, Bender CF, Widenhoefer RA (2007) J Am Chem Soc 129:14148
228. Ajamian A, Gleason JL (2004) Angew Chem Int Ed 43:3754
229. Fogg DE, dos Santos EN (2004) Coord Chem Rev 248:2365
230. Eilbracht P, Barfacker L, Buss C, Hollmann C, Kitsos-Rzychon BE, Kranemann CL, Rische T, Roggenbuck R, Schmidt A (1999) Chem Rev 99:3329
231. Eilbracht P, Schmidt AM (2006) Top Organomet Chem 18:65
232. Müller TE (2006) Top Organomet Chem 19:149
233. Heutling A, Pohlki F, Bytschkov I, Doye S (2005) Angew Chem Int Ed 44:2951
234. Mujahidin D, Doye S (2005) Eur J Org Chem 2689
235. Zhai H, Borzenko A, Lau YY, Ahn SH, Schafer LL (2012) Angew Chem Int Ed 51:12219
236. Liu X-Y, Che C-M (2009) Org Lett 11:4204
237. Han Z-Y, Xiao H, Chen X-H, Gong L-Z (2009) J Am Chem Soc 131:9182
238. Noyori R (2002) Angew Chem Int Ed 41:2008
239. Kim H, Lim W, Im D, Kim D-G, Rhee YH (2012) Angew Chem Int Ed 51:12055

Top Curr Chem (2014) 343: 261–282
DOI: 10.1007/128_2013_484

Published online: 26 October 2013

Asymmetric Reductive Amination

Chao Wang and Jianliang Xiao

Abstract Asymmetric reductive amination (ARA) affords synthetically valuable chiral amines straightforwardly. This chapter reviews the recent advances made in the area, focusing on ARA by hydrogenation, transfer hydrogenation, organocatalytic reduction, and biocatalytic reduction.

Keywords Asymmetric catalysis · Reductive amination · Hydrogenation · Transfer hydrogenation · Organocatalysis

Contents

C. Wang (✉)
Key Laboratory of Applied Surface and Colloid Chemistry of Ministry of Education, and Department of Chemistry and Chemical Engineering, Shaanxi Normal University, Xi'an 710062, China
e-mail: c.wang@snnu.edu.cn

J. Xiao (✉)
Key Laboratory of Applied Surface and Colloid Chemistry of Ministry of Education, and Department of Chemistry and Chemical Engineering, Shaanxi Normal University, Xi'an 710062, China

Department of Chemistry, University of Liverpool, Liverpool L69 7ZD, UK
e-mail: jxiao@liv.ac.uk

1 Introduction

Amines are widely found in natural products, agrochemicals, and pharmaceuticals. As a result, a great deal of attention has been drawn to the development of efficient and economic methods for producing chiral amines [1–24]. One effective method is the reduction of imino C=N bonds, which are most conveniently obtained from the condensation of carbonyl compounds with amines [17, 20, 23–25]. However, imines are not always easy to synthesize and have limited stability. Reductive amination (RA) exploits imines generated in situ from carbonyl compounds and amines, alleviating the problematic imine isolation. Tremendous efforts have been made to develop efficient and selective RA reactions. The progress is rather slow, however, probably due to the following issues. (1) The carbonyl group used in RA is reducible itself, giving rise to an issue of chemoselectivity. (2) The reaction of the carbonyl with the amine results in an equilibrium, which usually disfavors the imine product, unless water is removed. (3) Various reducible intermediates, such as hemiaminals, aminals, enamines, and iminium ions, may appear during the RA reaction, complicating the reaction. (4) The amine substrate, imine intermediate, and amine product may poison the catalyst, particularly metal complex catalysts. (5) The acyclic imine intermediate has *E*/*Z* isomers, which makes stereoselective reduction difficult. Indeed, successful RA systems are sparse and asymmetric versions are even fewer. In this chapter, recent advances of ARA from areas of organometallic catalysis, organocatalysis, and biocatalysis are described, aiming to show the state-of-the-art ARA reactions. Stoichiometric reduction using borohydrides is not discussed [26, 27].

2 Organometallic Catalysis

Organometallic catalysis is the major driving force in the general area of asymmetric catalysis. This is also seen in ARA, where organometallic catalysts dominate the scene.

2.1 Metal Catalyzed Hydrogenation

Metal catalyzed hydrogenation is one of the most successful asymmetric catalytic reactions [28]. Using hydrogen gas as hydrogen source is desirable both economically and environmentally, owning to the 100% atom efficiency for the reduction and the low cost of H_2. Ruthenium, rhodium, and iridium complexes are the most widely used catalysts for hydrogenation [17, 20, 23, 24]. The key to controlling the stereoselectivity rests on the ligands, which are mostly phosphines. New concepts, such as cooperative catalysis, have also been explored in ARA.

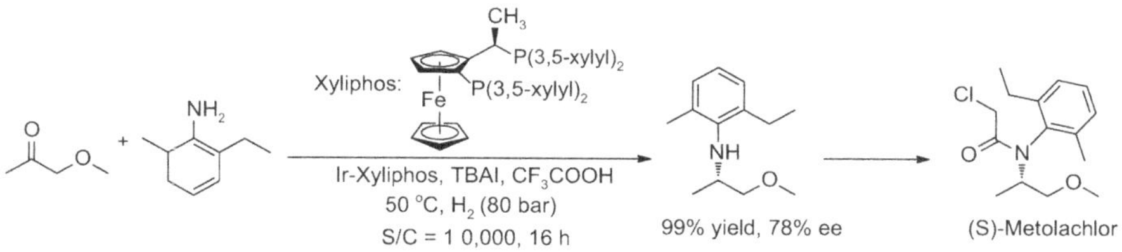

Scheme 1 First example of ARA of a ketone

The first ARA reaction by hydrogenation was reported by Blaser and co-workers [29]. In the production of the grass herbicide Metolachlor, a highly efficient imine reduction process was developed using a catalyst generated in situ from $[Ir(COD)Cl]_2$ (COD = cycloocta-1,5-diene) and the ferrocenyldiphosphine ligand Xyliphos (Scheme 1). Approximately 80% enantioselectivity was obtained at a substrate to catalyst ratio (S/C) of >1,000,000, with initial TOF up to 1,800,000 h^{-1}. A one-pot process, without the isolation of the unstable imine intermediate, was attempted. It turned out that the one-pot procedure can provide a similar *ee* (78%), but with much slower reaction rate. The best activity was observed at an S/C of 10,000 for 14 h, and the addition of iodide ions and acid was necessary. It should be noted that the ketone substrate used is an aliphatic one, which is a difficult class of substrates for obtaining high *ee*s by RA.

Despite the success of the Blaser system, highly enantioselective and active catalysts with broad substrate scope were lacking. In 2001, Zhang and co-workers reported a novel ligand, f-binaphane, which demonstrates good activity and enantioselectivity in iridium-catalyzed hydrogenation of imines derived from aryl-alkyl ketones and aromatic amines [30]. The one-pot ARA with this ligand was also studied (Scheme 2) [31]. Since the imine-formation step was found to be the rate-limiting step, various acids were used to accelerate the imine formations. The Lewis acid $Ti(O\textit{i}Pr)_4$ was found to be an effective additive. Further, the addition of 10% of I_2 was vital for the reaction to proceed; no reaction was detected without it. With 1 mol% of the in situ formed catalyst from $[Ir(COD)Cl]_2$ and (*S*,*S*)-f-binaphane, various aryl-alkyl ketones reacted with *p*-anisidine, affording amines with excellent yields and *ee*s.

In 2000, Borner and co-workers reported that cationic Rh(I) complexes $[Rh(dppb)(COD)]BF_4$ [dppb = 1,4-bis(diphenylphosphino)butane and $[Rh(dpoe)(COD)]BF_4$ [dpoe = 1,2-bis(diphenylphosphinito)ethane] catalyzed hydrogenative RA of aldehydes and ketones [32]. Various aldehydes were aminated to afford amines at S/C of 500, although the selectivity between amine and alcohol products was not satisfactory. When the achiral ligands dppb and dpoe were replaced with a chiral ligand **1**, the RA between an α-keto acid and benzylamine afforded 59% yield and 38% *ee* (Table 1).

Subsequently in 2003, the same group disclosed their search for chiral hydrogenation catalysts aimed at ARA [33]. High throughput screening technology was deployed in their research. After screening 96 chiral ligands with $[Rh(COD)_2]BF_4$ and $[Rh(COD)Cl]_2$, respectively, Norphos and Deguphos were identified as

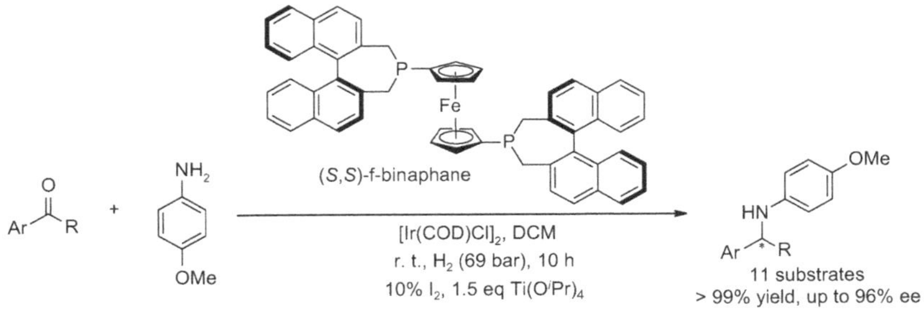

Scheme 2 ARA catalyzed by an Ir-f-binaphane catalyst

Table 1 Reductive amination catalyzed by chiral Rh-diphosphine catalysts

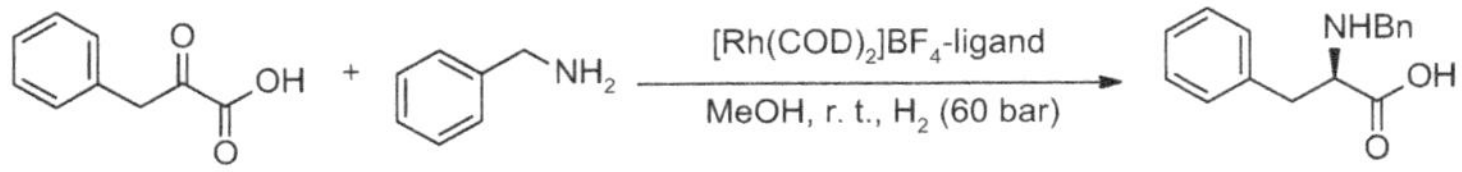

Entry	Ligand	Yield (%)	*ee* (%)
1	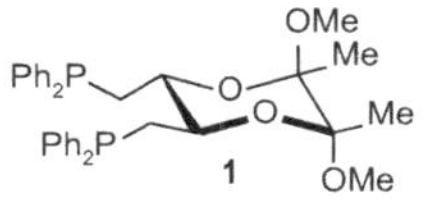	59[a]	38
2	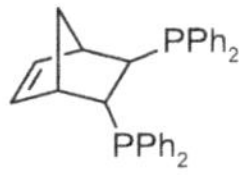 (*R*,*R*)-Norphos	99	95
3	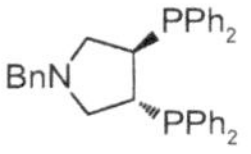 (*R*,*R*)-Deguphos	98 (99)[b]	92 (98)[b]

[a]50 bar H_2 pressure
[b]Isolated {Rh[(*R*,*R*)-Deguphos](COD)}BF_4 was used as catalyst for results in brackets

good ligands for the ARA of phenylpyruvic acid with benzylamine (Table 1). Using 1 mol% of in situ formed catalyst, *ee*s over 90% were observed with both Norphos and Deguphos. Up to 99% of yield and 98% of *ee* were obtained for phenylanaline by using isolated {Rh[(*R*,*R*)-Deguphos](COD)}BF_4 catalyst. However, only three substrates gave over 80% *ee* with this catalytic system.

Inspired by the work of Hsiao and co-workers on Rh catalyzed asymmetric hydrogenation of unprotected β-enamine esters and amides [34], Bunlaksananusorn and co-worker reported a Ru-catalyzed RA of β-keto esters with NH_4OAc to produce chiral β-amino esters in 2005 [35]. Using 1 mol% of Ru complex **2** as catalyst and trifluoroethanol (TFE) as solvent, both aryl and alkyl β-keto esters could be aminated into chiral β-amino esters with excellent chemo- and enantio-selectivities at 30 bar H_2 and 80 °C (Scheme 3). For example, when $R = $ Me, the desired product could be obtained in 80% yield and 96% *ee*.

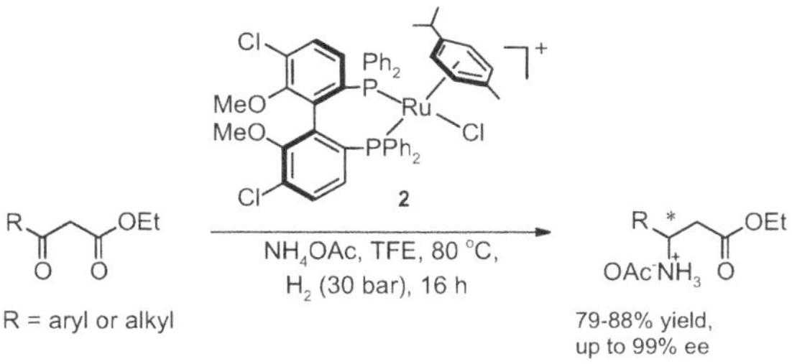

Scheme 3 Ru catalyzed ARA of β-keto esters

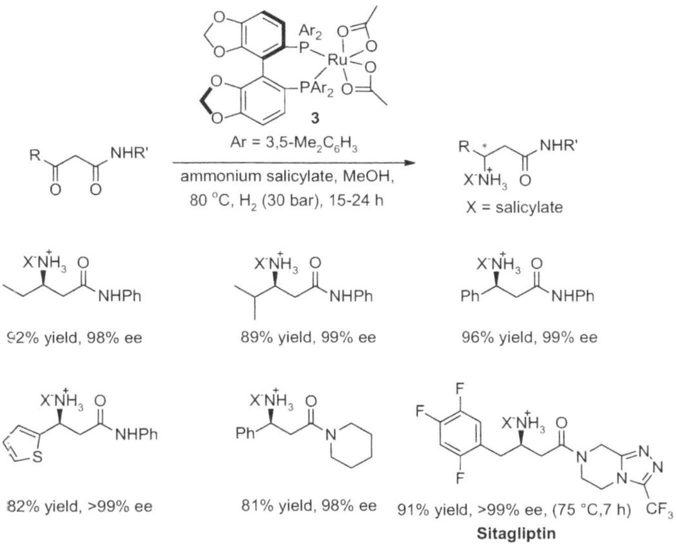

Scheme 4 Ru-catalyzed ARA of β-keto amides

Recently, research groups from Merck and Takasago reported RA of β-keto amides catalyzed by Ru-diphosphine complexes [36]. In this case, ammonium salicylate was used as the amine source and MeOH turned out to be the best solvent screened (Scheme 4). Excellent yields and *ee*s were obtained for all the substrates reported. Impressively, this method was applied to the synthesis of Sitagliptin, a potent DPP-IV inhibitor for the treatment of type II diabetes. With 1 mol% of catalyst **3**, Sitagliptin was obtained in 91% yield and >99% *ee* from its corresponding ketone. For the RA of β-keto esters and amides, imines were believed to be the intermediate that was reduced, although enamines are the more stable intermediate.

Since the first reported hydrogenative RA reaction, attention has been focused on the discovery of new diphosphine ligands. Development of phosphine-free catalytic systems with broad substrate scope and high enantioselectivities is still a challenge. In 2009, Xiao and co-workers reported a novel iridium catalyst for

Scheme 5 Hydrogenation of imines via cooperative catalysis

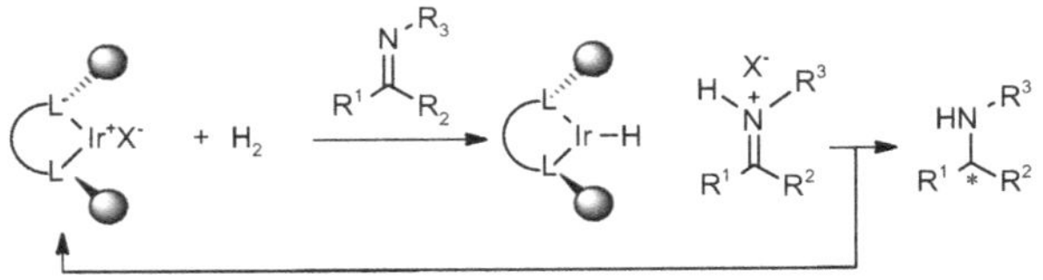

Table 2 Match/mismatch and anion structure effect on cooperative catalysis

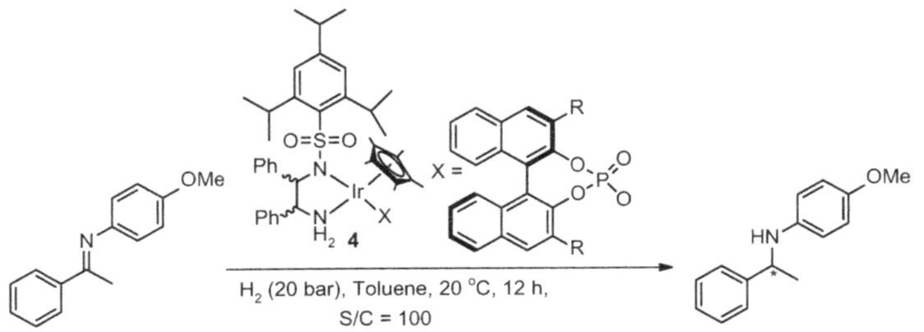

Entry	Configuration	*R*	Conversion (%)	*ee* (%)
1	(*R*,*R*)	2,4,6-(2-$C_3H_7)_3C_6H_2$	47	38 (*R*)
2	(*S*,*S*)	2,4,6-(2-$C_3H_7)_3C_6H_2$	60	97 (*S*)
3	(*S*,*S*)	H	53	17 (*R*)
4	(*S*,*S*)	Ph	57	26 (*R*)
5	(*S*,*S*)	3,5-$(CF_3)_2C_6H_3$	40	20 (*S*)
6	(*S*,*S*)	1-Naphthyl	43	38 (*S*)

asymmetric hydrogenation of acyclic imines, which is a departure from the usual catalyst development paradigm [37]. This catalyst uses a chiral diamine ligand instead of the commonly used diphosphine ligand and bears a chiral anion, which has a significant influence on the stereo outcome of the hydrogenation. Drawing inspiration from studies in organocatalytic imino reduction with Hantzsch esters, a chiral iridium catalyst with a chiral phosphate was devised. The latter was expected to ion-pair with the iminium cation resulting from deprotonation of the Ir-H_2 intermediate and thereby affects the stereoselectivity of the Ir-H hydride (Scheme 5) [38].

The strategy turned out to be successful. Good enantioselectivities required a match in chirality between the metal cation and its counteranion, however. The enantioselectivity increased from 38 (*R*) to 97% (*S*), when the configuration of the diamine ligand changed from *R*, *R* to *S*, *S* (Table 2, entry 1 vs 2). The substituents at the 3 and 3′ position of the phosphate also play an important role in the enantioselection. Without substitution or with a simple phenyl ring at the 3 and 3′ positions, poor *ee*s were obtained. Surprisingly, increasing the bulkiness of the substituent affords a product of not only higher *ee* but also opposite configuration (Table 2, entries 3 and 4 vs 2, 5, and 6). These results suggest that both the metal cation and its counteranion are involved in the enantioselectivity determining step [38].

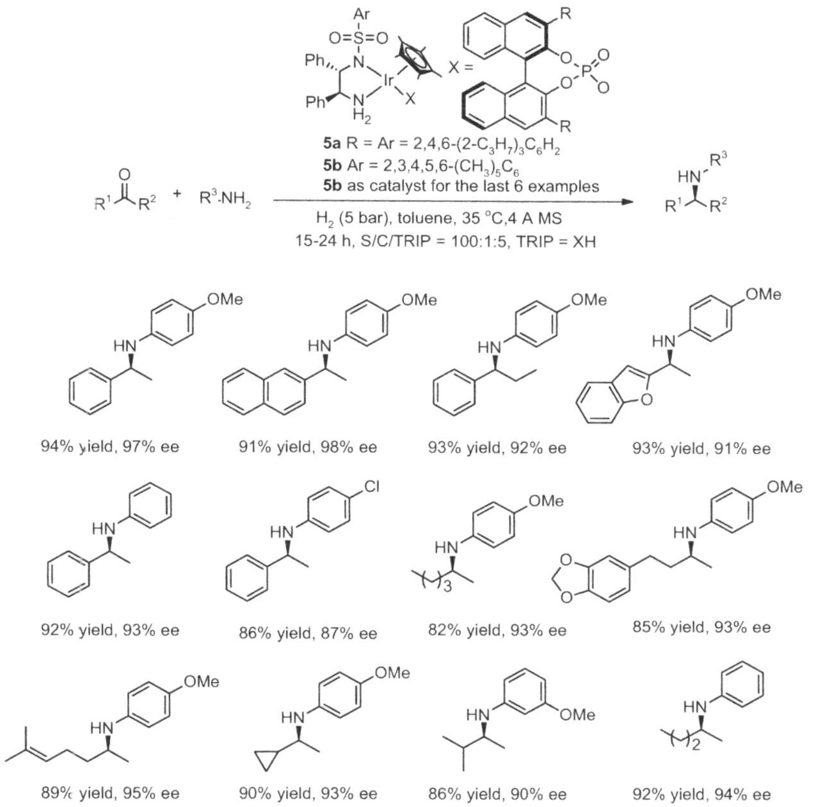

Scheme 6 ARA of aryl and aliphatic ketones via cooperative catalysis

This catalytic system was very stereoselective in the hydrogenation of various imines at S/C of 100. A one-pot ARA was later developed based on this system with an even broader substrate scope, thanks to the obviation of isolation of imine intermediates [39]. With 1 mol% of catalyst **5a** and 5 mol% of the phosphoric acid, various ketones could be aminated with aromatic amines to afford chiral amines under 5 bar of hydrogen pressure at 35 °C (Scheme 6). Impressively, aliphatic ketones reacted well to give amines with high yields and enantioselectivities with **5b** as catalyst. This is the first ARA system to have such a broad substrate scope.

The metals used in asymmetric hydrogenative RA are usually ruthenium, rhodium, and iridium. In 2009, Rubio-Pérez and co-workers reported that a chiral palladium diphosphine complex catalyzes hydrogenative ARA [40]. Interestingly, the Pd-BINAP catalyst gave better results for aliphatic ketones than for aromatic ones. While less than 50% of *ee* was obtained for aromatic ketones, over 90% *ee* was observed for aliphatic ketones. The optimal results were obtained with 2.5% of catalyst under 55 bar hydrogen pressure at 70 °C in $CHCl_3$ in the presence of 5-Å molecular sieves (Scheme 7).

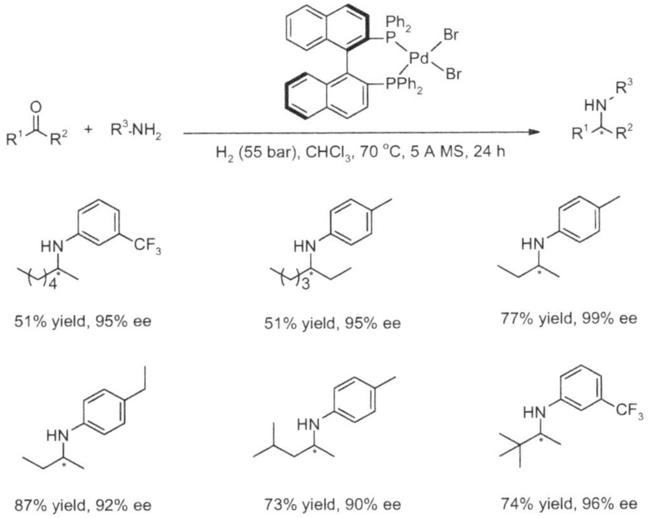

Scheme 7 Pd-BINAP complex catalyzed ARA

2.2 *Metal Catalyzed Transfer Hydrogenation*

Transfer hydrogenation, which uses hydrogen sources other than hydrogen gas, is an alternative way of reducing unsaturated bonds. The use of small organic molecules, such as alcohols, HCOOH, etc., as hydrogen sources avoids the use of hazardous hydrogen gas and high pressure apparatus. Due to its operational simplicity and versatility, metal catalyzed transfer hydrogen has attracted a great deal of attention and made substantial progress in recent years, particularly in the reduction of carbonyl groups [41–53]. However, the development of metal catalyzed transfer hydrogenation systems for reduction of C=N bonds lags behind that for carbonyl groups, and transfer hydrogenative RA reactions are even rarer. Only three examples of asymmetric transfer hydrogenative RA have been reported to date.

The first and only example of intermolecular asymmetric transfer hydrogenative RA was reported in 2003 by Kadyrov and co-workers [54]. After screening a series of Ru, Rh, and Ir catalysts, Ru catalysts with BINAP or tol-BINAP ligand gave the best enantioselectivities for the RA of acetophenone with ammonium formate in MeOH (Scheme 8). Addition of 15–20% of ammonia accelerated the reaction but decreased the enantioselectivity. A range of aromatic ketones could be aminated, affording excellent enantioselectivities. However, the reaction gave a mixture of free amine and its *N*-formylated product even under optimized conditions, although the free amine could be obtained by acidic hydrolysis of the reaction mixture. The catalyst was not good for aliphatic ketones, giving poor yields and enantioselectivities.

In the same year, Wills and co-workers reported the first intramolecular transfer hydrogenative ARA [55]. They applied the Noyori transfer hydrogenation system

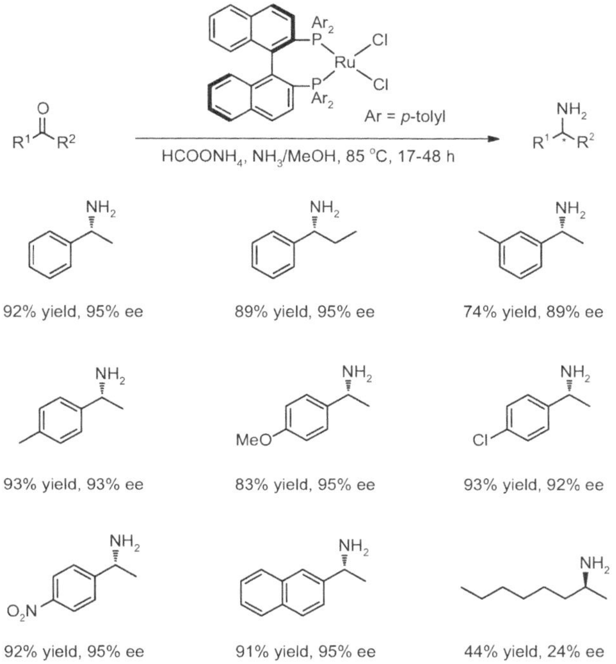

Scheme 8 Ru-tol-BINAP catalyzed transfer hydrogenative ARA

[56], which is highly effective for cyclic imine reduction, to the RA of substrates containing both a carbonyl and amino group, with the latter being Boc-protected. To perform the Ru catalyzed transfer hydrogenative RA, the Boc group had to be removed in pure formic acid first, and the reaction mixture was adjusted to be less acidic with Et_3N. Although many substrates afforded good conversions, enantioselectivity was observed only for one substrate. Compound **6** was converted to a chiral tetrahydroisoquinoline compound in 85% yield and 88% *ee* (Scheme 9). The authors believe that the configuration of the imine intermediate is crucial for the enantioselective step.

Strotman and co-workers carried out a detailed study of the Noyori's transfer hydrogenation system for the intramolecular RA of dialkyl ketones [57]. By using a sterically bulky ligand derived from DPEN, they achieved the first highly enantioselective intramolecular RA of dialkyl ketones. The carbon dioxide produced during decarboxylation of HCOOH was found to be detrimental to the reaction, decreasing the reaction rate by affecting the Ru hydride and Ru formate equilibrium and lowering the yield by forming carbamate with the product. Purging of the carbon dioxide in the system led to improved rate and isolated yield. Under optimized conditions, Suvorexant, a potent dual orexin antagonist, could be obtained with 97% yield and 95% *ee* (Scheme 10).

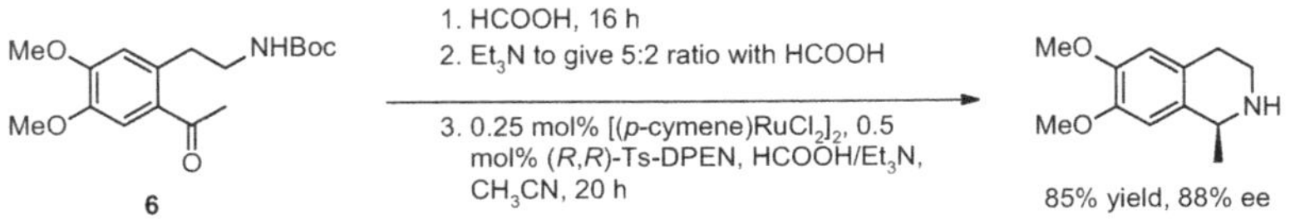

Scheme 9 Intramolecular transfer hydrogenative ARA

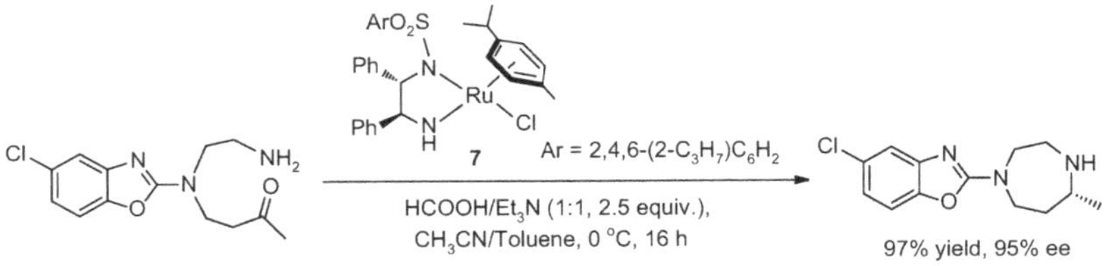

Scheme 10 Intramolecular transfer hydrogenative ARA of dialkyl ketone

3 Organocatalysis

Since 2000, organocatalysis has emerged as a powerful alternative to metal catalysis. Various organocatalytic asymmetric reduction systems have been developed, particularly for imine reduction. There are excellent reviews, which summarize the development of organocatalytic transfer hydrogenation reactions [19, 58]. In the following sections, ARA catalyzed by organocatalysts will be discussed.

The hydrogen sources used in organocatalytic ARA are usually hydrosilanes or Hantzsch esters, which require Lewis base and phosphoric acid catalysts to activate them, respectively. Reactions with hydrosilanes as hydrogen source are presented first, followed by those using Hantzsch esters.

3.1 Hydrosilanes as Hydrogen Source

In organocatalytic ARA using hydrosilanes, the catalysts normally possess Lewis basic centers and hydrogen bond donors, with the former activating the silane reagents and the latter interacting with the imine intermediate.

The asymmetric reduction of isolated imines is simpler than ARA so the reduction of imines was studied earlier than that of ARA. In 2001, Matsumura and co-workers reported the first example of imine reduction with trichlorosilane. Using an *N*-formylpyrrolidine catalyst, good yields and moderate enantioselectivities were achieved [59]. Further development of this type of system for asymmetric imine reduction was undertaken by Malkov and Kočovský [60–62], Sun [63], Jones [64] and their co-workers.

In 2007, the first ARA with hydrosilanes catalyzed by chiral Lewis base catalysts was reported by Malkov, Kočovský, and co-workers [65]. α-Chloroketones were

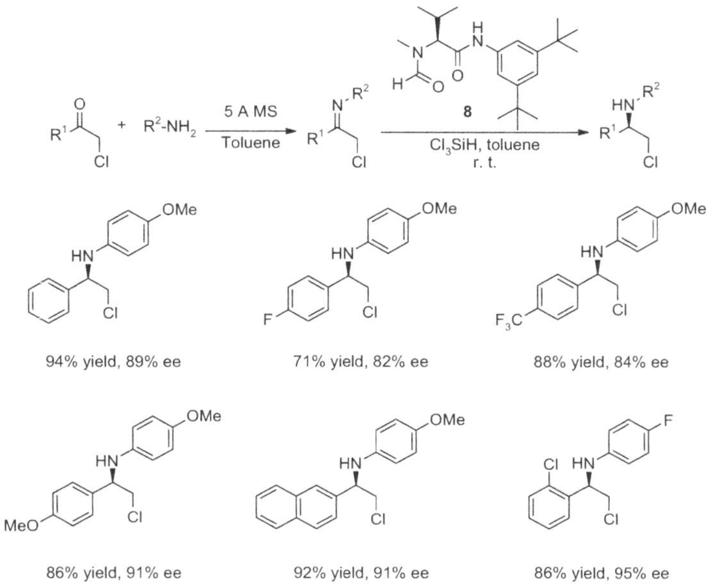

Scheme 11 ARA of α-chloroketones

condensed with amines to form imines at RT for 24 h, which were subsequently reduced without isolation at 0 °C. With 5 mol% of **8**, a range of chiral α-chloroamines could be obtained in high enantioselectivities and yields, which could be further transformed into chiral aziridines (Scheme 11).

In more recent studies, Benaglia and co-workers screened a series of organocatalysts derived from chiral amino alcohols for imine reduction [66]. After identifying the best catalyst, one-pot RA was also examined and shown to work well. For example, methoxylacetone could react with an aromatic amine to form the key intermediate for Metolachlor in 86% yield and 70% *ee* (Scheme 12).

Further development along this line was made by Jones and co-workers. An imidazole-based organocatalyst **10** was found to catalyze the reduction of imines as well as ARA [67]. In the ARA reaction, yields were moderate and low in some cases, due to the slow formation of imine intermediates. Using a two-step-one-pot procedure, the imine formation was accelerated by microwave heating, leading to improved amine yields. The system was applied to the synthesis of a calcimimetic (*R*)-(+)-NPS R-568, affording it in 67% yield and 89% *ee* (Scheme 13).

3.2 Hantzsch Esters as Hydrogen Source

ARA with Hantzsch esters as hydrogen sources and chiral phosphoric acids as catalysts constitutes an important class of RA reactions. Hantzsch esters are mimics of nature's reducing agent – NADH, the hydride source used by enzymes for

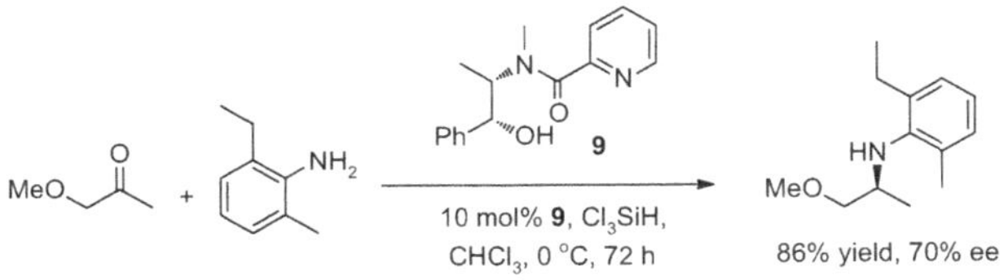

Scheme 12 Organocatalytic ARA of an aliphatic ketone

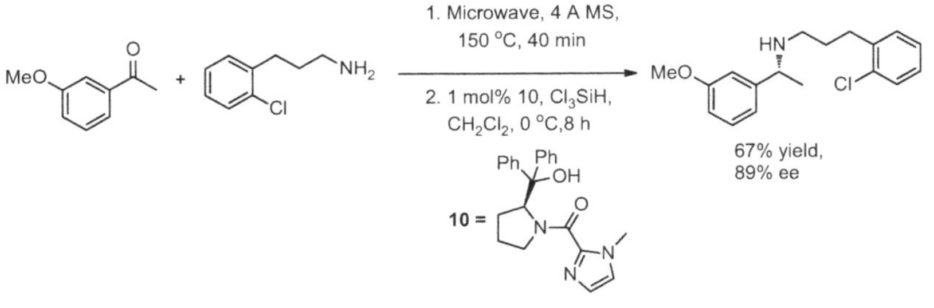

Scheme 13 Synthesis of (*R*)-(+)-NPS R-568

reduction reactions. In organocatalytic ARA with the chiral phosphoric acid/ Hantzsch ester system, it is believed that the acid activates the imine intermediate through hydrogen bonding or protonation by the proton from the OH group. The Hantzsch ester is also activated through hydrogen bonding between its NH unit and the P=O group of the phosphoric acid. This model of activation is somewhat similar to the aforementioned chiral Lewis base/hydrosilane system.

Again, asymmetric imine reduction was first explored. Early in 1989, the first asymmetric organocatalytic reduction of imines using Hantzsch ester was reported [68]. However, no further development of this protocol was reported until 2005. Probably inspired by the emergence of the concept of organocatalysis, several papers on organocatalytic imine reduction and ARA with Hantzsch esters were reported nearly simultaneously from the groups of Rueping [69], List [70], and Macmillan [71].

Almost in parallel, Rueping [69] and List [70] and their co-workers reported the reduction of imines derived from aromatic ketones and aromatic amines. Chiral phosphoric acids, introduced into asymmetric reactions by Akiyama [72] and Terada [73], were used as catalysts with Hantzsch ester as the hydrogen source. In the List paper an example of one-pot RA was presented [70]. A comprehensive study by MacMillan and co-workers was then followed, disclosing the ARA of aromatic ketones with aromatic amines [71]. Interestingly, all three groups used similar chiral phosphoric acid (**11a**–**11c**) and the same hydrogen source (Scheme 14). From these three studies it is evident that increasing the bulkiness of the substituents on 3 and 3′ positions of the phosphoric acid improves the enantioselectivity. **11a** and **11b** were used for imine reduction, while **11c** was employed for ARA. The catalyst loading was 10 mol% for **11b** and **11c** and 20 mol% for **11a**, with the reduction run at 35, 60,

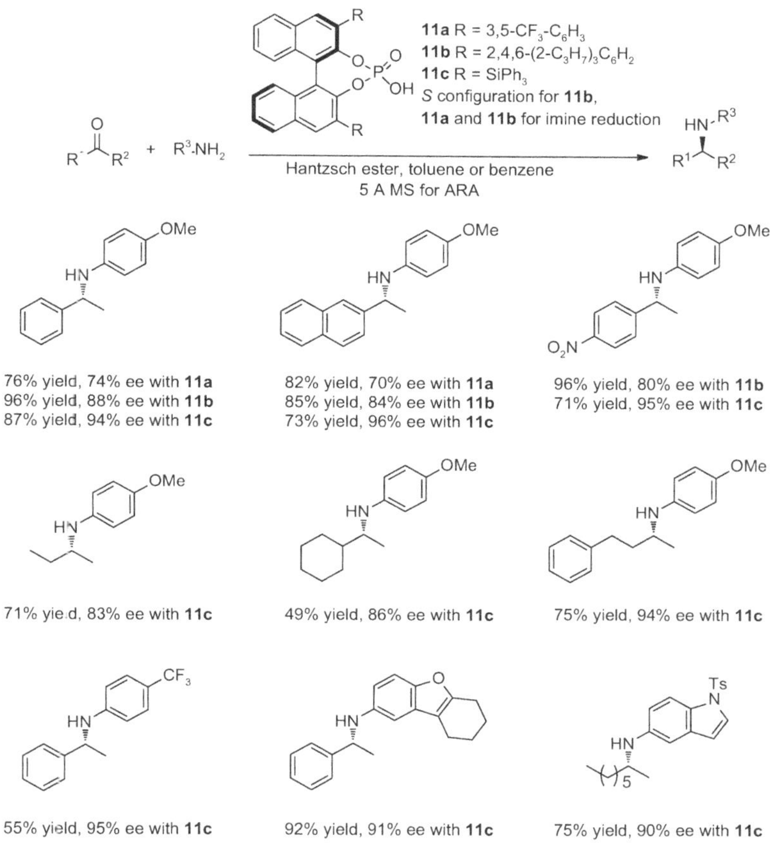

Scheme 14 Chiral phosphoric acids catalyzed imine reduction and ARA with Hantzsch ester as hydrogen source

and 40 °C using **11a**, **11b**, and **11c**, respectively [69–71]. In the ARA catalyzed by **11c**, 5-Å molecular sieves were used to promote imine formation. A broad substrate scope was observed in the ARA catalyzed by **11c**, including aliphatic ketones and aromatic amines with different substituents. Examples are shown in Scheme 14 [69–71]. A single crystal X-ray structure of **11c**-bound aryl imine was obtained, shedding light on the origin of the enantioselectivity observed [71].

The scope of the ketone partner for the organocatalytic ARA was extended to α-keto esters. Antilla and co-workers reported the asymmetric reduction of imines derived from α-keto esters as well as their ARA using catalyst **12** [74]. The yields of ARA were generally 10–20% lower than imine reduction, but the *ee*s were identical. Scheme 15 shows examples of ARA.

The amine partners in the organocatalytic examples above are all aromatic ones. Aliphatic amines seem to be challenging substrates for ARA. Recently, List reported that benzylamine could be used for ARA reactions [75]. A low pressure

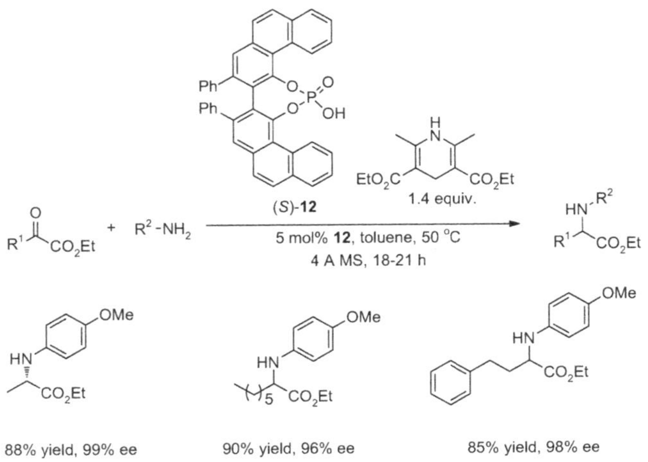

Scheme 15 Organocatalytic ARA of α-keto esters

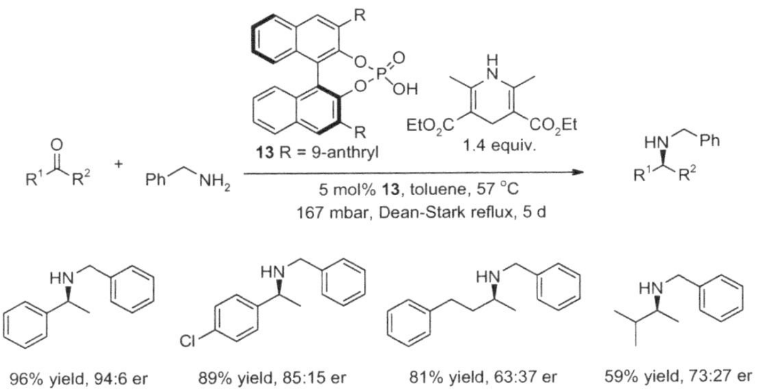

Scheme 16 Organocatalytic ARA of benzylamine

Dean-Stark trap was used to remove the water generated from the imine formation step. Aromatic ketones gave better *ee*s than aliphatic ones (Scheme 16).

Further application of the chiral phosphoric acid/Hantzsch ester system was explored by List and co-workers. α-Branched aldehydes were reductively aminated to afford chiral β-branched chiral amines via a dynamic kinetic resolution process (Scheme 17) [76]. In this case, **11b** turned out to be the best catalyst, and it was necessary to modify the structure of Hantzsch ester to ensure high enantioselectivity. Selected examples are presented in Scheme 18.

Another elegant application coming from the List group involves an aldolization-dehydration-conjugate reduction-reductive amination cascade process catalyzed by a single chiral phosphoric acid [77]. Starting from diketones, chiral cyclic amines could be obtained with good yields and high diastereo- and enantioselectivities

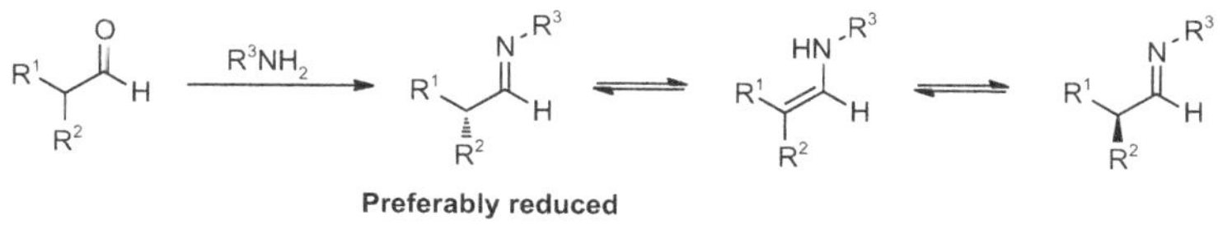

Scheme 17 ARA of α-branched aldehydes via dynamic kinetic resolution

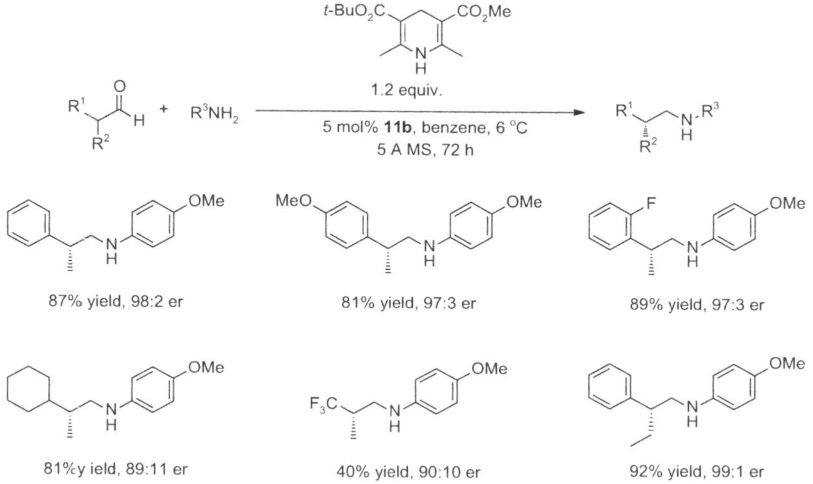

Scheme 18 Selected examples of ARA of α-branched aldehydes

(Scheme 19). Their best catalyst **11b** (TRIP) was again employed. Other development of the chiral phosphoric acid-catalyzed imine reduction or ARA includes modification of the structure of chiral acids [78] or exploration of hydrogen sources of properties similar to Hantzsch esters [79].

4 Biocatalysis

Biocatalysis is an important complement to chemical catalysis for chiral amine synthesis, as it often gives products that are difficult to access by chemical means. Many reviews have appeared, summarizing the progress in biocatalytic reduction to access chiral amines [80–86]. In the following sections, selected examples of ARA from ketones and amines catalyzed by enzymes are presented. Reactions that involve kinetic resolution as well as dynamic kinetic resolution will not be covered here. There are two main types of enzymes that have been used to transform carbonyl groups into amino groups – amino acid dehydrogenases and transaminases.

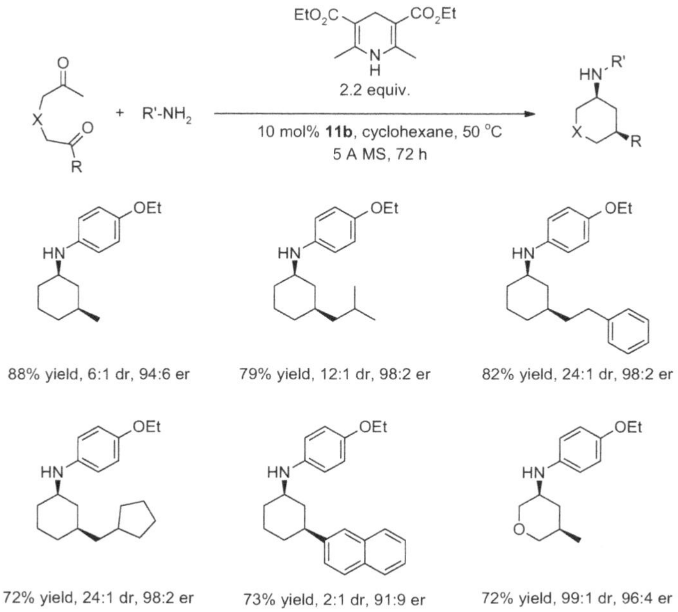

Scheme 19 TRIP-catalyzed cascade amination

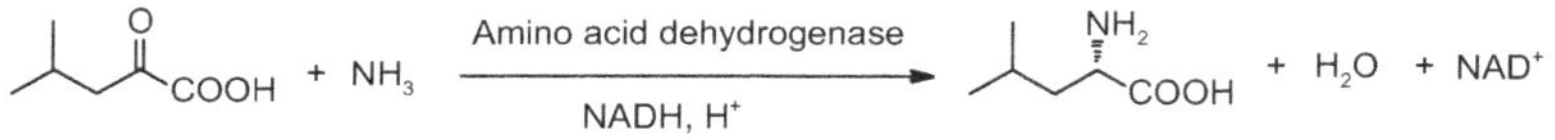

Scheme 20 Reductive amination catalyzed by amino acid dehydrogenase

4.1 ARA with Amino Acid Dehydrogenases

Amino acid dehydrogenases can catalyze the amination of carbonyl compounds, usually α-keto acids/esters, with NADH as the hydrogen source. As early as in 1961, an example of RA catalyzed by an amino acid dehydrogenase (AaDH) to produce a chiral amino acid was reported (Scheme 20) [87].

The biocatalytic ARA with NAD(P)H as hydrogen source can be practically useful only if NAD(P)H is regenerated. The NAD(P)H regeneration system has thus been developed and used together with AaDH to effect ARA of keto acids with ammonium as amine source. The NAD(P)H regeneration system normally uses enzymes, e.g., formate dehydrogenase (FDH) or glucose dehydrogenase (GluDH), which convert $NAD(P)^+$ to NAD(P)H with small molecule hydrogen sources, e.g., ammonium formate or glucose. Large scale productions of amino acids have been possible with these systems. Some examples of biocatalytic ARA using AaDH coupled with FDH are found in Scheme 21 [88–93].

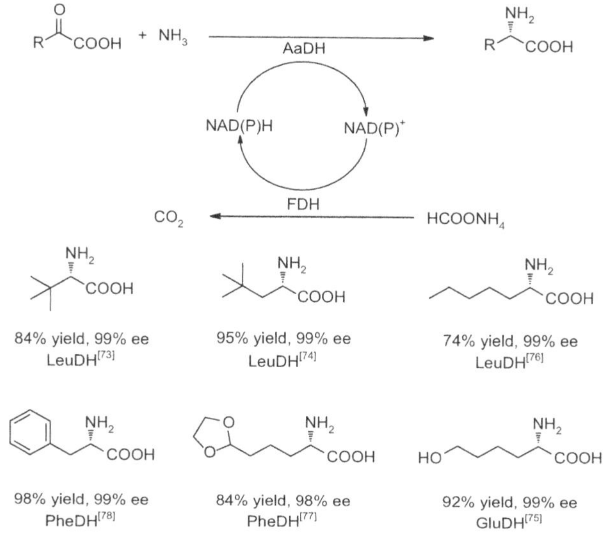

Scheme 21 Examples of biocatalytic ARA

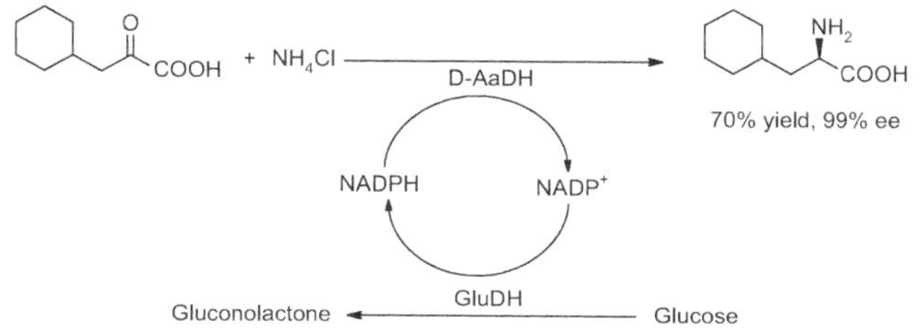

Scheme 22 Production of D-amino acid with engineered enzyme on a gram scale

Most AaDHs are natural enzymes, which only selectively produce L-amino acids. In order to obtain D-enantiomers or unnatural amino acids, the enzymes have to be engineered. In 2006, Novick and co-workers reported the first D-amino acid dehydrogenase by directed evolution of an existing enzyme [94]. The engineered enzyme was capable of producing D-amino acids via ARA of keto acids with ammonia. For example, D-cyclohexylaniline was produced on a gram scale using the engineered enzyme coupled with a NADPH regeneration system (Scheme 22).

In the examples described above, isolated enzymes and NAD(P)H were employed. However, whole cell catalysts for ARA of α-keto acids are known [95], and further development in the direction has been reported [91, 96].

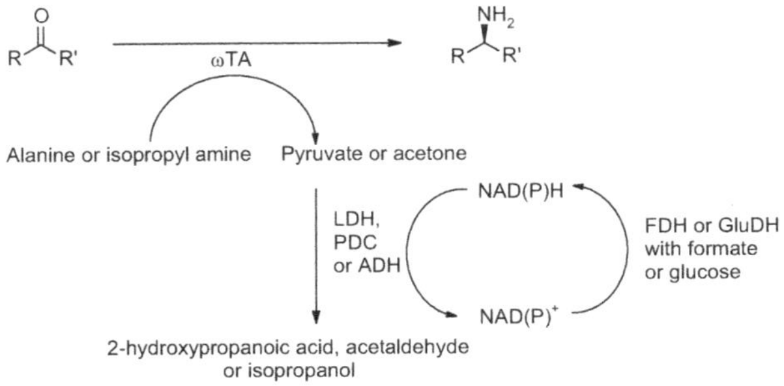

Scheme 23 ω-Transaminases catalyzed ARA

4.2 *ARA with ω-Transaminases*

The major drawback of AaDHs is that they can only catalyse the ARA of α-keto acids to produce α-amino acids. This limitation was overcome by the use of ω-transaminases (ωTAs). ωTAs catalyze the transfer of amino group from cosubstrates to carbonyl compounds to form new chiral amines. A problem faced by ωTAs for practical applications is that new carbonyl compounds are generated after transferring amino groups from the amine source, and the newly formed carbonyl compounds would compete with the carbonyl substrates for transamination. It is thus crucial to remove the co-produced carbonyls in order for the desired reaction to go to completion. Alanine and isopropyl amine are the two most often used amine sources. The by-product from alanine is pyruvate, which could be removed from the system using pyruvate dehydrogenase or pyruvate decarboxylase (Scheme 23) [97, 98]. In addition, the acetone produced from isopropyl amine could be removed by distillation or selective alcohol dehydrogenase [99, 100]. During the by-product removal process, NAD(P)H is used as the co-factor for the enzymes, which can be recycled with the previously described regeneration systems using FDH or GluDH.

An elegant multiple enzymes cascade system was designed by Kroutil and co-workers in 2008 [101]. Rather than removing pyruvate generated from transamination of alanine, an amino acid dehydrogenase was employed to regenerate alanine from pyruvate via another RA using ammonia as the amine source. Thus, the true amine source is ammonia in this case. The strategy allows for a wide range of ketones to be transformed into chiral amines with high enantioselectivities (Scheme 24).

The substrate scope of ω-transaminases catalyzed ARA has been broadened by engineering the enzymes. For example, by a combination of in silico design and directed evolution of ω-transaminase, Savil, Janey, and co-workers developed an enzymatic system showing different substrate tolerance from natural enzymes. The utility of the enzyme has been demonstrated in the manufacture of the important pharmaceutical product, Sitagliptin, with high enantioselectivity (Scheme 25) [102, 103].

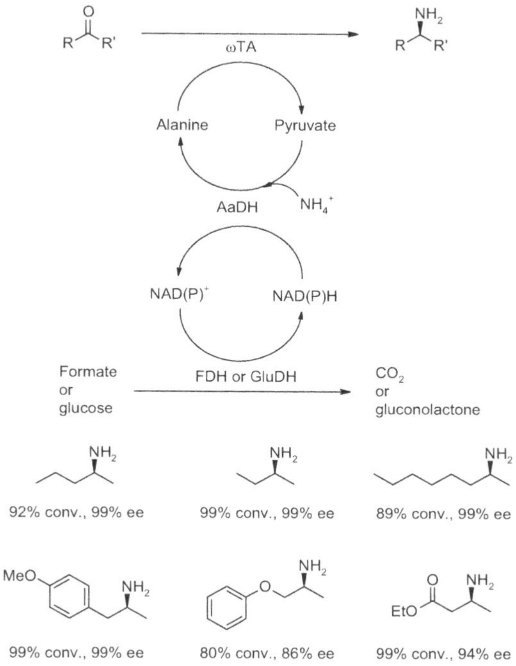

Scheme 24 Ammonia as amine source in ω-transaminases catalyzed ARA

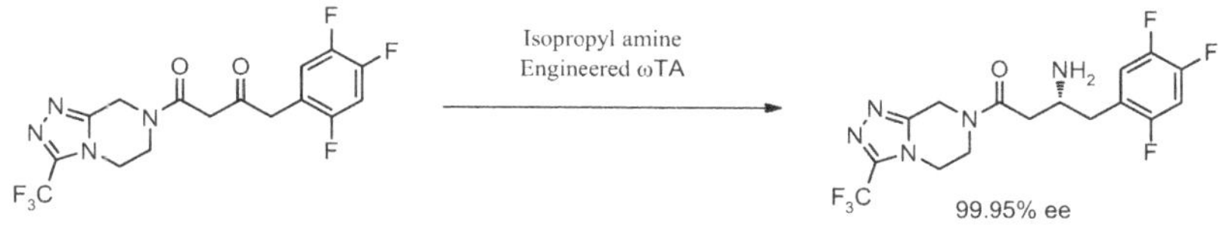

Scheme 25 Manufacture of Sitagliptin with engineered ω-transaminase

5 Summary and Outlook

Recent progress in the area of ARA has been summarized. From the examples shown, we can see that apart from the traditional metal catalyzed ARA, organocatalytic and biocatalytic methods have emerged as powerful alternatives. In metal catalyzed ARA, transfer hydrogenation with isopropyl or formic acid as hydrogen source still lags behind hydrogenation with hydrogen gas in terms of substrate scope, catalyst activity, and productivity. New transfer hydrogenation systems may provide opportunities to enable more efficient transfer hydrogenative ARA. Organocatalytic ARA requires improvement in activity and productivity before practical use is possible. Large scale applications of ARA have been seen in biocatalysis. Its limitation in substrate scope may be tackled by enzyme engineering.

References

1. Kobayashi S, Ishitani H (1999) Chem Rev 99:1069–1094
2. Ellman JA, Owens TD, Tang TP (2002) Acc Chem Res 35:984–995
3. Gomez S, Peters JA, Maschmeyer T (2002) Adv Synth Catal 344:1037–1057
4. Besson M, Pinel C (2003) Top Catal 25:43–61
5. Blaser H-U, Malan C, Pugin B, Spindler F, Steiner H, Studer M (2003) Adv Synth Catal 345:103–151
6. Ellman JA (2003) Pure Appl Chem 75:39–46
7. Kukula P, Prins R (2003) Top Catal 25:29–42
8. Breuer M, Ditrich K, Habicher T, Hauer B, Keßeler M, Stürmer R, Zelinski T (2004) Angew Chem Int Ed 43:788–824
9. Friestad GK (2005) Eur J Org Chem 2005:3157–3172
10. Tararov VI, Börner A (2005) Synlett 2005:203–211
11. Abdel-Magid AF, Mehrman SJ (2006) Org Proc Res Dev 10:971–1031
12. Morton D, Stockman RA (2006) Tetrahedron 62:8869–8905
13. Ramachandran PV, Burghardt TE (2006) Pure Appl Chem 78:1397–1406
14. Friestad GK, Mathies AK (2007) Tetrahedron 63:2541–2569
15. Kizirian J-C (2007) Chem Rev 108:140–205
16. Minnaard AJ, Feringa BL, Lefort L, de Vries JG (2007) Acc Chem Res 40:1267–1277
17. Petrini M, Torregiani E (2007) Synthesis 2007:159–186
18. Skucas E, Ngai M-Y, Komanduri V, Krische MJ (2007) Acc Chem Res 40:1394–1401
19. You SL (2007) Chem Asian J 2:820–827
20. Spindler F, Blaser HU (2008) Enatioselective hydrogenation of C=N functions and enamines. In: The handbook of homogeneous hydrogenation. Wiley-VCH GmbH, p 1193
21. Muller TE, Hultzsch KC, Yus M, Foubelo F, Tada M (2008) Chem Rev 108:3795–3892
22. Tripathi RP, Verma SS, Pandey J, Tiwari VK (2008) Curr Org Chem 12:1093–1115
23. Nugent TC, El-Shazly M (2010) Adv Synth Catal 352:753–819
24. Wang C, Villa-Marcos B, Xiao JL (2011) Chem Commun 47:9773–9785
25. Abdel-Magid AF, Carson KG, Harris BD, Maryanoff CA, Shah RD (1996) J Org Chem 61:3849–3862
26. Burkhardt ER, Matos K (2006) Chem Rev 106:2617–2650
27. Gribble GW (2006) Org Proc Res Dev 10:1062–1075
28. Noyori R, Ohkuma T (2001) Angew Chem Int Ed 40:40–73
29. Blaser HU, Buser HP, Jalett HP, Pugin B, Spindler F (1999) Synlett 867–868
30. Xiao D, Zhang XM (2001) Angew Chem Int Ed 40:3425–3428
31. Chi YX, Zhou YG, Zhang XM (2003) J Org Chem 68:4120–4122
32. Tararov VI, Kadyrov R, Riermeier TH, Borner A (2000) Chem Commun 1867–1868
33. Kadyrov R, Riermeier TH, Dingerdissen U, Tararov V, Borner A (2003) J Org Chem 68:4067–4070
34. Hsiao Y, Rivera NR, Rosner T, Krska SW, Njolito E, Wang F, Sun Y, Armstrong JD, Grabowski EJJ, Tillyer RD, Spindler F, Malan C (2004) J Am Chem Soc 126:9918–9919
35. Bunlaksananusorn T, Rampf F (2005) Synlett 17:2682–2684
36. Steinhuebel D, Sun Y, Matsumura K, Sayo N, Saito T (2009) J Am Chem Soc 131:11316–11317
37. Li CQ, Wang C, Villa-Marcos B, Xiao JL (2008) J Am Chem Soc 130:14450–14451
38. Tang WJ, Johnston S, Iggo JA, Berry NG, Phelan M, Lian L, Bacsa J, Xiao JL (2013) Angew Chem Int Ed 52:1668–1672
39. Li C, Villa-Marcos B, Xiao JL (2009) J Am Chem Soc 131:6967–6969
40. Rubio-Pérez L, Pérez-Flores FJ, Sharma P, Velasco L, Cabrera A (2009) Org Lett 11:265–268
41. Zassinovich G, Mestroni G, Gladiali S (1992) Chem Rev 92:1051–1069
42. de Graauw CF, Peters JA, van Bekkum H, Huskens J (1994) Synthesis 1994:1007–1017

43. Noyori R, Hashiguchi S (1997) Acc Chem Res 30:97–102
44. Palmer MJ, Wills M (1999) Tetrahedron Asymmetry 10:2045–2061
45. Wills M, Palmer M, Smith A, Kenny J, Walsgrove T (2000) Molecules 5:4–18
46. Everaere K, Mortreux A, Carpentier JF (2003) Adv Synth Catal 345:67–77
47. Clapham SE, Hadzovic A, Morris RH (2004) Coord Chem Rev 248:2201–2237
48. Gladiali S, Alberico E (2006) Chem Soc Rev 35:226–236
49. Samec JSM, Backvall JE, Andersson PG, Brandt P (2006) Chem Soc Rev 35:237–248
50. Ikariya T. Blacker AJ (2007) Acc Chem Res 40:1300–1308
51. Wu XF, Xiao JL (2007) Chem Commun 2449–2466
52. Wang C, Wu XF, Xiao JL (2008) Chem Asian J 3:1750–1770
53. Wu XF, Wang C, Xiao JL (2010) Platinum Met Rev 54:3–19
54. Kadyrov R, Riermeier TH (2003) Angew Chem Int Ed 42:5472–5474
55. Williams GD, Pike RA, Wade CE, Wills M (2003) Org Lett 5:4227–4230
56. Uematsu N, Fujii A, Hashiguchi S, Ikariya T, Noyori R (1996) J Am Chem Soc 118:4916–4917
57. Strotman NA, Baxter CA, Brands KMJ, Cleator E, Krska SW, Reamer RA, Wallace DJ, Wright TJ (2011) J Am Chem Soc 133:8362–8371
58. Ouellet SG, Walji AM, Macmillan DWC (2007) Acc Chem Res 40:1327–1339
59. Iwasaki F, Onomura O, Mishima K, Kanematsu T, Maki T, Matsumura Y (2001) Tetrahedron Lett 42:2525–2527
60. Malkov AV, Mariani A, MacDougall KN, Kocovský P (2004) Org Lett 6:2253–2256
61. Malkov AV, Stončius S, MacDougall KN, Mariani A, McGeoch GD, Kočovský P (2006) Tetrahedron 62:264–284
62. Malkov AV, Vranková K, Stončius S, Kočovský P (2009) J Org Chem 74:5839–5849
63. Wang ZY, Ye XX, Wei SY, Wu PC, Zhang AJ, Sun J (2006) Org Lett 8:999–1001
64. Gautier F-M, Jones S, Martin SJ (2009) Org Biomol Chem 7:229–231
65. Malkov AV, Stončius S, Kočovský P (2007) Angew Chem Int Ed 46:3722–3724
66. Guizzetti S, Benaglia M, Cozzi F, Annunziata R (2009) Tetrahedron 65:6354–6363
67. Gautier F-M, Jones S, Li X, Martin SJ (2011) Org Biomol Chem 9:7860–7868
68. Singh S, Batra UK (1989) Indian J Chem B 28:1
69. Rueping M, Sugiono E, Azap C, Theissmann T, Bolte M (2005) Org Lett 7:3781–3783
70. Hoffmann S, Seayad AM, List B (2005) Angew Chem Int Ed 44:7424–7427
71. Storer RI, Carrera DE, Ni Y, MacMillan DWC (2006) J Am Chem Soc 128:84–86
72. Akiyama T, Itoh J, Yokota K, Fuchibe K (2004) Angew Chem Int Ed 43:1566–1568
73. Uraguchi D, Terada M (2004) J Am Chem Soc 126:5356–5357
74. Li G, Liang Y, Antilla JC (2007) J Am Chem Soc 129:5830–5831
75. Wakchaure VN, Nicoletti M, Ratjen L, List B (2010) Synlett 2010:2708–2710
76. Hoffmann S, Nicoletti M, List B (2006) J Am Chem Soc 128:13074–13075
77. Zhou J, List B (2007) J Am Chem Soc 129:7498–7499
78. Kumar A, Sharma S, Maurya RA (2010) Adv Synth Catal 352:2227–2232
79. Zhu C, Akiyama T (2010) Adv Synth Catal 352:1846–1850
80. Hummel W, Kula M-R (1989) Eur J Biochem 184:1–13
81. Bommarius AS, Schwarm M, Stingl K, Kottenhahn M, Huthmacher K, Drauz K (1995) Tetrahedron Asymmetry 6:2851–2888
82. Hall M, Bommarius AS (2011) Chem Rev 111:4088–4110
83. Hollmann F, Arends IWCE, Holtmann D (2011) Green Chem 13:2285–2314
84. Ricca E, Brucher B, Schrittwieser JH (2011) Adv Synth Catal 353:2239–2262
85. Patel RN (2011) ACS Catal 1:1056–1074
86. Wildeman SMAD, Sonke T, Schoemaker HE, May O (2007) Acc Chem Res 40:1260–1266
87. Sanwal BD, Zink MW (1961) Arch Biochem Biophys 94:430–435
88. Bommarius AS, Drauz K, Gunther K, Knaup G, Schwarm M (1997) Tetrahedron Asymmetry 8:3197–3200
89. Krix G, Bommarius AS, Drauz K, Kottenhahn M, Schwarm M, Kula MR (1997) J Biotechnol 53:29–39

90. Hanson RL, Schwinden MD, Banerjee A, Brzozowski DB, Chen B-C, Patel BP, McNamee CG, Kodersha GA, Kronenthal DR, Patel RN, Szarka LJ (1999) Bioorg Med Chem 7:2247–2252
91. Gröger H, May O, Werner H, Menzel A, Altenbuchner J (2006) Org Proc Res Dev 10:666–669
92. Hanson RL, Howell JM, LaPorte TL, Donovan MJ, Cazzulino DL, Zannella V, Montana MA, Nanduri VB, Schwarz SR, Eiring RF, Durand SC, Wasylyk JM, Parker WL, Liu MS, Okuniewicz FJ, Chen B-C, Harris JC, Natalie KJ Jr, Ramig K, Swaminathan S, Rosso VW, Pack SK, Lotz BT, Bernot PJ, Rusowicz A, Lust DA, Tse KS, Venit JJ, Szarka LJ, Patel RN (2000) Enzyme Microb Technol 26:348–358
93. Cainelli G, Engel PC, Galletti P, Giacomini D, Gualandi A, Paradisi F (2005) Org Biomol Chem 3:4316–4320
94. Vedha-Peters K, Gunawardana M, Rozzell JD, Novick SJ (2006) J Am Chem Soc 128:10923–10929
95. Galkin A, Kulakova L, Yoshimura T, Soda K, Esaki N (1997) Appl Environ Microbiol 63:4651–4656
96. Menzel A, Werner H, Altenbuchner J, Gröger H (2004) Eng Life Sci 4:573–576
97. Höhne M, Kühl S, Robins K, Bornscheuer UT (2008) Chembiochem 9:363–365
98. Fuchs M, Koszelewski D, Tauber K, Kroutil W, Faber K (2010) Chem Commun 46:5500–5502
99. Cassimjee KE, Branneby C, Abedi V, Wells A, Berglund P (2010) Chem Commun 46:5569–5571
100. Truppo MD, Rozzell JD, Turner NJ (2009) Org Proc Res Dev 14:234–237
101. Koszelewski D, Lavandera I, Clay D, Guebitz GM, Rozzell D, Kroutil W (2008) Angew Chem Int Ed 47:9337–9340
102. Desai AA (2011) Angew Chem Int Ed 50:1974–1976
103. Savile CK, Janey JM, Mundorff EC, Moore JC, Tam S, Jarvis WR, Colbeck JC, Krebber A, Fleitz FJ, Brands J, Devine PN, Huisman GW, Hughes GJ (2010) Science 329:305–309

Index

The manufacturer's authorised representative in the EU is Springer Nature Customer Service Centre GmbH, Europaplatz 3, 69115 Heidelberg, Germany. If you have any concerns regarding our products, please contact ProductSafety@springernature.com

Printed and bound by CPI Group (UK) Ltd, Croydon, CR0 4YY

30/09/2026

02225602-0008